Map Worlds

Map Worlds

A HISTORY OF WOMEN IN CARTOGRAPHY

WILL C. VAN DEN HOONAARD

WILFRID LAURIER
UNIVERSITY PRESS

This book has been published with the help of a grant from the Canadian Federation for the Humanities and Social Sciences, through the Awards to Scholarly Publications Program, using funds provided by the Social Sciences and Humanities Research Council of Canada. Wilfrid Laurier University Press acknowledges the support of the Canada Council for the Arts for our publishing program. We acknowledge the financial support of the Government of Canada through the Canada Book Fund for our publishing activities.

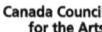  

LIBRARY AND ARCHIVES CANADA CATALOGUING IN PUBLICATION

Van den Hoonaard, Will C. (Willy Carl), 1942–, author
    Map worlds : a history of women in cartography / W.C. van den Hoonaard.

Includes bibliographical references and index.
Issued in print and electronic formats.
ISBN 978-1-77112-126-2 (paper)—ISBN 978-1-55458-933-3 (pdf). —
ISBN 978-1-55458-934-0 (epub)

    1. Women cartographers—History. 2. Cartography—History. I. Title.

GA203.V35 2013   526.082   C2013-903608-3   C2013-903609-1

Cover design by Sandra Friesen. Front-cover image: detail of *Prima pars Brabantiae cuius caput Lovanium* (Amsterdam, 1650), by W. Blaeu. Image provided by Antiquariat Reinhold Berg, www.alte-landkarten.de. Text design by Sandra Friesen.

© 2013 Wilfrid Laurier University Press
Waterloo, Ontario, Canada
www.wlupress.wlu.ca

Every reasonable effort has been made to acquire permission for copyright material used in this text, and to acknowledge all such indebtedness accurately. Any errors and omissions called to the publisher's attention will be corrected in future printings.

No part of this publication may be reproduced, stored in a retrieval system, or transmitted, in any form or by any means, without the prior written consent of the publisher or a licence from the Ca-nadian Copyright Licensing Agency (Access Copyright). For an Access Copyright licence, visit http://www.accesscopyright.ca or call toll free to 1-800-893-5777.

Dedicated to
Lisa-Jo van den Scott
L. Cheryl Power
Jordan (Jay) van den Hoonaard

Contents

LIST OF FIGURES, TABLES, AND CHARTS	ix
PREFACE	xi
ACKNOWLEDGEMENTS	xiii

1	Introduction: The Strands through Map Worlds	1
2	Who Is a Cartographer?	15
3	The Thirteenth to Seventeenth Centuries	29
4	The Eighteenth and Early Nineteenth Centuries (1666 to 1850)	45
5	Cartography from the Margins: From the Early Twentieth Century to World War II	75
6	Mid- to Late-Twentieth-Century Pioneers and Advancers in North America	93
7	Late-Twentieth-Century Pioneers and Advancers in Europe, Asia, and Latin America	139
8	"Getting There without Aiming at It": Women's Experiences in Becoming Cartographers	169
9	"We Are Good Ghosts!": Orientations and Expectations of Women Cartographers	185
10	Educational Opportunities and Obstacles	205
11	The Gendered Social Organization	217
12	Female Pathways through the Present-Day Map World	241
13	Gender Shifts	269

CONTENTS

APPENDICES

A	Methodology	285
B	Topics Covered in an In-Depth Interview	297
C	Overview of Twenty-Eight Women Pioneers in Cartography	301

NOTES	305
REFERENCES	323
COPYRIGHT ACKNOWLEDGEMENTS	353
INDEX	357

List of Figures, Tables, and Charts

Fig. 3.1	Herrade of Hohenbourg	31
Fig. 3.2	Ebstorf Map	32
Fig. 3.3	The Engravers' Map Atelier	36
Fig. 4.1	Embroidered Map of New York	48
Fig. 4.2	Finishing Room	54
Fig. 4.3	Founder's Room	55
Fig. 4.4	Composing Room	56
Fig. 4.5	Map Seller's Shop	57
Fig. 4.6	Rocque Title Page	58
Fig. 4.7	Globe by Elizabeth Mount	62
Fig. 4.8	Colban Map	64
Fig. 4.9	Perspective Map of Kabelvåg by Colban	65
Fig. 4.10	Church of Vågan	66
Fig. 4.11	Portrait of Shanawdithit, Newfoundland	70
Fig. 4.12	One of Shanawdithit's maps: Red Indian Lake, Newfoundland	71
Fig. 5.1	Thematic Map Created by Kelley	78
Fig. 5.2	Photo of Mina Hubbard in Labrador	83
Fig. 5.3	The Official Lunar Map for the International Astronomical Union by Mary A. Blagg	86
Fig. 6.1	Marie Tharp	96
Fig. 6.2	Ocean Map by Marie Tharp	97
Table 6.1	Membership in Selected Cartographic Organizations, 1987	104
Fig. 7.1	Photo of Regina de Almeida	165
Table 11.1	The Participation of Women Cartographers in Cartography and Related Fields	219
Table 11.2	Gender Composition of Selected Events at ICA 2011	223
Table 11.3	Chairs of ICA Commissions, by Gender, 1999–2011	224
Table 11.4	Map Librarians, Archivists, and Related Positions by Proportions of Gender, 1993	237
Table 12.1	Characteristics of Members of the Pathways through (In)equality	248
Table 12.2	Percentage of Active Women in the Surveying and Mapping World at All Educational Levels: The World by Regions, 1980 and 1990	265
Table 12.3	Gender Composition of Cartographers at ICA 2011	266

LIST OF FIGURES, TABLES, AND CHARTS

Table 12.4 Occupational Niches of Women Cartographers, Surveyors, and GIS Staff in Developed and Developing Countries, 1995   267
Table A.1 Source Regions of (Near) Contemporary Women Vignettes and Interview Participants, 1999–2012   293

Preface

Map Worlds plots a journey of discovery through the world of women map-makers. The journey starts in the "golden age of cartography" in the sixteenth-century Low Lands and ends with tactile maps in contemporary Brazil. As developers of resources that allowed early map ateliers to flourish through marital liaisons, women had an unmistakable role. Others, working from the margins, produced maps to record painful tribal memories or sought to remedy social injustices in the nineteenth century. In contemporary times, one woman produced a revolution in the way we think about continents, likened to the Copernican revolution. Still others created order and wonder about the lunar landscape, while others turned the art and science of making maps inside out, exposing the hidden, unconscious, and subliminal "text" of maps. Promoting social justice and making maps work for the betterment of humanity are goals shared by all these outstanding map-makers.

The enthusiasm for the topic grew from my meeting Dr. Eva Siekierska of Ottawa in 1993. During a coffee break at a committee meeting (unrelated to cartography), she told me about her excitement at chairing the International Cartographic Association's newly established Commission on Gender and Cartography. Her enthusiasm aroused my curiosity about the connection between gender and cartography, and for four years thereafter,

PREFACE

I reflected on its significance as a topic of research, especially as I am an active promoter of the equality of women and men. During these four years, I sought to understand maps as hegemonic devices constructed by makers of maps, trying to see patterns in city maps, for example, that pushed women and children to the margins either by ignoring them altogether in maps or representing them in secondary ways (van den Hoonaard, 1994).

By 2003, I had finished the first draft of *Map Worlds*. Writing across disciplines is never an easy task, as my book manuscript vividly demonstrated. Diligent reviewers critiqued the draft and offered many suggestions to improve it. With other writing projects on the go, I had to forsake attempts to improve the book manuscript until much later. When I was elected Chair of the Historical Cartography Interest Group of the Canadian Cartographic Association in 2008, I decided to rewrite the book entirely. In that connection, I began assembling some new material, first to be published in the Association's newsletter, *Cartouche*, and then to be incorporated into a greatly revised book manuscript.

The next phase reinforced a new approach: one-third of the book would highlight the twenty-eight women pioneers in cartography, both dead and living. This particular emphasis produced a round of correspondence and emails with those who knew the women who had died and with those who were the subject of the vignettes themselves.

Acknowledgements

A major portion of the research would have remained a pie in the sky if it were not for the nearly seventy interview participants who offered me a portion of their very busy schedules. I cannot mention most of their names in light of the anonymity they were entitled to in this research, but they may recognize their story.

Foremost among those I wish to recognize are Dr. Eva Siekierska, the first Chair of the Commission of Gender and Cartography of the International Cartographic Association; Dr. Monica Rieger of the University of Calgary; Dr. Susan Nichols of the University of New Brunswick; and Dr. Suzan Ilcan of the University of Windsor. Dr. Janice Monk has been an unfailing guide and help in my work. I should especially mention Dr. Alberta Auringer Wood (formerly at Memorial University in Newfoundland and Labrador), whose support for my research has been unfailing, uncompromising, and untiring throughout all fifteen years of my research and writing. I should not omit mentioning Dr. Cliff Wood, the gentle scholarly giant in cartography.

I owe a particular debt to Dr. John McLaughlin, President Emeritus of the University of New Brunswick; Ms. Alice Hudson at the New York Public Library; Ms. Joanne M. Perry of the University of Pennsylvania; Ms. Mary Ritzlin of Ritzlin Maps; Ms. Elizabeth Hamilton of the University of New

ACKNOWLEDGEMENTS

Brunswick; and Dr. Judith Tyner of California State University–Long Beach. I owe a special debt to all those who wished to have their stories included in this book; some have agreed to be openly acknowledged as pioneers in the field and have thus contributed much to their own story.

I must also acknowledge the two anonymous reviewers of a very early draft of the book (2003) who so diligently and constructively offered important critiques. The current version of the book is, I hope, a dramatic revision of that early draft. Ms. Samara Young, an undergraduate at St. Thomas University, was instrumental in pulling together the relevant sources of information for the score of women cartographers profiled in *Map Worlds*. She possesses that unique combination of intelligence, a practical talent for finding new information, and an enthusiasm and conviction that the work can be done in a compelled and compelling time. I thank her in particular for helping me with the research. Ms. Gill Steeves, also a student at St. Thomas University, also provided assistance, especially in relation to her discovery of Mina Hubbard, the accidental cartographer of Labrador.

Along the way, research projects have a habit of drawing in students: Ms. Natt Foster (who became a Ph.D. student in sociology at McGill), Dr. Mele Rakai (formerly a Ph.D. student in the Department of Geodesy and Geomatics Engineering at the University of New Brunswick), Ms. Lenora Sleep, Dr. Dawne Clarke, Dr. Linda Caissie, Mr. Truls Christian Waage, and Ms. Linda Hanford. I am also grateful to Dr. Margriet Hoogvliet, Dr. Ute Dymon, the staff of the International Institute for Aerospace Survey and Earth Sciences in Enschede (Netherlands), Ms. Joanne Schweik, Ms. Wendy Straight, Dr. Harry Steward, Mr. Patrik Ottoson, Mr. Hartmann, Mr. Ian Fussey, Mr. Frank Blakeway, Ms. Valerie Shoffey, Ms. Susan Pugh, Mr. Jake Kean, Ms. Grace Farrell, Mr. Dave Doyle, Dr. Barbara Buttenfield, Dr. Roger Wheaton, Dr. Anne Furlong, Mr. Jeffrey Bursey, Dr. Valerie Traub, Dr. Wladyslaw Cichocki, Ms. A. Suzanne Hill, Ms. Joanne Costello, Mr. Michael Meade, and Mr. Samandar Hajizadeh. I also wish to acknowledge the help of the Canadian Department of National Defence in Ottawa for providing information on handkerchief maps. I should not forget the help I received from Dr. Ewen A. Whitaker of Arizona, who was deeply involved in the matter of lunar nomenclature and who brought to my attention the name of an additional woman cartographer. Ms. Lisa Quinn, Acquisitions Editor at Wilfrid Laurier University Press, has been as helpful and encouraging as anyone can dream about. Mr. Doug Hildebrand at University of Toronto Press helped move the manuscript to the next stage of its development. I owe both of them deep thanks.

ACKNOWLEDGEMENTS

The Social Sciences and Humanities Research Council of Canada awarded in April 1997 a grant (no. 410-97-0219) in support of this program of research, to which I am deeply indebted. I also received a grant from the University of New Brunswick Retirees Research Fund. Throughout the process, the University of New Brunswick Research Ethics Board has been unfailingly supportive of this work. I owe particular thanks to the Canadian Cartographic Association, which accepted me in their midst.

While it has become customary to end acknowledgements with a healthy sense of gratitude to members of one's immediate family, it is by no means a perfunctory exercise. On the contrary. The true intellectual in the family will always be Deborah, my spouse, who stood wholeheartedly behind the project, including the essential pre-final draft task of editing the work. There is no other conceivable way of supporting a spouse's intellectual endeavours as she has. For my children: when my first book was published in 1991, Lisa-Jo was fourteen years old, Lynn Cheryl had just turned twelve, and Jordan was about to become eight. They are now thirty-five, thirty-three, and thirty.

CHAPTER 1

Introduction: The Strands through Map Worlds

Behind some eight thousand contemporary women from around the world stand not only more than five hundred years of history, but also one of the most popular cultural productions in the world: maps. The world of mapmakers is somewhat known to us, but we have only a microscopic knowledge about the involvement of women in map-making. *Map Worlds* has set itself the task of recovering these women from history. No less significantly, it also recounts the experiences of women with contemporary cartography. Oftentimes, the world of women cartographers seems to be hidden, much like the so-called dark side of the moon, but as every thinking person knows, the invisible side of the moon bathes in the sunlight just as much as the one that faces us does.

Few fields have changed as dramatically as cartography. In the thirty-six years between my making my last etch as a cartographic editor at Falk-Plan (a European map company) in 1966 and my re-entry into the world of cartographers—this time as a social scientist—the field has become unrecognizable. This awakening was a compelling one, for it demonstrated the earthquake-like technological shifts within cartography beyond anyone's imagination. As a social scientist, I began wondering about the social organization of cartographers—their world, their culture, and their habits. Had these changed drastically as well? As a modern individual, I wanted

CHAPTER 1

to apply the contemporary pincer of analysis: What role does gender play in all of this? I had originally envisioned *Atlas Shrugged* as a title for this book, but had Atlas truly shrugged? Was it a momentous shrug, signifying a radical departure of the old ways? Or was it a shrug of the kind that said, "So what?" This is one of the themes of this book: Have the technological changes that captivated cartography over the past thirty-six years also been reverberating through the organization and lives of women who are drawn to this field?

It is only through the recent passage of time that map librarians, cartographers, map collectors, and historical cartographers have begun to consider more explicitly the role of women in map-making. Part of the awakening process involves fresh research on women who have made important or interesting contributions to cartography (see, for example, Steward, 2001).[1]

I have identified three strands of research interests. One strand explores pockets of cartography where women have been particularly active and have contributed significantly to the field in the historical sense. This strand represents a recuperative history of women in cartography.

The second strand concerns itself with contemporary women pioneers in cartography starting around 1880 and up to the present. This strand offers vignettes (brief biographical sketches) of twenty-eight women who were (or are) pioneers or major advancers of cartography. The vignettes describe their parentage, education, careers, contributions to cartography, and anything else that would explain the circumstances of their place and time in cartography.

The third strand deals primarily with the experiences, problems, and obstacles contemporary women face in cartography. This strand also looks behind the social organization of cartography as the backdrop of those experiences.

THE SOCIOLOGICAL SENSE

There is, however, more than the historical and experiential study of women in cartography that makes it a subject of scholarly interest, especially from a sociological perspective. Sociologists are committed to the study of societal norms and of the breaching of those norms. The breaching of norms becomes the genesis of new norms and social forms, and such breaching of norms attracts the gaze of sociologists. When occupations believed to be traditionally filled with men are punctured by the emergence of women (or when women have become historically visible), it is an invitation to study the phenomenon. The significance of such a study rests, then, on discovering the new dynamism that has entered the occupation and its potential for, and nature of, change. Naturally, too, such a sociologist would discover the individual attributes that might attend those who breach the norms, as well as those who welcome those breaches.

Sociologists look at the world of cartography from the perspective of the social arrangements that create the map world, revealing the many taken-for-granted attitudes and behaviours that seem to reinforce that world. In that light, I assembled statements of women cartographers about that world, how they entered it, how they achieved their identity as a cartographer, how they developed professional relations with their colleagues, and what they thought their challenges and contributions might be. The "sociological imagination" (a term coined by C. Wright Mills to describe the inevitable link between personal biography and social structure) draws the sociologist's attention to (cartographic) organizations and wider social trends that shape the lives of women cartographers.

When a sociologist explores the research about women in cartography or about such breaches specifically, he or she discovers an imbalance in the breadth of coverage. On one hand, there is a growing body of lists and articles that pertain to the historical aspects of women in cartography (Ritzlin, 1986, 1989, 1993; Hudson, 1989, 1999a, 1999b, 2000; Miller, 1993; Tooley, 1978; Tyner, 1997),[2] but relatively few devoted to either contemporary women or to the social organization of cartography. On the other hand, there is an increasing number of studies that speak to the presence of women and to gender imbalances in fields allied to cartography: namely, geodesy and geomatics (formerly known as surveying engineering).[3] Taken together, there are over 110 articles (and one book) that discuss women in historical cartography and contemporary women in other fields. There are, in addition, at least half a dozen professional journals devoted to women in geodesy and geomatics, geography, and so on, with special issues on women published occasionally by trade magazines and professional journals.[4] The sociologist would conclude that not too much has been written about the recent crop of women in cartography. Still, the sociologist might find some useful reference points about the study of women in science, geography, and geodesy.

Women of Science (Kass-Simon, Farnes, and Nash, 1993) constitutes an early exploration of women in a number of scientific fields (archaeology, geology, astronomy, mathematics, engineering, physics, biology, medicine, chemistry, and crystallography). While that work aims at righting the record of the contributions of women to the sciences, it also underscores the

CHAPTER 1

persistent obstacles that women face. These obstacles come in the form of stereotypes. The twentieth-century stereotypes are "subtle replays" of the nineteenth-century ones: premenstrual syndrome the "othering" of pregnancy by medical and technological means, and hormonal deficiencies (1993: 386). Ruth Woodfield's study of women, work, and computing (2000) offers the prospect that a genderless future world of computers is well-nigh impossible. The world of computer work mirrors the gender imbalances of the wider world, despite the initial wave of optimism in the 1970s. Suzanne Le-May Sheffield's *Women and Science* (2006) still brings a discouraging message: despite the passing of the United States' *Equal Opportunity Employment Act* and affirmation-action policies in the 1960s, discrimination against women continues. Sheffield has found that "Men and women in positions of power keep women out, and in turn, women themselves internalize the social norms against women's participation in science, creating a self-fulfilling prophecy" (2006: 201). Joyce Tang, in *Scientific Pioneers: Women Succeeding in Science* (2006: 65), focuses on the stamina it took to withstand barriers, the defiance of cultural norms regarding the equality of women and men, the positive attitudes to setbacks, and coping with cumulative disadvantages in education and careers. However, sociologists are not as interested in attributes of individuals as in discovering how the social context reaffirms, rejects, or shapes those attributes. The interplay between personal attributes and social structure makes for a keen study. A more recent work is Ruth Watts's *Women in Science: A Social and Cultural History* (2007), which underscores the continuing "persistent biases" frustrating women in science (2007: 201) and the masculinization of formerly female-dominated areas. Her study of women in science from antiquity to modern times "illustrate[s] both the way that notions disadvantaging women in science keep recurring and the ways that women circumvent them" (201). With women scientists "progressively dropping out of academia and industry," Watts avers that we should analyze "the underlying structural power relationships" (203). Disenfranchisement of women is one consequence of exclusion of women from science, she states (202).

Within geography, the road to equality was rather bumpy. Mildred Berman (1984) describes her personal experiences whereby she had to endure the numerical superiority of male graduate students, the silencing of women students, and the foreclosure of paths for positions in universities. There are scores of individual biographies on women geographers, usually in article format, about Millicent Todd Bingham—a human geographer (Berman, 1980)—and about Ellen Churchill Semple (Berman, 1974).

INTRODUCTION: THE STRANDS THROUGH MAP WORLDS

Surveys have documented the early, inadequate wakening of equality of women and men among American geographers.

Aside from exploring the issue of gender in the context of the life of professional geography organizations and individual experiences of women geographers (whether as autobiographies or biographies), researchers are lavishing considerable attention on women as primary sources of geographic information. Researchers such as Marianna Pavlovskaya (2009) and Kevin St. Martin and Pavlovskaya (2009) have recently explored the wider theoretical and conceptual contours of women and feminism in the allied fields of geography and geographic information systems (GIS). Janice Monk (e.g., 1990, 1998, 2004) conducted interviews with at least sixty women geographers. The earliest women in her collection of interviews were in graduate school in the 1930s, including Mary Arizona Baber, whom Monk calls an "exceptional social activist" in the University of Chicago faculty, founding the Chicago Geographic Society (Janice Monk, email to author, 13 September 2011). Monk's other extensive research on women geographers, linking biography, culture, and political contexts, deals with their changing expectations in the 1970s (Monk, 2006) and their participation in the American Geographical Society (2003). Several other works stand out. *Complex Locations* by Avril Maddrell (2009b) provides a rather fine detailed account of women in UK geography, noting relevant historical, institutional, social, and economic contexts. In the final analysis, she found that gendered thought and practice exist in geography. First, she found that the language of the history of geography is couched in gender. There is, for example, no women's equivalent term to "founding fathers," and the text is strongly gendered in any case. Second, Maddrell alerts us to how women geographers refer to themselves as authors, designating themselves with more ambivalence than is the case for men authors. Third, the processes associated with the refereeing and reviewing of articles (as well as establishing one's reputation) are highly gendered, and only recently have begun to abate. Fourth, the vocabulary of memorializing the accomplishments of women is suffused with a particular subtext that plays a part in "undermining any past scholarly accomplishments." Women's ongoing work in geography well into their retirement years has left them being charged as "oddballs" (Maddrell, 2009b: 329–337).

Within geodesy, Clara Greed's *Surveying Sisters* (1991) is the most detailed of all these studies and comes closest in its attempt to portray an in-depth perspective of women in a field allied to cartography. Her research (conducted between 1986 and 1988) highlights the problems women surveyors face in the organizational structure, interpersonal relationships, and ethos of British surveying. She describes the "getting by" attitude of women who face a relatively undisturbed subculture in which they largely assume ancillary functions and must tread warily; they become socially fragmented and isolated, face taking "leftover" surveying tasks, move into dead-end jobs (even though their initial rise was rapid), and are overtaken by younger, inexperienced men (Greed, 1991: 181–184). Greed's work highlights the problematic aspects of gender in surveying more than the perspectives of women themselves, however. Clara Greed herself suggests that in the future, "ethnographic and more open-ended approaches to research may be the way to achieve [a "total explanation"]" that might offer more room for women's experiences than the "closed" approaches of surveys (Greed, 1991: 187). This strand in *Map Worlds* displays the results of interviews with thirty-eight women.

As strange as it may seem, there is a close empirical and theoretical comparison between women in cartography and women in architecture. Annmarie Adams and Peta Tancred, in *"Designing Women": Gender and the Architectural Profession* (2000), refer to the "gendered substructure" of the profession of architecture (123) whereby licensed women architects severed ties with their profession and branched out into "unofficial" architectural practices (118). In these new avenues of work, women fostered innovation in the field, on top of fusing their public and private work.

The scholarly attention on women in cartography-related fields over the past several decades summoned up new insights about the wider issue of power, spatial politics, and representation. Earlier work, such as by Shirley Ardener (1981) and Janet Siltanen and Michelle Stanworth (1984), elaborated on gendered private and public spaces, reflecting patriarchal structures. Private space symbolized restrictions placed on women, while the public signified freedoms belonging to men. There is, in fact, a large literature that resorts to spatial metaphors in today's critical writing. As Catherine Nash reminds us, many of the mapping terms (such as "site," "space," "terrain," and "map") are rightly used to describe theoretical and ideological spaces—they are "potent metaphors" for masculinist and colonial usages (Nash, 1994: 228, 229). According to Nikolas H. Huffman (1997) and other feminist geographers (van Ee, 2001), the masculinist vision of geography systematically excluded women in maps as images of power, order, and world views. Feminist geographers have been reluctant to use maps as sources (see, for example, Domosh and Seager, 2001; Nelson and Seager, 2005; Massey and Allen, 1984).

Others have more broadly extended the spatial imagery in social analyses, speaking of the "periphery" and the "core" as representing not only space, but also inequalities of power. Blunt and Rose (1994: 5), for example, refer to Gloria Anzaldúa's work on the "borderlands" between Mexico and the United States. In these wider contexts, it is knowledge (and thus power) that defines space. The power of space is evident in all societies. Take Canada as one example. Aboriginal (First Nations) reserves speak to the issue of the "politics of location" (Blunt and Rose, 1994: 7) in several ways: mainstream and First Nations societies may well perceive such space as "negative" or "positive" space. Do reserves reinforce marginalization or do they serve to preserve culture? As evidenced by feminist analysis of these and other spaces, it is clear that minorities and peoples inhabiting these spaces would more quickly recognize the critical dimensions of space as emblems of power or powerlessness. Being at the centre involves unexamined assumptions and a taken-for-grantedness that minorities cannot afford to entertain.

But let us locate women cartographers within their occupation as a whole. Their location can best be understood when one considers the larger parameters of cartography marked by "map worlds" that *Map Worlds: A History of Women in Cartography* sees as a sufficiently compelling concept as to warrant special consideration.

MAP WORLDS

The concept of map worlds embraces the totality of relationships, norms, practices, and technologies that shape and constitute the world of mapmakers. While "map trades" is about the retail of and products associated with maps per se, map worlds is about the wider context of cartography (in which "map trades" are located), suggesting that many more elements contribute to the field than what one normally thinks of as map-making. The map world is an explicit recognition that cartography is multi-faceted; there are no margins in this conception of map worlds, where boundaries are contiguous. All kinds of relations, practices, and ideas occur on the borderline of the map world involving powerful forms of knowledge, struggles, and tensions, invoking change and interchange.

Map-makers constitute a community, but such a community is not an instantaneous phenomenon. Rather, it is created through the inevitable passage of time. The cartographic community has been building over the past five hundred years, but has undergone dramatic transformations heralded by technological innovations and social and political trends. Map worlds[5] paralleled these transformations. They have seen the death knell of many cartographic occupations, ranging from woodcutters, engravers, colourists, Letraset producers, and many others, to the birth of new ones, such as GIS specialists. Relationships and patterns of social and economic activity in map worlds have also changed and realigned themselves accordingly. Map worlds fluctuate constantly, pushed or pulled by what seems to matter most at that time. For example, the state, the military, and corporations all now parlay more prominently into the map world than ever before, but so do democratic processes actuated by computers and the Internet.

Those who study map worlds see the necessity to discover the building blocks that make up that community, such as the professions and occupations that grow and wither and the norms, values, meanings, symbols, and terminology that guide the community (or establish its deviants). Technology, on top of that, also constitutes, defines, and unites (and divides) its subculture, and comprises a social organization whose going concern melds all these elements. On the surface, a community shares norms and values, although on closer inspection, these are more heterogeneous than we assume. Conflict is part of any community—conflict that may involve protocols, procedures, ways of doing things, production, products, and what constitutes creativity in any given field. These points of conflict define boundaries. A sociologist will profit from the study of these boundaries because they become the means of figuring out what constitutes the shared values and norms of the community.

What makes community, or collective action, possible? Collective action refers to a social organization that accomplishes any human activity. Howard S. Becker, in his study on the creation, production, and distribution of art (1982), adapts a term used by artists to refer to the whole borderless community that makes art possible, namely "art worlds." The social organization of art worlds (not unlike our map worlds) requires a division of labour, cooperative links, conventions, mobilizing resources of all kinds, patronage, sales by dealers, agents, culture industries, education or training, accreditation, and so on. It is a world where the initiative and work of an artist is linked in many tangible and intangible ways to a wide variety of things that make his or her art possible, from someone's making a particular colour of chalk to the organization of an art gallery. We can extend the idea of "worlds" to other areas of human endeavour, whether they are music, schooling, plumbing, nursing, or the making of maps—cartography.

The map world is not static, but dynamic. The issues of gender follow this dynamic expression of the map world. *Map Worlds* suggests that the role of women (and men) in cartography is predicated on the structure of the changing map world. The salience of gender is thus varied and, historically, cannot be measured along a single line. Rather, each period in cartography unfolds in terms of its own map world, which has shaped the gender relations of its time and created a structure of gender relations that is different from the previous or succeeding map worlds. Relationships among production processes, occupational niches, the prevalence of certain technologies, fashions in map design, public tastes for particular maps, and inventions all play into the map world, and are constantly reshaping themselves. For example, from the mid-nineteenth century on, the growing internationalization of the map world has resulted in an increasing order of homogeneity within that world that prevails in most countries around the globe now. However, there are instances where issues of gender are still unbent by the passage of time.

A short walk to a library to pick up an atlas can yield an example of how women have participated in a special branch of cartography: the creation of atlases. The inside of the title page tells us there are at least thirty tasks that signify collaboration in producing an atlas, in addition to seventeen specific editorial tasks. From an examination of twenty-seven atlases, Natt Foster (1997) found that the number of women involved in producing world atlases has increased between 1944 and 1992 and that such involvement has moved women from being research assistants or compilers to positions of greater responsibility. Women are often associated with the "gathering" of knowledge, such as research, compilation, assistance, and preparation of map materials. Project directors, chief cartographers, presidents, contributors, or consultants are more likely to be men than women. Even in the physical layout of the tasks, we find women at the lower end of the task chain.

In terms of school atlases, other examples of gender and map-making abound. The cover of the historical atlas for primary schools in a central European country (Papp-Váry, 1998) showed gladiators; while the display of portraits inside the atlas was indeed quite rich (and the maps were wonderfully laid out), the page devoted to eighty-two scientists and composers of the nineteenth and twentieth centuries included only one woman: Marie Curie (24). Another primary school atlas, this time devoted to the history of Hungary (Papp-Váry, 2000), features a front cover displaying eighteen figures, including two women, where the face of one woman is partially hidden by the face of another woman! And while here, too, the maps and illustrations are lively and colourful, the only page where there are marginally more women than men is devoted to the theatre: of the eleven images,

CHAPTER 1

six are of women (23). These few examples (namely, the gendered division of tasks and the themes found in a school atlas) give credence to the idea that gender makes a difference in the way a map is designed, but we need to gather more substantive data about the ideological nature of maps and whether that has changed significantly in contemporary times.

It is noteworthy to mention that particularly since 2000, women were directly involved in the production of atlases. Seager and Olson created *Women in the World: An International Atlas* (1986). This atlas is now in its fourth edition as *The Penguin Atlas of Women in the World* (2009). In 2000, Raji, Atkins, and Townsend produced *Atlas of Women and Men in India*. Elisabeth Buehler of Switzerland, in 2001, created *Frauen und Gleichstellungs Atlas Schweiz*, as well as its translation, in 2002, as *Atlas Suisse des femmes et l'égalité*. More recently, Huq-Hussain, Khan, and Momsen created *Gender Atlas of Bangladesh* 2006. In Japan, Takeda, Kinoshita, and Nakazawa produced *Gender Atlas of Japan* (in Japanese) in 2007.

STRAND ONE: WOMEN IN THE HISTORY OF CARTOGRAPHY
I am making the argument that any study of women in cartography should invariably involve an inclusive, historical approach. There is a tendency for sociologists to undertake ahistorical work, although there is an explicit need to anchor the work in historical roots, or even in its tendrils. The scholarly contributions of Hudson and Ritzlin (2000a, 2000b) are singularly important because, for the first time, we now have a list of women cartographers in history. What surprised these researchers was the large number of pre-twentieth-century women cartographers. By 1989, Alice C. Hudson and Penny Barckley reported that they had found 150 pre-twentieth-century women map-makers (Hudson, 1989: 2); ten years later, this list expanded to two hundred names (Hudson, 1998, personal communication), and today it includes nearly three hundred (Hudson and Ritzlin, 2000b). The two map librarians scoured the personal name indices of 1,100 volumes on the history of cartography (Hudson, 1989: 4). In the earliest stage of their project, they had anticipated finding five to ten names.

Increasingly, thus, an awareness is beginning to extend to historical cartography. Norman J.W. Thrower's books (1972, 1996) indicate the extent of this growing interest. The index of his 1972 publication, *Maps and Man*, contained no reference to "women."[6] His 1996 revision of the same book, now entitled *Maps and Civilization*, contains two references to women in the index (although the reference to women's playing an important role in map-making in China [Thrower, 1996: 265] is not referenced in the index).

INTRODUCTION: THE STRANDS THROUGH MAP WORLDS

What is also missing in the index are references to daughters of various map-making families, such as Colleta van den Keere in the Netherlands, and Virginia Farrar and Lady Lovelace in Britain (Thrower, 1996: 279, 295). Interestingly, Thrower's reference in the earlier book to a younger sister of a Chinese prime minister (1972: 23) has now grown to assigning a larger role to women in Chinese map-making (Thrower, 1996: 265). Both publications rely on Joseph Needham's classic four-volume work, *The Shorter Science and Civilization in China* (Ronan, 1981), which makes only one reference to a "girl cartographer."

While the history of cartography extends across numerous centuries (and across many regions of the world), it is unfair to abstract that history as if it is one unbroken history. As Harley and Woodward convincingly point out in their comprehensive oeuvre, *Cartography in the Traditional East and Southeast Asian Societies* (1994), we simply cannot assume that the history of cartography is a unilinear process, culminating somehow in a world cartography based on European standards of measurement and symbology (see also King, 1996: 36). The only significance that one can attach to recounting such a history would be to illustrate the kind of impact technology can exercise on cartography in a manner that has also shaped the involvement of women in cartography—the theme of *Map Worlds*.

However, these records have not documented much in the way cartography was organized or undertaken, let alone plotting and memorializing the participation of women in such an enterprise. There is very little known about this map world, given both the relative paucity of the historical record and the decimation of mapping tools.[7] Literary and scholarly works were usually confined to the priestly, scholarly, and ruling classes, much like isolated islands in an ocean of people who sometimes left the task of such matters to monks, the literati, and officials.

The early periods of cartography (before the nineteenth century) were no less enmeshed in the larger cultural and social framework than is the case today. Chapters 3 and 4 in *Map Worlds* illustrate the gendered contexts that were predicated on a particular division of labour associated with fifteenth- to eighteenth-century map workshops and on the rules and functions of intermarriage that made the workshops economically and socially viable. Chapter 4 concerns itself with women cartographers active in the nineteenth century.

CHAPTER 1

STRAND TWO: (NEAR) CONTEMPORARY WOMEN AS PIONEERS IN CARTOGRAPHY
Few, if any, would go as far as claiming that, historically speaking, maps are simply the production of the male perspective of the world. By all accounts, that would be an extreme perspective: maps are far too complex in their conception, design, production, and intended interpretation to warrant such a perspective. However, in light of J. Brian Harley's theories about maps as social creations (to be referred to later), it is not an exaggeration to claim that gender has, or might have, shaped the cultural production of maps. With the influx of more women into cartography, have those social assumptions in maps changed at all? This is one of the areas *Map Worlds* hopes to explore.

Chapter 5 delves into the lives of five early-twentieth-century women map-makers. The vignettes of these women intend to heighten our awareness of the silent struggles and loud victories at a time when the only research on early-twentieth-century map-makers includes work on female Army map-makers, the so-called Mapping Maids or Millie the Mapper (McPherson, 1993; Tyner, 1999; Anon., 1993: 1–2). Chapters 6 and 7 offer vignettes of twenty-three women pioneers in cartography as a means to learn about the connections between personal biography and the workings of the larger structure—whether these impede or encourage the participation of women in cartography. They are the stories of courage, sacrifice, disappointments, resistance, and, ultimately, the melting of patriarchal attitudes that had stood in the way of the strivings of women. Virtually each vignette is worthy of a movie script.

STRAND THREE: CONTEMPORARY EXPERIENCES AND SOCIAL ORGANIZATION
No sociological study of mapping would be complete without examining facets from the everyday world of cartographers in relation to the larger aspects of fads and fashions in cartography. That everyday world consists of social interaction among those who participate in it. Social interaction consists of getting involved in that world, being recruited, acquiring the cartographic perspective, learning meanings from one's peers, achieving identity as a cartographer, developing a professional self, "doing" cartography, experiencing professional relationships such as bonds and networks, and defining deviant or conformist behaviour (Dietz, Prus, and Shaffir, 1994). Chapters 8 and 9 convey my findings of interviews with thirty-eight women.

INTRODUCTION: THE STRANDS THROUGH MAP WORLDS

A sociologist like myself, however, does not explore the individual characteristics of cartographers. Yes, a sociologist does interview individuals and trace their individual careers. But what is central to all sociological work is finding patterns shared by all occupational incumbents. The sociologist finds the general in the particular (Macionis and Gerber, 1999: 4–5). A sociologist abstracts the culture of cartographers from individual experience. In some respects, a sociologist seeks to find quasi-"tribal" elements in a group that has undergone tremendous change. In short, a sociologist is interested in studying the everyday life of cartographers (i.e., social interaction), their community (i.e., values and norms), and the wider sets of connections ("map world") of everything that has a bearing on cartography, including social organization, technology and their suppliers, techniques of map-making, universities, education, and the like. One could even be persuaded to include the gender breakdown of the CartoPhilatelic Society—collectors of stamps illustrating maps—that shows that 13.8 percent (N=18) of its 130 members are women (Sulkowski, email to Rabbi Yosef Goldman, Fredericton, NB, Canada, 14 March 2012).

Chapter 10 brings us up to date on the education of women cartographers, including issues of mentorship, while Chapter 11 focuses on the International Cartographic Association (ICA) as the principal social organization that features the struggles and accomplishments of women at the highest level. In this context, I should briefly mention the pivotal studies since 1991, which include the survey undertaken by the ICA Task Force on Women in Cartography (1989) that identified the barriers standing in the way of women's participation in ICA activities (Siekierska, 1995). The findings by Carol Beaver (1993) and Brandenberger (1997) are not dissimilar in this respect, and point to the low proportion of women in the field and to remedial steps that can be undertaken. The findings of *Map Worlds* thus far form the basis of Chapter 12, "Female Pathways," which identifies both mainstream and alternate ways that women cartographers move through the formal social organizational aspects of cartographic associations. The concluding chapter (13) permits us to take the findings to a higher level and analyze them in terms of feminist approaches.

CHAPTER 1

METHODOLOGICAL FRAME OF *MAP WORLDS*

As Appendix A goes into sufficient depth outlining my research strategy, there is no need to dwell on those details in this chapter, except to distill the generalities of my strategy. It emerged in the course of my research, but did not follow the chronological sequence of *Map Worlds*.

Aside from the historical and documented accounts of early women cartographers, the principal theoretical direction I used relies on symbolic interactionism. It is the everyday experience of the thirty-eight interviewed women cartographers that is the empirical focus of that part. Symbolic interactionism and Grounded Theory are theoretical perspectives appropriate for this project. The first highlights the meanings that people assign to the things they say and do; while the second offers a systematic way of moving constantly between data collection and theoretical constructs, a dialectic between evidence and ideas. Grounded Theory occupies a special niche in qualitative research; developed in the 1960s by Glaser and Strauss (1967), it depends on a systematic form of data analysis, with extensive use of coding, theoretical samples, the constant-comparative method, and theoretical saturation.

Deeply embedded in the social thought of George H. Mead and Herbert Blumer, symbolic interactionism stresses the importance of meanings as they emerge through social interaction. Social interaction becomes problematic when participants do not share a definition of the situation, as might be the case of women cartographers, many of whom still see themselves as participating in a "man's world." Unlike psychologists, who seek "motives," symbolic interactionists are interested in "how" social processes unfold (Becker, 1998).

From the micro level to the macro level, symbolic interactionism seems an appropriate tool to understand the world of cartography. If we, for example, consider the alleged gender differences in map reading (which, by the way, is not a topic in this book) from a symbolic interactionist perspective, we would claim that map reading involves patterns of expectations that are built into social interaction—it is not merely a psychological "state of mind." Men are expected not to ask for directions (women are); women are not expected to be able to read maps (Hamicz, 2001). What is crucial in understanding these expectations is that when men and women enter social interaction involving asking for directions or map reading, all the parties bring these assumptions into social interaction.

CHAPTER 2

Who Is a Cartographer?

With such a vast array of fields enshrined in cartography (science, geography, geodesy, computer science, geomatics, map libraries, and map archives, to name a few), one is justified in asking who is a cartographer.[1] The question requires a complex answer. Many types of occupational incumbents inhabit(ed) the map world, but over time the shape of the map world itself has undergone fundamental changes. These changes involved major shifts in the way the incumbents relate to cartography and to each other: some tasks disappeared entirely, while others experienced intrinsic changes as they faced new opportunities and challenges. While technology drove these shifts, it was not the only element. Political and social trends, demography, and even family structures realigned and dictated cartographic tasks. For instance, family ateliers in some regions of Europe were at the core of cartography during the sixteenth century, but became less viable or relevant as time wore on. Map archivists arose in the nineteenth century, but were hitherto unknown. Currently, as one scholar noted (Anon., 2012), the rapid development of "locationally referenced mobile devices and the emergence of Google Map, Google Earth, and a number of other map databases have seen the emergence of location-based social networks, an increasing amount and variety of volunteered geographic information (Crowdsourced information), and the application of mapping to a wide variety of topics not usually mapped." These developments have particularly fanned the interest of writers, artists, historians, regional planners, and real estate agents, among others, who are now active cartographers. Most recently, the Open Street Map over the past five years has democratized cartography.

The analytic idea of map worlds is an attractive one because it acknowledges these fundamental shifts and realignments. A static definition of who is a cartographer traps the definition in one moment of history; a more fluid definition would serve our analytic purposes much better. While some see drawing maps as the core task of an incumbent in the map world, many others are engaged in tasks that allow others to create maps. In contemporary times, one could visualize some surveyors inhabiting that world, sharing the social space with others, such as geographic information systems (GIS) specialists and hands-on cartographers, even though each is trained in widely divergent university departments and settings. Surveyors are trained in departments of geomatics and geodesy, GIS specialists have found their calling through departments of mathematics or the social sciences, and cartographers might have received their training in cartography laboratories. In some respects, there is little overlap among them; in other respects, there can be considerable overlap, such as when GIS and applied cartography intertwine.

Similar varying intensities accompany, for example, applied cartographers and map librarians. Neither may have received formal training, but were drawn to their work through love of maps and fell into their respective tasks through opportunity or chance. Still, applied cartographers can draw on fine examples of maps dating from earlier epochs saved by map librarians or map archivists. Both groups might serve on the same executive of an association devoted to promoting the many facets of cartography. Academic cartographers might have found their niche in university departments of history or geography, but may well have written the fundamental text on how to go about making maps. Map archivists and map librarians sometimes have interchangeable roles in preserving and collecting maps, along with educating the public and scholars about maps. Moving to the outer edges of the map world, one might also include traders and sellers of antique maps.

As perplexing as these categories may be, many incumbents might identify themselves in a manner that eludes any categorization at all. Fragments of the map world are, no doubt, contested spaces. Viewed from afar, it looks bizarre; from the inside, it makes almost perfect sense. If history, technology, politics, social trends, demography, and even family structures pervade the map world, redefining and reshaping cartographic tasks and relations among the incumbents, we have a problem of defining who is a cartographer. What is a sociologist to do?

First, one would include just about anyone who defines him- or herself as a cartographer. This is not a pell-mell approach, because once a practitioner inhabits the map world and self-defines as a "cartographer," it would be impossible for an outsider to claim otherwise. Would it therefore be sufficient to take anyone whose name appears on the membership roster of a professional cartographic society who thus sees him- or herself as contributing to, or being part of, cartography? It is not uncommon to find historians, geographers, applied cartographers, academic cartographers, GIS specialists, and map librarians enlisted in such a society. A sociologist would, furthermore, take great pains in finding out what constitutes a cartographic event and seeing who participates in it. These three approaches do not coalesce perfectly, but do contribute to our answering the complex question of who is a cartographer. If nothing else, the map world consists of diverse groups of incumbents.

Today's occupational diversity reflects the technological impulses induced by innovation and coalescing of fields and disciplines that hitherto were remote and unaffiliated with cartography. There is no reason to assume that these impulses will stop in the foreseeable future. In the past, the map world has witnessed radical technological changes with the arrival of the printing press in the fifteenth century, land-surveying instruments in the seventeenth, aerial photography in the early twentieth, and, more recently, satellites and computers—to name a few innovations that have given shape to a new map world in each instance. However, the current transformation is an opportune one for a sociologist like me to look at cartographers in terms of social interaction, the community, and norms and values that make up the map world. In the context of *Map Worlds*, we take great interest in the current technological transformation in the map world that appears to coincide with the opening up of cartography to the wider participation of women.

The participation of women in the field is not new, as Alice Hudson (2000) and several others point out. Women have participated in the map world in a wide variety of ways, whether as workers, colourists, business owners in sixteenth-century map ateliers in Europe, nineteenth-century geography teachers in the United States (Ritzlin, 1990c), or engravers, explorers, or American "mapping maids" during World War II (McPherson, 1993). The contemporary difference, however, within the map world is that the cartographic space occupied by women is less niche oriented than was the case in the past.

With the disappearance of exclusive niches, space becomes contested. The GIS people boldly claim more of cartography than applied cartographers are comfortable with. One is not even very sure who is a cartographer anymore, indicating changing self-definitions: can those computer experts really know something about the aesthetics of maps? Occupational boundaries within the map world are shifting and contested.

The context of our study vividly reflects these shifting realities. The range of incumbents in contemporary cartography who appear in *Map Worlds* is both impressive and troublesome. On one hand, they represent at least twenty-five allied occupations in cartography, displaying a vast array of specializations. On the other hand, such a long list of occupations is problematic when one asks, "Who is a cartographer?"[2] The book contains stories of individuals who once practised applied, traditional cartography who now have deliberately chosen to work with GIS. We also include a software engineer whose purpose was to develop computer programs, but who now invests more time, energy, and resources in understanding the aesthetics of maps and is drifting closer to cartography. Still others have moved completely away from cartography, but have left their mark on the field. These shifting occupational directions are not uncommon in the map world. Even if someone has stayed faithful to one's original "calling," he or she is still immersed in the map world and should therefore be part of our study.

This chapter demonstrates that the membrane that constitutes the map world varies in thickness: in some areas, it is quite cohesive; in others, less so. The overlaps among affiliated fields might be overt or covert. One needs to acknowledge that complexity and admit that it is no easy matter to identify who is or is not a cartographer. Throughout the study, two questions have guided my decisions about who should be considered a cartographer. In interview situations, I directly asked, "Do you see yourself as a cartographer?" or "Whom do you consider a cartographer?" Both questions ought to answer the question of who is a cartographer—at least for the bulk of incumbents referred to in *Map Worlds*.

At the heart of the map world, the practitioners of cartography seem to agree, as Denis Cosgrove (1999: 1) says, that mapping involves "acts of visualizing, conceptualizing, recording, representing, and creating spaces graphically." Even if one does not wholly depend on the formal interactions among all of these fields and incumbents, one can still imagine how a mapmaker will benefit from seeing what a map archivist has preserved about a given area, both to fill in the gaps and to be creative about how to incorporate new data need into a revised map. The collection of a map librarian might inspire someone else in the map world to create maps with particular themes or cartouches. Academic cartographers abandon their desks and immerse themselves in finding out about how colour-blind people might respond to particular maps. A practising map-maker might focus all her attention on the teaching of map reading to children and be the impetus for a worldwide competition for children's making of maps. Sometimes, the overlaps reside within one individual, or among several individuals who inspire each other. When a map librarian converts data from Darwin's diaries into data sets and creates a dynamic map showing Darwin's travel around the world, should we regard her as a cartographer?

Running ahead of my data in *Map Worlds,* I wish to offer the considered opinions of some interview participants in my research. From the perspective of some, graphic spatiality stands at the core of cartographic work, according to certain conventional or traditional conceptions of cartography. Others offer shades of varying opinions about what constitutes cartography and how their definitions extend beyond the traditional ones. It is still not easy to draw lines among these definitions, for they are not mutually exclusive.

One set of conventional (or traditional) definitions evokes a heterogeneity of meanings, which the interviewees ascribe to the work they do, involving universalistic, specific, landscape-related, and general definitions. It is clear, however, that even within these definitions, there is no widely accepted agreement that, as it turns out, ranges from very wide to very narrow definitions. In a posting on the maphist listserv (maphist@geo.uu.nl, 14 April 2008), Judith Tyner makes the point that one could get into quite a few arguments about what is meant by "cartographer." Should one consider a woman who embroidered a map on silk in the third century A.D. (to make the map more permanent) a cartographer? Should one consider globes embroidered by schoolgirls in Pennsylvania to be made by cartographers? Or are they simply engaged in school projects?

FROM THE PERSPECTIVE OF INTERVIEW PARTICIPANTS

As mentioned earlier, one of the means available to find out who is a cartographer is to pose the question directly to the thirty-eight interview participants in the research.

"Kristina" (2),[3] for example, is vitally interested in tackling social problems with her maps and believes that as a generalist, she should not only know "the field of cartography and geography but also have a big interest in other fields, such as history, biology, environment, literature, and so on." This *universalistic* definition seems particularly appropriate for pre-GIS cartographers who are steeped in wider ways of knowing the space they document on maps.

On a specific, limited note, a number of interviewees confined their definition to the actual process of *making maps*. "Gabrijela," an atlas cartographer, defines a cartographer as someone "who takes an active part of making maps" (Gabrijela, email to author, 2 April 2001), not unlike "Shpela," who describes a cartographer as someone "who also knows all criteria or procedures of map production, generalization, and light and colour representation" (7). Similar definitions come from atlas makers like "Natalie" (6) and "Zia" (5), and a "tabletop"[4] cartographer such as "Lynn" (7).

For others, a cartographer takes her cue from the *landscape*, the surface, and the use of her imagination in converting the landscape into a graphical representation—a "projection of the imagination" (Zia: 3). She added: "I see it as someone who can get ... the most fascinating representation of the surface, with several dimensions that they could be the sounds ... [*digital?*] ... yes ... the senses ... [*yes*] ... all the human dimensions that we can put in a place ... [*yes*] ... that a map could give us those sensations" (Zia: 3).

"Helle," who works in a cartographic firm, defines a cartographer as someone who has the best way of "describing the landscape" (Helle: 2). "Ramona," who teaches cartographic literacy to children, emphasizes that a cartographer organizes "space through maps" (Ramona: 2), while "Sawa," of the Indian subcontinent, specifically denotes the importance of the scale and "lots of generalization" (Sawa: 5) in defining a cartographer.

On a more general level, a cartographer could be anyone who is associated with maps, whether as someone who actually "makes" maps, someone without any cartographic training (for example, someone who uses mapping software), as an academic cartographer, as a map collector, as a map historian, or as a GIS specialist (Loukie: 2–3; Helena: 3). In several countries, such as the Netherlands, Britain, and Denmark, it was not uncommon for map librarians to have been under the organized aeges of national cartographic societies (Wallis, 1979b: 108, 109). Even so, every occupation within cartography underscores an ethos about the production and maintenance of particular forms of knowledge that target regions, places, or populations for particular purposes, spatializing people's lives according to these transformative efforts: that is, making people's relations to particular spaces more relevant ideologically and materially than their relations to other spaces (Ilcan, 2002). The question that comes to mind is this: what is the relationship between map-making and the production of knowledge? How do these articulate with each other and through what means, processes, and configurations? How does this relation underscore the ethos of cartography? How are such relations predicated on technological innovation?

In contrast to the (above) conventional or traditional definitions, the newer definitions echo technological advances. It is not the imaginative use of light, colour, landscape interpretation, or other fields that result in maps that "give us those sensations" (Zia: 3); rather, the focus is on the process of translating "data sets" into "visualization." A human agency has interpreted the surface features into data sets that are, in turn, interpreted into visualization. This form of visualization is two steps removed from spatiality, which allows the "data" (and hence the represented space) to become objectified and, as a result, less subjective or personal. Finding one's way through data sets becomes an art, as opposed to finding one's way through space.

"Els," for example, defines a cartographer as someone who "visualizes geographical data and translates these into information" (Els, 2–3). "Visualization" and "spatial information management" are cornerstone concepts, the hinge that defines who is a cartographer (e.g., Myra: 2; Leslie: 1; Ettie, email to author, 1 May 2001). "Interfaces," "procedures," "management of data," and "software" are nodal concepts in GIS-based definitions.

With both traditional and GIS-based definitions holding their own, contemporary cartography has evolved to a point where there is considerable room for a variety of cartographic practitioners. Cartography has been invested with a dynamism that can embrace its various "traditions," some ancient, some ultra-modern. Regardless of the tradition one adheres to, there runs a strong element that is shared among these practitioners: namely, that maps should be designed with the user in mind, hopefully and possibly incorporating aesthetic qualities. Technological innovations and transformations account for many of the changes in map worlds.

TECHNOLOGICAL TRANSFORMATIONS

For the purpose of illustration, this section demonstrates how technological innovation has, in the past, transformed the map world. The availability of technology, the tools to produce maps, and technological procedures associated with printing are some of the basic change agents in the map world that invariably influenced the position or role women played in the map world.

CHAPTER 2

AVAILABILITY OF TECHNOLOGY

What animates the shape of the map world can be something as basic as whether the map is drawn by hand, machine, or software program and geodata. The transitions from phase to phase, over the past seven hundred years, were either very rapid or extremely slow. As if the interplay of all of these technological innovations is not complex enough in shaping the map world, one also notes distinctive cultural preferences and the availability of particular technologies that insinuate themselves into the map world, whether it is the Bavarian or Swabian preference for woodcutting, the Italian insistence on copper engraving (Woodward, 1975: 40), or the Belgian love for photolithography (Ristow, 1975: 103) at another time that multiplies or diminishes interconnecting links. Lithography took twelve years after its invention—from 1796 to 1808—to take hold in cartography (Ristow, 1975: 79); copper engraving lasted from the sixteenth century well into the nineteenth century (Robinson 1975: 21–22; Koeman, 1975: 149–150), with remnants still being practised until the 1940s.

There are, however, not many accounts about technological processes that could offer us an intimate glimpse of past map worlds. The few we have were written much after the technological process occurred in map-making. The first account of woodcutting came out in 1766, more than three centuries after the fact, as Robinson acknowledges (Robinson, 1975: 2). Woodcut maps seem to have been the only available style favoured north of the Alps, especially in Germany. The maps could be executed only in lines and dots. The maps were drafted on oil paper that could be transposed, like a transparency, on a wood block. The actual cutting of the block would occur after such transposition. Copper plates soon replaced wood blocks, and within a relatively short time, much detailed knowledge about the woodblock process was lost (Grenacher, 1970). Quoting Twyman, a contemporary authority on lithographic processes, Ristow notes that "we know very little about the earliest lithographic presses and next to nothing about their actual appearance" (Ristow, 1975: 79), although David Woodward (1975) is able to rely on numerous books from the sixteenth to the nineteenth centuries carrying illustrations of people at work in various related trades. One does, thus, get a reasonable sense of what that work entailed and the division of labour that it required, leaving space, perhaps, for the involvement of women.

TOOLS TO PRODUCE MAPS

There is an astonishing array of material tools that are called into play when making maps—each one involving makers, buyers, sellers, intermediaries, craftspeople, conventions, inventions, and habits, creating a social world unto itself. As David Woodward (1977) informs us, the printing of maps requires "special attention by the printer" that involves "format, size, number of impressions, level of precision, degree of color, line and consistency in tone, ease of revision, reproducibility of fine detail, the frequent need for color, and the need to combine lettering and line work ... [these] characteristics of the map have always posed the unique combination of problems for the printer" (Woodward, 1977: 4).

We learn, for example, that copper engraving in the seventeenth century entailed at least sixty-three different artifacts and processes.[5] Creating photographic plates entailed twenty-five processes and items, with some processes ranging in use from diamonds to distortion-free lenses, with each process or item governed by rules and conventions.[6] As another example of the complex processes that define the map world, one could look at a time when acetate sheets were used (in the 1950s and 1960s) and letters had to be cut from a plastic sheet. A dizzying number of items and processes were involved even before the first letter could be pressed on a map design![7] What interests us here is not so much the material artifacts and processes required for map-making, although that can be a subject of its own study, but the many interactions—the social organization of the map world that interlaces issues of gender in various means, at different levels. Elizabeth Harris informs us that the success of a cartographic effort was due to the "superb organization and workmanship" (Harris, 1975: 113) rather than to any particular technical process.

TECHNOLOGICAL PROCEDURES ASSOCIATED WITH PRINTING

The nature of the printing enterprise added another layer or set of relationships to the map world, such as the use of wood, copper, stone, film negatives, laser printers, or digitalized output, for example. These relationships were reflected in labour-force participation, the role of women, unions, attempts to increase (or curtail) the number of workers, training, and owner-management relations. The introduction of film negatives diminished the expertise of copperplate engravers (which hitherto entailed the participation of women), resulting in shifts in the map world. The nineteenth century, for example, was noted for its "very considerable confusion" in the relationship between map-maker and map printer (Robinson, 1975: 15), because, within a space of 150 years, there was photography, wax engraving or cerotyping, electrotyping, line engraving, wood engraving, and half-tone screen, which, "like a musical medley, would come forward and flourish for a while, only to fade out or blend into yet another" (Robinson, 1975: 15).

We learn that the occupational links between cartographer, printer, medieval guilds, and so on were intricate and delicate at times (Robinson, 1975: 18). Within fifty years, printing moved from hand-cut blocks, with two hundred impressions an hour, to fast steam presses and automatic inking, producing thousands of prints an hour (Harris, 1975: 113).

For over fifteen years (since 1997) I have been exploring the vast technological changes in cartography and wanted to understand their relationship to the emergence of the active participation of contemporary women in cartography. Have these technological shifts and participation of women altered the face of cartography? However, there is also a realization among some cartographers since the 1970s that it is not only technology that stands behind changes in the map world. The infusion of change comes from an entirely different quarter: namely, the realization by cartographers that maps have a purpose cultivated at the behest of agendas, intentions, and biases of cartographers themselves.

THE PURPOSES OF MAPS

Maps have complexes purposes. Commonly, one defines the purpose of maps in terms of their primary users, such as navigators, the state, the military, corporations, or the public, to name a few of the more obvious ones. The purposes of maps also guide their creators, however. In this instance, both overt and covert assumptions serve the intended purpose of maps, whether as a mechanism to build a political state, to create a beneficial interpretation of the past (or the future), or to move the public to accept or reject political trends. The belief that the goal of maps must reflect accurately the dimensions of space persisted generally well into the 1970s. The purposes intended for the users and those of the creators are often intermingled. Road maps that serve to underscore the presence of brand-name gas stations also serve as a river guide for the canoe enthusiast. Map creators have at their disposal a large array of techniques (such as scale, projections, and symbols) to achieve the purpose of their maps. However, users can easily bend or even suffocate those purposes. For example, I have in hand a flat-lying image of an icosahedron (twenty-sided) world map. Its creators, the United States National Geophysical Data Center, want the purchaser to cut out the twenty triangles with the tabs and glue them in such a way as to create the effect of a globe. Instead, I use this world map as a means for students to configure the world in a manner that follows their own imagination: rather than creating a standard globe, I cut out the twenty triangles and give them randomly to each student. As soon as students realize that the triangles can fit in different ways, they realize the world map is, after all, a social construction. The original intent of the world map (as visualized by the National Geophysical Data Center) is to affirm the prevailing view of the world; my purpose, however, is to affirm an entirely different view of the world.

Some early lone voices raised the presence of the "subjective" in maps, such as Armin Lobeck in his *Things Maps Don't Tell Us* (1956). Judith Tyner (1974, 1982) uses the term "persuasive cartography" to underscore the social-constructivist and subjective nature of maps. Persuasive cartography refers to the social and political functions of maps. One can also explore the assumptions and ideologies that enter maps even before they are made. Other map thinkers, such as Mark Monmonier in *Maps, Distortion and Meaning* (1977) and J. Brian Harley in his latest *The New Nature of Maps* (2001), and Arthur H. Robinson and Barbara Bartz Petchenik in their *Nature of Maps* (1976), also started to speak out about the subjective nature of maps. These authors were ahead of their time, and it was not until the early 1980s that the ideology of maps became a more persistent theme in cartographic scholarship.

The idea that maps can or should be deconstructed is now common. John Pickles, in a parallel fashion, speaks about the social implications of maps and map-making, such as in his work on propaganda maps (1992) and his *Ground Truth* (1995). While Harley, Monmonier, and Pickles enjoy some measure of acceptance in the cartographic community of scholars, Denis Wood's work on the "power of maps" (1992) seems to have picked up on the idea of "subjective" maps, but has turned it on its head to make "power"ful statements about one's community. There is considerable public appreciation and interest in his attempts to localize maps with elements that are relevant and important to local populations.

Maps "inevitably distort reality" (King, 1996: 18). Such awareness, however, as indicated above, is recent. Cartographers have, until the recent past, believed that there is distortion, but that such distortion can be kept to a minimum. The "shadow" ("noise") between reality and the reality drawn on maps cannot be completely eradicated, but it could be weakened. As Geoff King avers, cartographers usually do not think about the "possibility that the map may affect understanding of the real world itself, radically destabilizing such a scheme" (King, 1996: 20). Perhaps colonial cartographic schemes to rename or name places of "discovery" are the most egregious examples of social and cultural distortions from the perspective of the "conquered."

What we have, then, is a double challenge. Not only do suppositions enter into the map-making process, but maps themselves shape people's perceptions of the world. As noted earlier, a number of thinkers have tried to come to grips with the "new" reality of cartography. Ironically, the debate is not so new after all, for it was already in the mind of one of the most renowned cartographers in the world, Gerardus Mercator in Flanders, who was already speaking in 1552 of his "new geography." The new geography would not simply place locations according to their distances (as one would make note of limbs on a dead corpse), but would enliven the "geographic corpse" by indicating the "mutual political relationships" of those locations (cited by Crane, 2002: 250). With the use of particular colours, whether or not particular coats of arms would appear on the fringes of maps, and with standardized symbols to indicate town size, persuasive cartography was vigorously born.

The immersion of cartography, especially Western cartography, into a scientific, rational, positivistic model led cartographers to consciously ignore the social or persuasive contexts of their maps. How close could they get to achieve both precise and accurate measurements of the space that would eventually appear as symbols on their maps? It seemed relatively easy to ignore the persuasive impulses of map-making in light of the increasing rational perspective of the world to which literature, inventions, notions of progress, exploration of new lands, and even art were then subject. If persuasive or subjective dimensions were thought of at all, they stayed on the margins of the map world.

This chapter has provided a complex answer to a simple question: "Who is a cartographer?" Given the vast amount of change in cartography, it makes no sense to provide a narrow definition. The concept of map world accommodates the idea that occupational incumbents of cartography take on many different forms, with shifting roles and tasks, in ways that can be unpredictable. While we acknowledge that one can trace the sources of some shifts to the political realm, demography, social trends, and family structure, we have been particularly insistent that technology is the leading motor of these shifts. Today, the general accessibility of technology to create maps, along with the realization that maps are indeed quite subjective in their conception and intent, has eliminated particular niches of cartographic activity traditionally occupied by women. No longer confined to niches, women can now engage in the map world on a wider scale. Will our data support the idea that this engagement is complete or still incomplete, full or partial, equal or unequal?

CHAPTER 3

The Thirteenth to Seventeenth Centuries

The period between the thirteenth and seventeenth centuries in the Western world coincided with huge transformations that deeply affected map worlds: the rise of the Church to a height of tremendous secular power and authority, the appearance of the Renaissance, the advancement of technology, world-encircling discoveries that brought cartography into its wake, and the early signposts of science and rational thought that would start challenging the hegemony of the Church itself. Initially, European maps had more symbolic and moral aims than practical ones, whereby the "T-O" maps served as "pictorial representations of Church dogma" (Tooley, 1978: 12).[1] The Renaissance loosened many cultural enterprises from the moorings of the Church and permitted map-makers to free themselves of the theological structures imposed by the "T-O" maps. It was a period when the slightest attempt to alter conventional map design (e.g., the use of non-Gothic script) might lead to prison, as it did for Mercator; at least one other cartographer was sent to his death (van den Hoonaard, 2010). Voyages of discovery stimulated awareness of areas hitherto unknown. Map ateliers were the new phenomenon that could create the demands of new technologies that further down the road allowed for the mass production of maps. The making of multiple exact copies of maps and their rapid dissemination become a reality.

CHAPTER 3

The engagement of women in the map world reflected these cultural shifts. The creation of the "T-O" maps in the medieval epoch were largely the outcome of work within monasteries and some cloisters; in the latter case, a number of women stepped into the map world to produce striking maps. These locations, hitherto infused with a theological spirit, eventually gave way to map ateliers that began to rely on the introduction of Ptolemy's *Geographia* and on new technology to produce maps on a larger scale (Thrower, 1996: 58). Run by households, the map ateliers became the cynosure of planning and organization, relying increasingly on women for the production, colouring, and selling of maps to fulfill the restless demand for ever-newer maps.

THE THIRTEENTH AND FOURTEENTH CENTURIES

Nancy G. Heller points out that most medieval women were married by fourteen years of age and were preoccupied with childbearing and childrearing, as well as "endless domestic chores" (Heller, 1991: 12). It comes as no surprise that because unmarried women (and especially women in cloisters) were free of these obligations, they "held a remarkable range of jobs, working as brewers and butchers, wool merchants and ironmongers" (Heller, 1991: 12). There are, however, only three claims of the involvement of women in map-making: Ende, Herrade Landberg, and the thirteenth-century Ebstorf map.

Ende, a tenth-century Spanish nun, made a world map, along with illuminating the manuscript of Beatus of Liebana (Heller, 1991: 11–13). Herrade of Hohenbourg (d. after 1196), the Abbess of Lorraine, authored *The Garden of Delights*, the first known encyclopedia written by a woman. The work contains many examples of symbolic representation of space. Her work indicates, as Hoogvliet suggests (1996), that women were not excluded from geography or cartography.

The Ebstorf map[2] overlays the known world on the body of Christ and was discovered quite by chance by the overseer nun (chanoiness) at the Ebstorf nunnery in the nineteenth century—around six centuries after its creation, possibly in 1239 or 1284.[3] The story of its discovery is nothing short of remarkable. In 1830, the chanoiness checked a storeroom of the former nunnery of Ebstorf, situated near Lunenburg and Brunswick, believing that she would recover religious artifacts. What she found was the original map in thirty parchment pieces. The map was brought over to Hanover where, in 1896, Konrad Miller made a facsimile of it, although heliotype copies had been made in 1888.[4] We are fortunate that he made a copy, for bombing

Figure 3.1 Herrade of Hohenbourg (1125–1195) became famous for authoring and illustrating the *Hortus Deliciarum* (*Garden of Delights*) (1159–1175), the first known encyclopedia by a woman. Source: http://figurationfeminine.blogspot.com/search/label/1125%20De%20Landsberg%20Herrade.

raids on Hanover in 1943 destroyed the original, which measured more than 12 square metres (see Figure 3.2).

The map is the largest known *mappaemundi*, a medieval term to refer to these "didactic and symbolic" maps that served "to present the faithful with moralized versions of Christian history from the Creation to the Last Judgment" (Harley and Woodward, 1987: 504). It conveys a deeply religious meaning, as the map is superimposed on the body of Christ. East, as was common in those days, is placed at the top of the map, with continents grouped around the Mediterranean; Jerusalem, in gold, is in the middle of the world. Although the sequence of towns is almost accurate, everything else seems "off." The map indicates some recently founded towns, such as Riga, founded in 1201. The reader is treated to a wealth of historical, biblical, ethnographic, and biological details. Alexander the Great appears several times, with conquests. Most of ethnographic description of peoples is on a small strip in the extreme south, with odd creatures who eat through a

CHAPTER 3

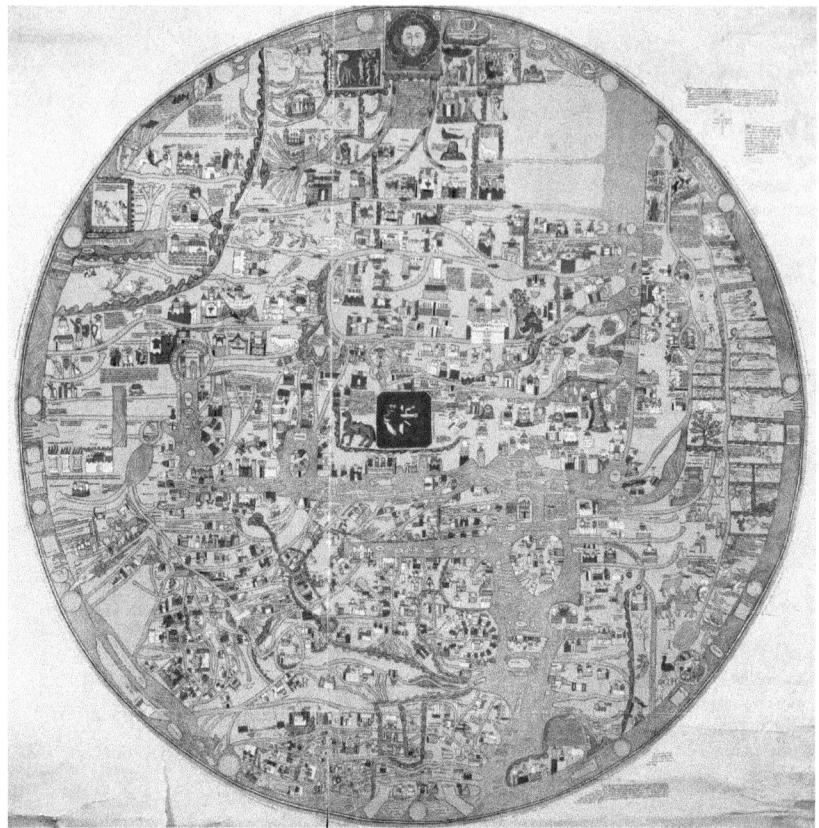

Figure 3.2 Konrad Miller's facsimile of the Ebstorf map, produced in the nunnery of that name in Germany, ca. 1239. *Source*: Hoogvliet (1996).

straw and people "who protect themselves against the burning sun by covering their head with lips." A bull leaves "burning excrement" after its rear end (Hoogvliet, 1996: 18).

There is no agreement among scholars that women created the Ebstorf map. According to Margriet Hoogvliet, "most scholars were inventing all kinds of odd arguments to avoid the conclusion that the nuns of Ebstorf could have made the map" (Hoogvliet, 1996 and 1998, personal communication). Walter Rosien (1952) found it "hard to believe that a woman's hand could have made the map." Geoff King, another scholar, holds the established position that the Ebstorf map was drawn by Gervase of Tilbury (1140–1220), an English priest, although the Ebstorf map had complexities that are absent in the known works by Gervase (1996: 32). Unlike in many other cloisters, these nuns came from nobility, were allowed to possess things, and were allowed to travel. The area of the cloister experienced an economic revival between 1290s and 1310s, with nuns involved in rebuilding and decorating Gothic churches, producing enormous carpets. The map fits well within this period of heightened devotional and artistic activity. Aside from this particular example, there is other evidence that women may have been involved in the map world.

THE FIFTEENTH CENTURY

The invention of movable type printing occasioned by Johannes Gutenberg in 1439 produced a wave of supplementary effects and had an immense impact on the creation and production of maps. The increased tempo of explorations and geographical discoveries by Europeans created both the need for maps that recognized a new world and a thirst on behalf of the population to acquire the latest maps. The quick succession of two technological advances in the sixteenth century—the printing press and the copper plate—was destined to transform map-making (Bagrow, 1964: 91). The "golden age of cartography" was born.

My raison d'être of selecting cartographic activities in the Low Countries starting with the fifteenth century lies in the fact that the period not only corresponded to the beginning of the golden age of cartography, but also coincided with the most accurate and detailed records about the participation of women in cartography. Tooley (1978: 29) characterizes the Low Countries as having "produced in some respects the greatest map-makers of the world," and for "accuracy, ... magnificence of presentation and richness of decoration,[5] the Dutch maps of this period have never been surpassed." An account of the participation of women reveals the diverse nature of that participation.

THE SIXTEENTH AND SEVENTEENTH CENTURIES

The advanced production techniques, moreover, entailed work in the family workshops that transformed into map ateliers.[6] The ateliers brought together the many talents required to compose, engrave, and publish maps, although the heads of ateliers sometimes outsourced the publication of maps. These ateliers became the heart of the map world. The solitary copyist or solo creator of maps became a thing of the past. Map ateliers formed the hub of activity for women as colourists.[7] The owners of map ateliers also sought to be affiliated to other ateliers by pragmatic marriages. A third gendered element stands with the map ateliers: namely, the preponderance of widows who continued the business; widowers, however, were more likely to remarry, often for pragmatic considerations in light of the needs of the business. The day-to-day business of the map trade included women (Thrower, 1996: 124, 279, n. 45).

THE LIFESPAN OF CARTOGRAPHERS

Families in the Low Countries that made maps together lived long and prosperous lives in the sixteenth and seventeenth centuries.[8] The average life span of affluent men and women was thirty-five years and thirty years, respectively. If you were less affluent, cut off ten years. Somehow, the three main sets of extended families (comprising seventeen or so smaller families) who engaged in cartography (and related fields, such as publishing maps) lived well beyond those limits. The well-populated Hondius line of map-makers lived, on average, fifty-two years. The Ortelius families stretched their life span to fifty-seven years, on average. The Blaeu line managed to live even longer, to sixty-nine years—twice the average of other Low Landers. Was it the ink that led them to live so long? The chemicals from the copper plates? A sedentary life? The truth might be closer than we think: wealth. As today, the best predictor of good health (and a long life) is high income. This was no less true in the sixteenth and seventeenth centuries, when cartography became a fashionable and highly lucrative industry as new areas of the globe opened up to discovery and exploration.

The accretion of wealth, however, was (and still is) contingent on families holding together. Nothing will potentially ruin wealth more than divorce, especially for women and children (except that the courts in the Low Lands in those times were favourably disposed to deserted women). If some marriages were unhappy, they were happy making maps and having some children to boot. The accumulation of wealth in the hands of a relatively few people was made possible only by marriage within the cartographic "tribe." One cannot but notice that the cartographic ateliers, related businesses, and stores were, in fact, in the hands of three extended families, sustained by a calculated system of intra-marriage. With few exceptions (such as Anna Bertius, the widow of Josse de Hondt/Hondius Jr., who sold some of her husband's plates to Willem Janszoon Blaeu), it was unusual for members in each set of families to cross the line into another family's "territory." The family "map" was well demarcated, especially between the Hondius and the Ortelius lines. When it came to the wealth of individuals and families involved with making maps, owners resorted to encouraging pragmatic marriages to perpetuate, expand, or rescue a map-making business.

The following sections convey the breadth of these extended families and what each household brought to the equation.

THE HOUSES OF CARTOGRAPHY

Some twenty-two families in the Low Countries (the Netherlands and Belgium) exercised an enormous influence on classical cartography.[9] A chronology of the golden age of cartography (in the sixteenth and seventeenth centuries) reveals that in 1575, eight people significantly controlled the cartographic ateliers, but there was important collaboration among the cartographers. Ortelius, one of the major cartographers of the day, himself acknowledged the contributions of eighty-seven different cartographers to his work. When land surveys became *de rigueur* in 1666, the number of cartographers associated with the Houses dropped considerably.

The largest of these Houses was Hondius, which comprised at least five families.[10] The Ortelius House of map-makers was comprised of six smaller families and just over twelve individuals. The same size applied equally to the famed House of Blaeu. There were also a number of other families and minor Houses.

Women laboured in these map ateliers. Diderot offers a lithograph of a copperplate engraver's atelier (Gillispie, 1959b) (see Figure 3.3). Although there is not a map in sight, one might imagine that the map atelier was set up in a similar fashion.

The historical record of the major map ateliers in Europe leaves no doubt that women participated as colourists. The use of colour and painting was thus one of several avenues by which women participated in the map world. According to Margriet Hoogvliet, numerous women were border painters (Hoogvliet, personal communication, 25 August 1998) and, according to Norman Thrower (1996: 60), "hand coloring of black-and-white copperplate prints became an important activity in European cartographic establishments, with women apparently playing an important role."

However, a number of sources bear out the fact that most colourists and engravers, both women and men, remained obscure. There are hundreds of thousands of unsigned engraved plates produced over the centuries. According to Ritzlin (email to author, 12 February 2011), people regarded this kind of work as unoriginal: the engraver merely followed the drawing of the artist or cartographer and the colourist carefully followed (in theory, anyway) a master print. Martin Hardie (1906: viii, 145) also comments on the rarity of engravers and colourists getting credit lines; in the case of coloured lithographs, one fails to find the names of any engravers in a survey of

CHAPTER 3

Figure 3.3 The Engravers' Map Atelier. This picture shows a woman etching the design into the copperplate with a burin, a very sharp engraving tool of hardest steel. *Source*: *A Diderot Pictorial Encyclopedia of Trades and Industry*, Vol. 2., New York: Dover Publications (1959b), Plate 380.

books with colour plates published in England from the time printing was introduced to that country. On the subject of a colourist's receiving credit, Mary Ritzlin mentions that "almost the only time I've seen a colourist credited on an antique print was for high-end nineteenth-century French fashion plates. I'm afraid for every Anna Beeck [publisher and colourist] or Elizabeth Haussard there are probably scores of unknown women who earned a living by means of their paintbrush or burin. And the total would probably be in the thousands if non-cartographic work were considered" (email to author, 12 February 2011).

John and Katherine Ebert (1974) reinforce the finding that colourists were held in low opinion. It was the "unsuccessful or student artists" who were paid a penny a print in the "production line" method of colouring. Hardie (1906: 90) relates an anecdote about author and antiquary F.W. Fairholt, who had claimed that he "was glad to earn ten shillings a week at the same mechanical task" in 1836. In a similar vein (Hudson and Ritzlin, 2000a: 4–5), colourists in London received less than a penny per map.

HOUSE OF HONDIUS

No doubt the largest extended family belonged to the Hondius (de Hondt) line, which comprised at least six large families and one loosely affiliated family (the Visschers), involving nearly forty individuals who were born between 1563 and 1678. If one were to include Gerardus Mercator (Kremer), whose legacy of maps was left in the hands of the Hondius family, and include Mercator's year of birth (1512), we would come to a span of 166 years. The family had two branches: one in The Hague, the other in Amsterdam (Keuning, 1948: 63). The Hondius House, which formed the centrepiece of the early Flemish cartographers, took over the Mercator atlas. The prolific businesses produced some of the most remarkable maps, better than even the Blaeu and Janssoon Houses, whose maps had achieved considerable fame in the sixteenth century.

Roger Stewart's "In the Family Way—Mapmakers" (2011) offers a six-page précis of all the intra- and intermarriages of the House of Hondius. Each new marriage partner brought with him or her knowledge about engraving, printing, colouring, geography, binding, publishing, or trading, ensuring the continuation of a cartographic dynasty. These intermarriages resulted in a vast concentration of cartographic acumen and business skills. Collectively, they included engravers, publishers, artists, text composers, a librarian, geographers, mathematicians, historians, a theologian, and booksellers—a heady mix of talent that ensured that the dynasty of cartographers would remain dynamic. The Hondius (de Hondt) line of cartographers intermarried with the Montanus (van den Berghe), Kaerius (van den Keere), Bertius (Bert), Goos, van Waesbergen, and Janssonius (Janszoon) families (in at least one case involving the marriage of two affinal cousins). This cluster perpetuated the work of these cartographers for at least another generation.

After the elder Jodocus (Josse) Hondius I's death (in 1612),[11] his widow, Coletta van den Keere (1565–1629)—and later, two of his sons, Jodocus II Hondius I (1595–1629) and Henricus (1597–1651)[12]—took over the business in 1612 (Duncker and Weiss, 1983: 46; Thrower, 1972: 279, n. 45).[13] Hondius II married the twenty-five-year-old Anna Staffmaecker (Keuning, 1948: 63). Later, as a widow, Anna married Petrus Kaerius (Pieter van den Keere) (1571–1646) in 1599. Anna's brother was Petrus Bertius (Peter Bert) (1565–1629), child of Johannes Bertius (Jan Jans Bert, 1615–1662), a widower who married Elisabeth, the widow of Henricus Kaerius (Hendrik van den Keere) and Petrus Kaerius (Pieter van den Keere) (1571–1646). Petrus Bertius (1565–1629), a professor of mathematics and a librarian at Leiden University who had fled his home in Flanders, was thus related by marriage to both Jodocus Hondius and Petrus Kaerius. However, Coletta continued publishing her husband's maps until 1623 at *den Wackeren Hondt* ("the Watchful Hound"), which she kept in full operation, including the publishing of an edition of the Ptolemy map in 1618, as well as several editions of Mercator's

Atlas (Ritzlin, 1986: 2712); according to van der Krogt (1996: 63), she held the monopoly on the *Atlas* in the world of atlases. In 1585, Jodocus's sister, Jacomina, who was an engraver and calligrapher, married Petrus Montanus (Pieter van de Berghe, 1560–1628) (Duncker and Weiss, 1983: 45).

In the second generation, Elisabeth Hondius (d. 1627), the daughter of Coletta (née Kaerius (van der Keere) and Jodocus Hondius, married Johannes Janssonius (Jan Janszoon, Jr) (1588–1664), who took over ownership of the atelier (Shirley, 1996: 17; WorldView Antique Maps, 2002), but when Janssonius died, his property passed over to their two daughters. Daughter Elizabeth married Jan van Waesberger (1651–1681), while the other daughter married Elizée Weyerstraet (Bagrow, 1966: 181). Their son, Jodocus Janszoon (1613/14–1665), married Francina van Offenberg in 1642.

The son of Pieter Goos (1615–1675), Abraham Goos, a noted engraver in Amsterdam who prepared plates for many maps, was already related to the Hondius family, by whom he was employed (WorldView Antique Maps, 2002). Abraham's son Pieter Goos II married Margareta van den Keere, but after Margareta became a widow, she sold the rights to the copper plates to Jacobus Robijn (ca. 1649–1707).

The Kaerius family in Ghent, consisting of the four siblings Petrus, Elisabeth, Margareta, and Coletta van der Keere, constituted an important publishing and engraving group (Brown, 1949: 165),[14] with affinal and agnatic ties to the Goos, Hondius, Bertus, and Janssonius families. In many respects, their marriages to both major and minor cartographic families made them the core of the business.

As mentioned in the introduction to this section, there was at least one other family that was connected to the House of Hondius, although not through marriage. Janszoon Visscher (1587–1637/1660?), a versatile engraver (Bagrow 1966: 181) and a student of Jodocus I Hondius, the patriarch of the House, managed to acquire the copper plates from the Kaerius family, whose matriarch, Coletta, Hondius I had married. Eventually, this acquisition greatly aided the fortunes of the Blaeu group (see below). Janszoon passed the plates on to his son Nicolaes Visscher I (1618–1679), who in turn gave them to his son Nicolaes Visscher II (1649–1692/1702?). Visscher II's wife, Elizabeth Verseyl (d. 1726), took over the business in Amsterdam (which specialized in war maps) after her husband's death; she had been married to him since 1680, leaving no children. She published several editions of *Atlas Minor*, *Atlas Maior*, and an atlas related to the War of Succession. Individual maps list her name, but it is not listed consistently on the title pages of atlases (Duncker and Weiss, 1983: 104; Koeman, 1967: vol.

3: 152–153; Ritzlin, 1986: 2712). After Elizabeth Verseyl's death in 1726, the business passed into the hands of Pieter Schenk, who was married into the Valck family, one of three families that bought the Blaeu copper plates at an auction at the Blaeu printing presses in 1672.[15]

HOUSE OF ORTELIUS

The Ortelius House constituted the second major family force in cartography. It was comprised of six smaller families and about a dozen individuals, whose dates of birth ranged from 1527 to possibly 1586—a range of fifty-nine years. It became the major competitor to the Hondius clan. Their intermarriages involved the Cool (pronounced "coal" in English) family and the Plantin (Plantijn)-Moretus (Morentdort) group, who printed many of the Ortelius maps (Cole, 1996: 91).

Abraham Ortelius (Ortel) (1527/8–1598) had two sisters, Anna and Elizabeth (ca. 1570–1594), both of whom received commissions to colour copies of *Theatrum Orbis Terrarum* (Hudson and Ritzlin, 2000b: 20). Elizabeth married Jacobus Colius I (Jacob Cool Sr.) (d. 1591). Their son, Jacobus Colius II (1563–1628), became the heart of the Ortelius House, especially after his maternal uncle, Abraham Ortelius (Ortel) (1527–1598), bestowed the name Ortelianus on Jacobus Colius II.

While the creation of maps involves the participation of a large number of people, the names of colourists did not appear on maps unless the colourist was particularly accomplished and was registered with the local St. Luke's Guild (Hudson and Ritzlin, 2000a: 4). As mentionned before, the colourists generally were not held in high esteem. The purchasers of these maps might ask the map atelier to colour the map, or the family of the purchaser might want to do it themselves, much along the lines of the paint-by-number routine that many readers still remember. It was not uncommon for women to do other work than colouring. When Abraham Ortelius, who was a member of the guild of St Luke as an illuminator, wanted to increase his earnings, he began, according to Brown (1949: 160), buying maps on the side. His aforementioned sisters, the colourists Anna and Elizabeth, "mounted them on linen," working as *kaartafzetters* (Potter, 1999: 40), whereupon he sold them at fairs, after having them painted (Ritzlin, 1986: 2711).[16]

This publishing house employed a large number of colourists around 1580: namely, Lynken (Lisken) née Seegers (also Seghers, Zegers), who was the wife Abraham Verhoeven (Hudson and Ritzlin, 2000b: 20, 22). The Verhoevens were employees in the Plantijn-Moretus printing press in the 1580s. Lynken and Abraham's daughter Mynken (ca. 1540–1594) also

CHAPTER 3

contributed to the printing of the maps. After Mynken's marriage to Liefrik, we also find Liefrick working there, as well as their daughter Marguerite and their son-in-law being active as colourists (Hudson and Ritzlin, 2000b: 15, 17).

The Plantijn-Moretus publishing house, which was so intimately tied to the Ortelius House, was itself the result of intermarriage. The daughter of Christoffel Plantijn (1555–1589), Martina (Maria) (1550–1616), married Johannes Moretus (Mortentdort) (1543–1610) in 1570. The cartographic tradition continued through their sons, Balthasar (1574–1641) and Jan II Moretus (1576–1618). Another household consisted of Pashina Jodocus, who married Gerard de Jode (1509–1591), who was affiliated with the Ortelius House.

All of these marriages added considerable talents and resources to the Ortelius House, including publishers, a mathematician, and colourists, and ensured that some family members received commissions to publish the famed *Theatrum Orbis*. In the end, the Vrients family (Jan Baptist Vrients, 1522–1612, and his wife, name unknown) became the successors to Abraham Ortelius and heirs to his maps.

HOUSE OF BLAEU

Comprised of six families, including just over a dozen individuals, the Blaeu House of cartographers combined a vast array of some of the best talents creating the most beautiful maps. However, intermarriage was not as frequent, which consequently interrupted the forward movement of the House. The dates of birth of individuals ranged from 1571 to 1725, covering 154 years. Their combined talents included astronomy, instrument making, and mezzotint engraving; publishing (of, for example, globes); selling prints, maps, atlases and architectural drawings; dealing as art dealers; and selling books. The Blaeus relied on business rather than family arrangements to pursue their goals. Blaeu preferred associating with nobility and merchants. In 1633, he was appointed map-maker of the Dutch East India Company. The success of his maps was partly due to his talents of promoting himself—and his maps.

The founder was the famed Guillermo Blaeu (Willem Janszoon Blaeu)[17] (1571–1638). His two sons, Joan (1596–1673) and Cornelis (d. 1642), carried the cartographic tradition forward. After the death of Joan Blaeu in 1673, the firm's surviving stocks of plates and maps were sold—a huge fire in 1672 had gutted the presses—some to Frederick de Wit (1630–1706), Pieter Schenk (1655–1718), and Gerard Valck (1651/2–1726). These arrangements included the sister of Gerard Valck, who married Pieter Schenk in 1687.

However, the strongest family link was probably through Frederick de Wit, who managed to convince Pieter Mortier (1661–1711) and his brother David Mortier (ca 1714), a prominent bookseller, to extend the life of the Blaeu maps through publishing them. De Wit's maps, according to Bagrow (1966: 183), were noted for their "fine workmanship, beauty, and accuracy." Cornelis Mortier, the son of Pieter Mortier, who was a bookseller in Amsterdam and founder of the firm, married Amelia 's-Gravensande (1666–1719), the sister of Johannes Covens, in 1685; this put the firm on a more secure footing. The conjoined firm of Mortier and Covens was such an important success that it lasted until the middle of the nineteenth century, producing "thousands of maps and dozens of atlases" (WorldView Antique Maps, 2002; van Egmond, 2002: 68).

It appears that Amelia Mortier may have issued the *Atlas* with the help of Pieter van der Aa (1659–1733). Intermarriage between these latest groups resulted in a publishing company that lasted for nearly two hundred years. Interestingly, some of the Hondius maps and plates were acquired by Pieter Schenk, who married into the Valck family, which also acquired a portion of the Blaeu maps. The flow of material from a major House (Hondius) to another one (Blaeu) was unusual, and spelled the end to the Blaeu House.

OTHER CARTOGRAPHIC HOUSES AND FAMILIES

Eight other cartographic families moved around the edges of the three significant Houses described above. The Ottens family consisted of Joachim (1663–1719), Josua (1704–1765), and Reiner (1698–1750)—three brothers who produced maps in the seventeenth and eighteenth centuries. The most active period of map publishing was concentrated in the years between 1720 and 1750, with an enormous collection of maps, some as large as fifteen volumes. This family had acquired maps from the Danckerts, whose founding member was Cornelis Danckerts (1603–1656) and later involved his son Carel Allard (1648–ca. 1709). The first two brothers benefited from Josua Otten's widow (from 1765 until 1780), who published a map of St. Eustatius in 1775. Of the remaining five families, the records mention only one woman—Anna van Westerstee Beeck (ca. 1697–1717)—whose husband, Barents Beeck, deserted her.

In France, social developments reflected, to a large extent, the gradual weakening of women's material dependence on men (Möbius, 1982: 8). Along with its discoveries in science, people saw the period as one emerging from an era of darkness and ignorance. In this light, more opportunities became available to women to emancipate themselves. Their

CHAPTER 3

involvement in map-making and engraving, in particular, speaks to these new opportunities.

DIVORCED, WIDOWED, AND DESERTED WOMEN

Despite the prevalence of pragmatic marital liaisons, divorce was not an uncommon occurrence. However, divorced women were able to hold on to their legitimate entitlements. For example, the courts awarded Anna van Westerstee Beeck (ca. 1697–1717) the business after the divorce from her husband Barents (Hudson and Ritzlin, 2000b: 9).

On a wider European commercial scale, widows often continued the map ateliers or businesses founded by their husbands, especially in the seventeenth century. Alice Hudson discovered, in 1989, that of 150 women map-makers, fifty-five had taken over their husbands' businesses (Hudson, 1989: 4). By 2000, Hudson (along with Ritzlin) came up with nearly 108 names of women who were clearly identified as widows in a field of three hundred women map-makers, more than one-third (Hudson and Ritzlin, 2000b). There are probably more widows than the ones indicated as such. The unambiguous assertion that the women were widows seems to suggest no reduced status of the woman.[18] In Holland and Flanders alone, the historical record leaves us with the mention of eight widows who, through their business efforts, were able to influence the activities and fate of at least ten cartography houses, some more prominent than others.[19]

Widows were sometimes courted by potential successors in the map-making business. Thanks to Hudson and Ritzlin (2000b), we have the names of at least twenty-seven women map publishers in the Low Countries. The following illustrate some of the salient ways in which they contributed to the family map trade. The three van Keulen widows—the nameless widow of Johannes (1654–1711), Ludwina Konst (widow of Gerard, 1678–1727), and Catherine Buys (widow of Jodocus II)—played a major part in continuing the business of their respective husbands. The nameless widow (ca. 1640–1648) of Joannis Cnobbbari was a publisher in Flanders and produced one of the evocative maps of the time, "Leo Belgicus" (Hudson and Ritzlin, 2000b: 12).

Johannes Bertius, who was first married to a woman whose name we do not have (the mother of Petrus Bertius and Anna Elisabeth Hondius), married Elisabeth, the widow of Henricus Kaerius. Margareta Kaerius, the widow of Abraham's son, Pieter Goos (1615–1675), a diamond cutter, sold the rights to copper plates to Jacobus Robijn (1649–1707). Coletta, the widow of patriarch Jodocus Hondius (1565–1627/9?) took over the business in 1612, until her death seventeen years later (Duncker and Weiss, 1983: 46; Thrower, 1972: 279, n. 45). Johannes Janssonius (1588–1664), who took over ownership of the atelier (Shirley, 1996: 17; WorldView Antique Maps, 2002) and who had been married to Elisabeth Hondius, married another woman after Elisabeth's death in 1627. Many other widows were engaged in the map trades, such as Albert Magnus's widow, who became a bookbinder and seller of maps, especially the Blaeu atlases; the widow of Jacob van Meurs; Catherine, the widow of Theodore Morentdort (1602–1667); and Amelia Mortier (1725–1763), widow of Pieter Mortier. Johanna Ottens, the widow of Josua Ottens (1704–1765), became active as a publisher right after her husband's death in 1765, as did Aaltje Fredericks van der Linden, the widow of Joachim Ottens (b. 1663) in 1719 (Hudson and Ritzlin, 2000b: 19, 20). The widow of Reiner Otten (1698–1750) continued to run her late husband's business (Bagrow, 1966: 184). The widow of Henrick van Peetersen also became publisher (1551). The widow of Joannis A. Someran produced maps of constellations in 1681.

Deserted women were also left to run the map business of their husbands. Anna Beeck "published and colored the maps she printed" in The Hague. Her husband, Barents, deserted her in 1693, after fifteen years of marriage and seven children. As noted above, she later divorced him, and local courts supported her running the family business. She provided the latest map reports of the War of Spanish Succession (1701–1713) and made some sixty maps and battle plans.

In the end, the best combination of enduring influence would involve endogamous marriage to create a sufficiently wide pool of talents that could serve the world of cartography in the most handsome terms.

While the thirteenth century saw as its business to copy the reintroduced and then well-known Ptolemy map, the sixteenth century witnessed a radical advance in creating and producing maps, so much so that it called for a complete reorganization of the production processes and subsequent gender relations in the map world. One such reorganization involved the transformation of family workshops into map ateliers. These ateliers represented the culmination of a tremendous evolution in the way maps were created and produced. The solitary copyist in nunneries or monasteries had to make way for the new maps, ushering in the golden age of cartography.

Women found numerous places in the map ateliers. Some were engaged as colourists, while others took an active role in producing maps and furthering the business interests of the ateliers. To maintain and increase their reservoir of cartographic talents, the cartographic Houses fostered pragmatic marriages. Many women who were widowed or deserted, and who did not seek pragmatic marriage liaisons, continued the business.

The latter part of the seventeenth century—invariably linked to the fundamental upheavals of the preceding century, declared the German observer Helga Möbius (1982: 8)—brought forth immense changes in cartography. The commercialization of cartography that had created such a demand for maps developed inside the map ateliers had become the distinguishing feature of map-making. The map world entailed a larger number of active participants, but the production and distribution of maps were initially in the hands of guilds. What pushed the advancement of cartography in particular toward the end of the seventeenth century was the rise of the State, requiring more advanced tools to measure the land mass and borders of national political entities, especially when embroiled in wars and conflict. Maps were becoming politicized, not just commercialized. The founding in 1666 of the Académie royale des sciences during the reign of Louis xiv led to accurate surveys of France involving triangulation and the charting of its coasts and harbours (Robinson, 1982: 19). No less significant, during the eighteenth and nineteenth centuries, as we shall see, was the impact of colonization on the general interest in maps, geography, and the creation of globes.

CHAPTER 4

The Eighteenth and Early Nineteenth Centuries (1666 to 1850)

By the seventeenth century, cartography had moved out of the ateliers. Increasingly, the State began to occupy a larger share of the map world. While it is true that the founding in 1666 of the Académie royale des sciences during the reign of Louis XIV was a forerunner of the zeitgeist of rational and scientific thought, its more immediate purpose was to assure the precise and accurate measurements of the borders of France. Other European countries began to emulate the mixed motivation of combining scientific advances with political aims. Maps were becoming national political tools, not just commercial products. No less significant during the eighteenth and early nineteenth centuries was the impact of colonization on the need for, and continuing interest in, maps, geography, and globes. Within two centuries (1666–1880), the purpose of maps had been radically changed to accommodate the political culture. This chapter explores the distinctive niches filled by women as part of the larger progression of society to encompass larger territories, nationally and internationally.

The era of topography and scientific cartography began with the establishment of scientific societies in the second half of the seventeenth century and continued well into the nineteenth century with the use of lithography by cartographers. This period roughly coincides with the Era of Enlightenment in Europe, which is said to have come to an end with the French

Revolution in 1789. However, for our purposes, while the map world was indeed shaped in this period by the Enlightenment and the "Age of Reason" (implying precise measurement), it ends somewhat later, in 1808, when a new reproduction technique, lithography, found its way into cartography. Invented in 1798 by Aloys Senefelder (1771–1834) in Munich, it resulted in less expensive ways of producing maps. On top of that, tonal gradation had become an achievable fact (Robinson, 1982: 57). One may wonder whether the reshaping of the map world (by politics and by technology) also reshaped the participation of women in cartography. There is a conspicuous lack of information about women in cartography, even in Robinson's otherwise very detailed 1982 history, *Early Thematic Mapping*.

With the arrival of the eighteenth century, the map world still saw the continuing rise of commercial interests, but the interests of the State prevailed more and more. For women, as Helga Möbius points out, two trends mark their changing status in society. Women were excluded from "intellectually demanding or skilled work, while at the same time, the number of women engaged in the unskilled labor force rose dramatically" (Möbius, 1982: 90). As Suzan Ilcan (2002) points out, women have always been involved in skilled work within households and household economies, but this kind of work has been undervalued and gone unrecognized because it was "unpaid" work "even though it made direct economic and social contributions" (Suzan Ilcan, email to author, 26 December 2002). As Stone (1977) avers, the extended family was being replaced by the nuclear family, effectively closing off the era of family firms in cartography.

THE MAP TRADES

Cartography had arrived as a science—the Parisian salons in women's homes from the latter part of the seventeenth to the early part of the nineteenth no doubt reflected this new-found enthusiasm for science, and spots opened up in the map world for women to render areas of service that were unimaginable even fifty years earlier, especially in the field of education. However, the commercial interest in promoting maps continued—there was still a public interest in them. Such a demand created the need to have handsome atlases and maps that were produced through colouring and design, although they were far less elaborate than before. A more powerful social influence was the rise of public education and knowledge, which grew not only out of the Enlightenment, but also out of the public fascination with new lands around the world.

The map trade, as we shall see in this chapter, entailed engravers, artists, colourists, map publishers, and sellers, but one normally does not consider the larger cultural and relational package attached to making maps. In addition to these activities, women participated as embroiderers, map sellers, teachers of maps, globe makers, and cartographers in France, Britain, the Low Countries, Norway, and North America. The fullest biographies of women cartographers belong to Virginia Farrar in England and Kirstine Colban of Norway.

COLOURISTS

Despite the promise of lithography, the printing of coloured maps continued to be very expensive. Each additional printed colour needed its own plate and a "separate trip through the printing process" (Robinson, 1982: 24). Moreover, it was difficult to position colours accurately on the plates. As a result, hand colouring of maps remained the best option. If anything, the demand for hand-coloured maps grew, safeguarding the role women had come to play in cartography.

Before 1850, lithographic techniques made it rare to see colour printing (Robinson, 1982: 193), but between 1850 and 1860, the colour printing process became more accessible and proved to be quite popular. By 1860, all obstacles to colour printing had been overcome, and the number of women who applied themselves in that area declined rapidly and eventually wholly disappeared.

EMBROIDERED MAPS

Embroidered maps straddled the eighteenth and nineteenth centuries, from 1770 to 1840 (Tyner, 2001: 37).[1] The interest in geography, spurred by discoveries and voyages, led to making geography lessons as interesting as possible. Toward the end of the eighteenth century, needlework was used to indicate "multiplication tables, family trees, 'almanacs'—and maps" (Tyner, 2001: 37). It was again John Spilburg (1739–1769) who, in addition to coming up with dissected maps (the forerunner of jigsaw puzzles), is credited for printing maps on fabric, including handkerchiefs.

Map publishers expanded their repertoire of things to sell and began producing maps on paper and fabric for needlework.[2] The London firms Laurie and Whittle, and Bowles and Carver, issued paper maps of the world and particular countries especially for use by young women.

Tyner observes (2001: 38) that embroidered maps were largely confined to English-speaking countries, and especially to England and the United States. She found only one example of such a map made by a French woman, Fanny le Gay of Rouen, in 1809. Only British publishers seemed interested in printing paper copies of such maps. A number of teachers in

CHAPTER 4

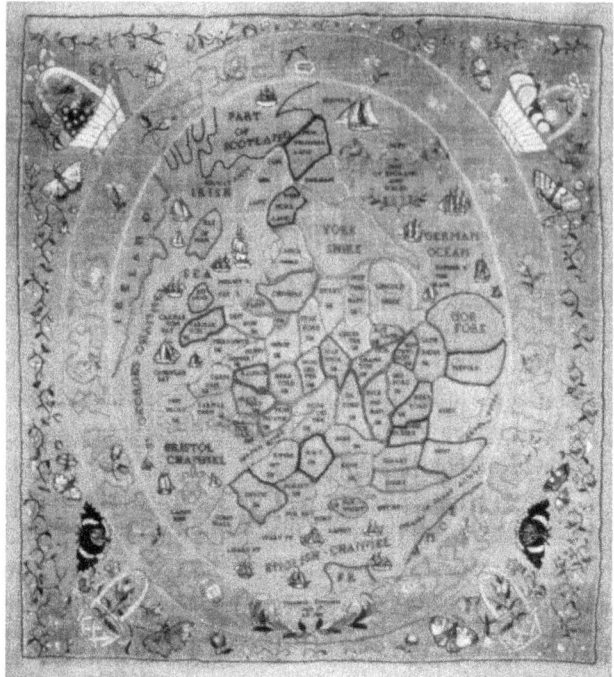

Figure 4.1 Elizabeth Ann Goldin (1829). *Source*: ART 440399.

the United States took up embroidering maps because they themselves were taught in British schools (see Figure 4.1 as a sample of such a map). The two Quaker schools that had a significant influence were the Mount School and the Ackworth Boarding School, both in York (Tyner, 2001: 40). The Mount School had ties with the Quaker (Oakwood) School in Poughkeepsie, New York, just two hours north of New York City. The School produced the largest number of embroidered maps in the United States.

The third place where maps were embroidered was also a Quaker school, this time in Maryland; they were based on Samuel Lewis's map of Maryland, dating back to between 1798 and 1800 (Tyner, 2001: 40–41). There are, however, not many examples coming from other schools, but Mrs. Rowson's Academy, a boarding school in Boston, produced a map, with ornate borders, as is typical of non-Quaker schools.

It was technology that accounted for the decline of embroidered maps: the sewing machine had been invented in 1829 by a French tailor (and the first lockstitch machine was devised around 1834 by an American).

THE EIGHTEENTH AND EARLY NINETEENTH CENTURIES

Decorative needlework as a requirement in home economics classes fell into disfavour (Tyner, 2001: 41). Today, embroidered maps are still being made, but by adults and not as part of any school curriculum. At the turn of the eighteenth and nineteenth centuries, however, "hundred of girls learned their geography by stitching maps on silks and linen" (Tyner, 2001: 41).[3]

ENGRAVERS

Technologically speaking, 1806 marked the adoption of lithographic printing, pushing the capacity of map publishers to exceed their press runs by several thousand copies and around seven times more cheaply than copperplate engraving (Blair, 1983: 2).[4] By 1818, within twenty years of its invention, lithography had reached the New World (Blair, 1983: 2).

Of all fields, engraving stood the closest to cartography. We learn from Hudson (1989: 2) that of the 150 women in map-making before the twentieth century, some twenty-one were engravers; most of these were eighteenth-century French women. Following her more recent work (Hudson and Ritzlin, 2000b) and studying the lists of engravers at the Berne City and University Library (Ryhiner Collection), we can narrow the number of women engravers in Europe in the eighteenth century to twenty-five, with twenty-one residing in France alone.

Jacqueline Panouse was an active engraver, employed by the great Nicholas deFer (1647–1720) (Ritzlin, 1986: 2711). A later family were the Jaillots, who re-engraved and reissued the Sanson maps in France in the early eighteenth century (Ross, 1996: 44). Hubert Alexis Jaillot (1632–1712), originally a sculptor, married the daughter of Bercy, a map colourist, and thereby came in contact with cartographic materials (Duncker and Weiss, 1983: 109). There was also a Jeanne Deny (1749–?), an engraver in Paris who was the sister of Martial Deny (1745–?), an active engraver (Ryhiner Collection). A collection of maps in the Berne City and University Library lists another fifteen or so women engravers in Paris in the eighteenth century.

While fewer in number than in France, the Low Lands also produced a number of notable women engravers of maps. Anna Catherine Brouwer, for example, engraved the 1170 plates for *De Nederlandsche Stad- en Dorpbeschryver [The Netherlands City and Town Guide]* (Hudson and Ritzlin, 2000b: 11). Somewhat later, around 1799–1807, we find Jeanne S. Maillart in Flanders engraving plates with her brother, Philippe Joseph Maillart, and Agnes Coentgen Schalk, a Dutch engraver, also of the eighteenth century.

Unless engravers were fine-art engravers, it was unusual for them to receive praise for their work. One compiler, Joseph Strutt, wrote condescendingly about them (Hudson and Ritzlin, 2000b: 5); if praise came their way, it was only after many years of service. Olga Herlin in Sweden is a good example. The topographic mapping of Sweden dates back to 1805. From the beginning, this was a purely military activity, and all the work was carried out by topographic officers. The maps were engraved on copper plates. After some decades, it became necessary to employ civil personnel for the engraving work, as it proved too difficult to recruit officers for this part of the work. But still only male personnel were used for fieldwork and engraving. No crack appears in this system until 1890. At that time, Olga Herlin was employed as the first female engraver, at the age of fifteen. She had already learned the work, since her father served as an engraver at the Survey Office. Her employment was of an occasional nature and she did not get a permanent appointment until 1927, some thirty-seven years after her initial employment. She stayed on in the office until 1940 and received a gold medal for her fifty years of "zealous and devoted service." She was a very skillful engraver and connoisseur, according to one veteran Swedish cartographer, and one could easily identify copper plates engraved by her. One reason she was able to continue her appointment at the Survey Office for such a long period was that she remained unmarried. At that time, women employed by governmental authorities were not allowed to keep their job if they married (Patrik Ottoson, email to author, 7 October 2002).

It is not easy to assess when the first attempts were made to engrave upon wood or metal in the United States. Those controlling the colonies forbade the establishment of printing presses for almost a century after the first settlement (Stauffer, 1907: xxi). There was, in fact, virtually no demand for printing or engraving until the wilderness had begun to be tamed, toward the end of the nineteenth century, and permanent communities and national parks (with new roadways) came into being (Louter, 2006). Interestingly, United States women associated with maps were cosmopolitan in their orientation, while the distinctive character of Canadian women linked with maps was associated with Aboriginal settings (the next chapter provides details about that Canadian side of cartography). The first evidence of an engraved work dates back to 1652, but the first engraver on a copper plate did so in 1702, while the first copperplate-engraved town plan or map (of Boston) dates as late as 1722 (Stauffer, 1907: xxi, xxiii). Eliza Coles (1776–1799) was, as far as we know, America's first map engraver. She may have learned her craft when she was just thirteen years old. Eliza was one of eleven children of Christopher and Anne Keough Coles. Christopher was

THE EIGHTEENTH AND EARLY NINETEENTH CENTURIES

an Irish engineer, involved with mapping and canal construction before migrating to the United States (Ristow, 1980; Ritzlin, 1986: 2711). Cornelius Tiebout was a young engraver who assisted Mr. Coles in his work; when he left the project early, thirteen-year-old Eliza took over the engraving work; she had watched Tiebout prepare plates for her father's surveys of roads in 1789. She contributed maps for *Survey of the Roads of the United States*, which was published that same year. Eliza contributed not only to Christopher Coles's *Geographical Ledger and Systemized Atlas* in 1794, but also to other projects of her father's. Yellow fever cut short Eliza's life at the age of twenty-three (Ritzlin, 1986: 2711).

In 1762, Mary Biddle (1709–1789) edited a map of Philadelphia by her father Nicholas Scull (Tyner, 1997: 47; Ritzlin, 1990a: 6; Hudson and Ritzlin, 2000b: 10). Biddle was also a publisher and map seller in Philadelphia.[5] The eldest child of Nicholas Scull and Abigail Heap, at the age of twenty-one she married William Biddle, a descendant of one of the original owners of New Jersey. After nineteen years of marriage, her husband (who had not much talent for business) made a number of poor financial investments, leaving his family nearly destitute; he died just months before the publication of his 1762 map of Philadelphia. It is not known how Mary Biddle managed to enter partnership with the man who co-edited the map with her (Matthew Clarkson), but it saved her and her family from further economic hardship.

The roster of women engravers in the United States was not plentiful. David Stauffer's list (1907) of seven hundred American engravers lists seven women. Even so, we do not know whether any of them engraved maps in particular—they are probably of the fine art variety.[6]

One writer on women's issues (in her *The Employments of Women: A Cyclopedia of Women's Work*) lists engraving as a possibility as women's work, with, however, few such opportunities outside the publishing world beyond Philadelphia and New York. This particular writer, Virginia Penny (1863), does speak about Europe and the fact that the "best map engraving in Paris is done by women. There were also women in London who were engaged with engraving maps" (Hudson and Ritzlin, 2000b: 5).

PUBLISHERS AND SELLERS OF MAPS

In the map publishing area, the participation of women, especially as widows, was impressive in France, Britain, and the Low Countries. During the seventeenth and eighteenth centuries, there were more than ninety known women publishers in these countries (see Hudson and Ritzlin, 2000a), with about one-third of this number residing in each of those areas. Other countries followed suit: Germany, Poland, Russia, Scotland, Ireland, Spain, Mexico, and the United States.

According to Hannah Barker (1997), a widow enjoyed a higher degree of independence than other women, as there was no shame attached to her status as widow. Some of the widows were sufficiently successful to maintain their royal warrants, and depended on the guild to allow them to carry on their husbands' businesses. It is a question, though, raised by Hudson and Ritzlin (2000b: 4), for example, if the guild would insist on their marrying a journeyman who could be given master status. A widow, like a single woman, had a wider range of opportunities than a married woman, including the freedom to choose her legal representative (Möbius, 1982: 54). N.H. Keeble also vests widowhood during this time with considerable status. Common law in Protestant England would leave the widow with a third of her husband's estate (Keeble, 1994: 252).

Still, some men cartographers were able to convince widows of cartographers to sell them the rights of their late husbands' works, as in the case of the widow of Johann Tobias Mayer of Göttingen. Maskelyne was thus able to continue the publication of the important *Nautical Almanac* (started in 1767), containing the relevant lunar tables made by Mayer (Brown, 1949: 202).

The rise of scientific cartography was an outgrowth of the Age of Enlightenment, where rationalism and empiricism held sway. This Age involved a massive cultural transformation. Printed materials were available more than ever before. Many women directed the *salons* of Paris and ventured into publishing. A few examples illustrate the significance of women in the map publishing business in France. By 1654, the Sansons were the chief map-makers in France. Nicholas Sanson became known as the "Founder of the French School of Geography." Marie Desmaretz, wife of Pierre du Val (1618–1683), the nephew of Nicholas Sanson, and one of their daughters (either Marie-Angelique or Michelle) actively participated in the map trade by 1687, operating the firm until 1704 (Pastoureau, 1984: 136, 137; Ritzlin, 1986: 2712). Upon Pierre's death in 1683, Madame du Val continued operating their prosperous concern, reissuing a number of maps, including *Amérique* in 1684 (Ritzlin, 1986: 2712).

Amelia Mortier, the widow of Pierre Mortier (1661–1728), whose work embraced French, English, and Dutch publishing interests, continued his business until her son, Cornelis, was able to take over. Lebreton *fils*, a firm that achieved masterpieces in making marbled paper in the seventeenth and eighteenth centuries in France (using the secret of mingling threads of gold and silver with coloured waves and veins in paper), was inherited by the widow of Lebreton, who had, however, "fallen into penury and distress" (Diderot, 1959b: Plate 384). This, and a few other examples, demonstrate that although a widow could in law succeed her husband in the running of a business, she could also end up "among the scores of beggars" if she owned a business that was difficult to maintain (Möbius, 1982: 65).

In Paris, the widow of Sebastian Hure, 1683, published an edition of Hennepin's *Louisiane*, which "had an enormous impact on the European view of the New World," and the widow of Ganeau issued Charlevoix's *Nouvelle France*, 1744, which included Bellin's important *Carte des Lacs du Canada* (Ritzlin, 1986: 2712).

A large number of women feature in the life of Nicholas deFer (1647–1720), a successful and famous cartographer, and his wife Anne Hus (or Huée), who had six daughters. Not only did four of their daughters marry engravers or men associated with the publishing trade, but they also brought their husbands into the family business. Bénard (Marie-Anne deFer's husband) kept his father-in-law's shop as well as deFer's logo—the *Sphère Royale*. Another son-in-law, Guillaume Danet (who had married Marguerite-Geneviève), had his own shop at the Pont Notre-Dame, but used the deFer sign concurrently with Bénard. Unrelated to the family, there was also Mrs. Hollar, wife of Wenceslaus Hollar, who assisted Nicolas deFer in his studio (Ritzlin, 1986: 2712).

Judging by the depictions in Diderot's *Encyclopedia of Trades and Industry*, paper making and printing were the trades that invited the active participation of women. France, according to Diderot (1959b: Plate 359), was the greatest producer of paper in the eighteenth century.[7] Women were engaged at many stages of producing paper and printing. The *Encyclopedia* identifies at least seven stages of making paper; two of those stages involved only women and one stage involved women and men working together. During the first stage, where fibres are mechanically separated, Diderot depicts women undertaking this separation between cellulose and non-cellulose matter (Plate 360). The depiction of the process of drying the sheets of paper shows two women (and one man), "an operation [that] required space, care, and patience, but not much skill" (Plate 367). It is in the finishing room (Plate 368) that we again see women (actually "little girls") inspecting, polishing, folding, and counting the sheets (see Figure 4.2).

Printing, like paper making, was also at a high level in the eighteenth century. The same can be said of maps printed during this period as of

CHAPTER 4

Figure 4.2 "Little girls" inspect, polish, fold, and count the finished sheet of paper, destined for printing, whether for literary or other purposes, such as printing maps. *Source: A Diderot Pictorial Encyclopedia of Trades and Industry,* Vol. 2 (New York: Dover Publications, 1959b), Plate 368.

published works in general: typefaces were graceful, well-designed, and pleasing to the eye, and "pages were admirably planned, perfectly squared, and clearly pressed" (Diderot, 1959b: Plate 369)—all clear characteristics of printing an artful and precise map. Diderot's *Encyclopedia of Trades and Industry* depicts at least six stages of printing, three of which show the participation of women in printing. The founder's stage produces the metal type, which is a low-melting alloy made by adding molten antimony (Plate 370) (see Figure 4.3).

We find the third stage of printing in the composing room, where a woman is depicted preparing a single line of text. She arranges the letters; the men later justify the margins and even out the surface (Plate 373). This work requires a high level of literacy, especially as the letters must be arranged in mirror form. Figure 4.4 offers a scene of a composing room.

London, England, seemed a particularly vibrant place for map distribution where between 1660 and 1720, there were fifty-five booksellers (Tyacke, 1978: 160). What is unique about British map sellers is that widows were allowed to retain guild membership in the Stationers' Company, even after remarriage (Ritzlin, 1986: 2711). Maps became so topical that map sellers' shops became quite common. Figure 4.5 clearly shows a woman in a map seller's shop, interested in purchasing maps.

Figure 4.3 Two women file burrs, along with other imperfections, off the letter that has just emerged from the mould. *Source: A Diderot Pictorial Encyclopedia of Trades and Industry*, Vol. 2 (New York: Dover Publications, 1959b), Plate 373.

Anna Seile comes to mind, as does the daughter of the French publisher Pierre Duval, who took over her husband's business around 1663 (Ross, personal communication, 18 August 1998). She published a map from Heylin's *Cosmography* and flourished in the London map trade for some five or ten years after her husband's death (Tyner, 1997: 47; Hudson, 1999a).

Ritzlin, using Tyacke (1978), offers an example of the very active businesswoman Anne Fitz (or Fitch) Lea, widow of Philip Lea, whom she had married in 1684. When Philip died in 1700, Anne received a third of the business; she and her brother held the remaining two-thirds in trust for the children of the marriage: two daughters (Ann and Deborah) and a son (Philip). Anne Lea was likely in her thirties when she inherited her husband's business, which she vigorously ran.

Around 1712, her son-in-law, Richard Glynne (or Glin), became her business partner. Anne had become less active by 1725 and died in 1730. Anne Lea's account does not seem to differ from those of other women map sellers.

Mary Senex's husband, John, was not only a member of the Stationers' Company, but a fellow of the Royal Society and Geographer to Queen Anne. He died in 1740, leaving his entire estate to Mary. Perhaps a more noteworthy example is Mary Ann Rocque (née Bew). Her husband, John, was a Huguenot surveyor, engraver, and landscape artist, appointed as

CHAPTER 4

Figure 4.4 The woman (on the right) composes the first line of text. *Source*: *A Diderot Pictorial Encyclopedia of Trades and Industry,* Vol. 2 (New York: Dover Publications, 1959b), Plate 373.

Topographer to the Prince of Wales. Mary Ann (who was his second wife) inherited the large establishment in 1762 after eleven years of marriage (Ritzlin, 1986: 2711) and continued to run the business. She published *A Set of Plans and Forts of America,* 1765 (Varley, 1953; Ritzlin, 1986: 2711). Figure 4.6 offers the title page of this work.

It is clear, however, that not all women decided to keep the business. Some sold their husband's copper plates to other map sellers and publishers (Ritzlin, 1986: 2712).

As far as the Low Countries were concerned, thanks to Hudson and Ritzlin (2000b), we have the names of at least twenty-seven women map publishers. The most active period of map publishing was concentrated in the years between 1720 and 1750, with enormous collections of maps, some as large as fifteen volumes[8] (Ritzlin, 1986: 2712).

In Ireland, however, there was reluctance to "letting a workshop be run by a woman [as it] was regarded as an interim solution," and "nobody wanted her to continue to do so." The general worsening economic situation forced guilds to keep the number of craftspeople low in an effort to guarantee their survival (Möbius, 1982: 68). Jane Grierson (ca. 1753–1759), a widow, published the *Atlas of the World* in Dublin; in Scotland, Mrs. Ainslie published British country maps around 1792 (Hudson and Ritzlin, 2000b: 9, 15).

Figure 4.5 The interior of an eighteenth-century map seller's shop. *Source*: Brown (1949: 175). *Original source*: Nicolas Sanson, *Atlas nouveau*, ca. 1730, Amsterdam.

Germany, too, had map-publishing houses, but such places as Württemberg (in 1701) forbade girls and women from entering into an apprenticeship, although the city did allow girls and women already so employed to continue their apprenticeship until their death. Guilds, however, were intent on driving women out of the trades (Möbius, 1982: 67). In the case of the Seutter-Probst-Lotter House, one of the more notable map publishing firms in Germany, women seemed to have played an ancillary role, primarily as daughters whose husbands were brought into the business. For example, Matthäus Seater, an engraver of maps in a variety of atlases, ranging from twenty to four hundred sheets, and "one of the most famous map publishers of Central Europe" (Ritter, 2001: 130, 132), relied on his daughter, Euphrosina (1709–1784), to engage her husband, Tobias Conrad Lotter, to become active in the business. Matthäus's other daughter, Anna Sabina (1731–1782), first married an engraver, divorced him, and then married another one, George Balthasar Probst. As it turns out, one of Matthäus's sons had married George Probst's sister, Jacobina (1726–1757). Other affinal arrangements were in place, but it was the husbands who became noted as engravers, not the women.

CHAPTER 4

Figure 4.6 Title page of Mary Ann Rocque, *A Set of Plans and Forts in America*, 1765. *Source*: Edward E. Ayer Collection, Newberry Library, Chicago.

THE EDUCATIONAL AREAS

The social organizational features of the map world in the nineteenth century made it possible to incorporate some new participants: schoolteachers. Until that time, education was not only confined to the elites, but also to men. The current method and system of education emerged in the first decades of the nineteenth century, due to industrialization and the requirement of abstract learning in various subjects that could no longer be obtained through traditional means (Giddens, 1991: 511–12). The new century opened up education to women (both as teachers and students), bringing cartography to the foreground in civic life. Women thus stimulated the practical and ideological values attached to maps, with positive repercussions throughout the map world.

Maps and globes became important tools for both boys and girls at this time (Thrower, 1996: 125): in an era when schools served to instill national and civic pride and a feeling of citizenship, maps became one of the chief ways to illustrate a nation's expansion and international interests. The depiction of far-off lands heightened ethnocentric thoughts and feelings as "civilized" explorers mapped lands of "uncivilized" peoples. The teaching of astronomy, surveying, and the drawing of elegant maps was part of any academy's curriculum, especially in young women's academies (Tyner, 2001: 41). The role of women as globe makers, authors of atlases, authors of geographies, and teachers (of geography) constituted some of the major contributions of women to schools and to society at large.

The rise of public education in the Western world coincided with the colonization of non-Western lands in the eighteenth and nineteenth centuries. As a consequence, geography about distant lands came to play a critical role in educating children as future citizens of colonizing powers. Some authors of geographies included maps as a matter of course. On one hand, maps in geography textbooks did indeed serve to educate these future citizens about other territories; on the other hand, maps were very effective in disenfranchising peoples through disinheriting these territories of indigenous names (and replacing them with Western toponyms) and through imposing unfamiliar, divisive boundaries in areas that had been culturally or historically united.[9] Aside from using globes to illustrate the wonders of the world, women authors and teachers of geography turned to using maps for displaying social and cultural phenomena.[10] Thus, teachers, most of whom were women, occupied the position of map users, instead of map decorators and creators, as they had been in the past. There are at least thirteen known women authors of atlases and maps in the United States alone, four in England, at least two in France, as well as several in Germany and Poland (information derived from Hudson and Ritzlin, 2000a, 2000b).

Emma Willard (1787–1870) developed teaching methods that placed emphasis on maps.[11] One of a family of seventeen children, she was a "bright and inquisitive child" who received much encouragement from her family. At twenty, she married a widower; when he suffered financial reversals, they opened a boarding school for girls in their home. After Emma's publication of *Plan for Improving Female Education* in 1818, the school grew in popularity. The work was so progressive that it was called the Magna Carta of women's education. The school relocated to Troy, New York, and was renamed the Troy Female Seminary, where Emma continued to develop particular teaching techniques, using maps in geography and history. Her techniques were so successful that she asked her talented student Elizabeth Sherrill to draw maps for a new book. Emma collaborated with William Channing Woodbridge to publish *A System of Universal Geography on the Principles of Comparison and Clarification* in 1822 or 1824. Emma Willard's contributions consisted of the sections on ancient geography, problems on globes,

CHAPTER 4

and rules for the "construction of maps." Their work "produced a revolution in the method of presenting geographical facts in schools" ("Woodbridge, William Channing," *Dictionary of American Biography*, 1936).

Just before her retirement from the Seminary in 1837, she published *A System of Universal History in Perspective, Accompanied by an Atlas, Exhibiting Chronology in a Picture of Nations and Progressive Geography in a Series of Maps*. If one were to follow her teaching method, a student would have to demonstrate "her understanding of the day's lessons by drawing, from memory, the appropriate map" (Ritzlin, 1990c) and "marking the paths of navigators and explorers and the march of armies" (Lutz, 1974: 40). Willard had an enormous impact on geographic education in America, even if we considered the use in schools of blank maps alone (Tyner, 1997: 48; Ritzlin, 1990c: 4; Hudson, 1999a). Her Troy Seminary was one of the very first institutions to prepare women for the teaching profession. After her retirement, she worked tirelessly to establish "normal schools," the equivalent of teachers' colleges. It is estimated that her own school trained over seven thousand young women.[12]

Mary Troy Seminary of Kinderhook, New York, was also a contributor to this genre in maps. Showing an ongoing interest in geography and the world, the Geography and Map Division has in its collections two maps by Van Schaack (Library of Congress, 2001), drawn twenty-two years apart. Her earlier work involves a hand-coloured manuscript map outlining the countries of the world with the most recent discoveries.

MAKING GLOBES

Teaching geography (and proper citizenship) with globes (and maps) was a fairly widespread phenomenon as well. In India, for example, wall maps and globes began to appear in the mid-nineteenth century in schools around the country. "In the absence of globes in the classroom," says Ramaswamy (2001: 99), "teachers were instructed to use an orange or a wood-apple to emphasize that the earth was spherical...." In France, it was Julie Bouréche who produced, in 1828, the *Atlas géographique et sphérique dessiné* for Miss Roullet's Academy for a competition held in Paris (*Imago Mundi*, 2001: 53, 176).

Hudson notes that twenty-three of the 150 women map-makers were nineteenth-century authors of atlases, including one who published a handbook on globes (Hudson, 1989: 4). As Hudson (1989) herself suggests, the interest in geography may reflect the opening up of schools to girls. Madame Coindé offers this 1813 introduction to her atlas: "[T]his map ... is particularly adapted for the use of young ladies, into whose hands very few books on [mythology] can be put with propriety, and, without some knowledge of this branch of history, how could they enjoy reading or theatrical representations of the best authors and poets ..., which by the assistance of this map, will become clear, amusing and instructive?" (cited by Hudson, 1989: 4).

Madame Coindé not only edited her atlas, *LeSage's Historical, Genealogical, Chronological, and Geographical Atlas*, but was also one of the twenty-five agents selling it, probably to her own students.[13]

Mary Ritzlin (1986: 2712) offers the following information on women in map-making in the United States: "In America, too, we find instances of widows running the family business. Roxanna Hamilton married surveyor-teacher John Farmer in 1824; he later founded a map business producing important maps of Michigan and other midwestern states. When John died in 1859, Roxanna continued to publish maps with the help of her daughter and sons under the name 'R. Fanner & Company.' Son Silas bought out his sister's (Esther) and brother's interest in 1864, and Roxanna's in 1868."[14]

Roxanna's husband died of a nervous breakdown, presumably caused by "his intensive work habits and his long hours of labor spent in preparing and engraving maps" (Ristow, 1985: 276).

Mrs. S.I. Watrous, 1824, was a Vermont illustrator (Hudson, 1999a) through whose efforts the 1821 and 1824 maps of Ebenezer Hutchinson of Hartford included a cartouche with a view of Montpellier (Ristow, 1985: 89); Sarah Sophia Cornell (1854–1870) produced numerous atlases (including twenty school atlases) and geographies for Appleton, including *Key to Cornell's Outline Maps*, in 1864 (Hudson and Ritzlin, 2000b). From today's perspective, Cornell's achievements are astonishing given the fact that she lived for only sixteen years. Another east coaster, Elisha Robinson, owned her own publishing firm, producing real estate atlases in New York City (Ristow, 1985: 260). According to Ristow (1985: 260), real estate atlases were introduced around 1880 in the United States.

Interestingly, there are remnants of the work done by schoolchildren. Fairly large manuscript maps of Connecticut and Maine were created by Jane Renwick (born in 1801 in Scotland) in 1813 as part of a school atlas (Hudson, 1999a; 1999b: 29–30). These maps were fairly typical classroom exercises in the nineteenth century, engaging American schoolgirls to make beautiful and accurate maps. Renwick was twelve years old when she drew her Connecticut map, but was thirty-five when she drew the Maine map around 1836. There were a large number of schoolgirl maps as geography

CHAPTER 4

Figure 4.7 Elizabeth Mount was only sixteen years old when she produced this globe in 1822. It features exceptional details. *Source:* Dymon and Kaye (1999: 7).

assignments. Many of these are still in existence. Like the Westtown globes, they are marginal to our study, although they do show the kind of training women had in geography and in globe- and map-making.

Globes have a particularly fascinating history in cartography, much of which has been recorded and analyzed by Dahl and Gauvin (2000). As Denis Cosgrove avers, globes were (and are still) used to teach not only geography but also astronomy. He says that "[t]hey were long regarded as especially suited to the education of women, whom men viewed as more capable of understanding through images and measure than by reading from a text" (Cosgrove, 2000: 61). Globe making, according to Judith Tyner (1997), was particularly popular during this time. There were some nine women known to have made globes, with a number receiving patents for their "globes, globe mounts, and other teaching apparatus" (Tyner, 1997: 48).

In the first decade of the nineteenth century, girls at Westtown, a Quaker School in Pennsylvania, made elegant hand-drawn maps and learned surveying technique as school exercises (Tyner, 1997: 48; 2001: 40)—a different category from commercial globes. The Westtown School in Chester County, Pennsylvania, was modelled after the Ackworth School in York, England, but did not create map samplers like its British counterpart. As early as 1804 students made silk globes, six years before James Wilson, usually considered America's first globe maker, made his commercial wooden globes (see Judd, 1974: 69).[15] No other schools were known to have produced such globes. Some twenty-eight of these globes have been discovered. The girls made the globes probably "in conjunction with classes in astronomy and mathematical geography" (Tyner, 2001: 40).

Many of the active women globe makers,[16] whether young or old, could be found in the United States. For example, in 1820, Elizabeth Mount of Long Island made a twenty-inch globe from paper gores glued onto papier-maché (see Figure 4.7). She was only sixteen years old when she made this exceptionally detailed globe. We can surmise that she made it for her pupils (Dymon and Kaye, 1999: 7). Elizabeth Oram was the first woman to obtain a patent for making a globe for teaching geography in New York in 1831 (Tyner, 1997: 48, 50). Ellen Eliza Fitz, in 1875, also obtained a patent for a new method of mounting and operating globes—one of best globes at the United States Centennial celebration in Philadelphia. She published her techniques in *Handbook of the Terrestrial Globe* (Tyner, 1997: 48).

The making of globes was not confined to the United States. In England, Miss Readhouse made a particularly interesting globe—a map of the moon, which she displayed at the Great Exhibition at Crystal Palace in 1857 (Hudson and Ritzlin, 2000b: 20).

CARTOGRAPHERS

In Britain, one finds in the mid-seventeenth century Virginia Farrer (or Ferrar), the daughter of cartographer John Farrer, who "issued later versions of her father's map [*Mapp of Virginia*, 1651–)] ... under her own name" (Thrower, 1972: 279, n. 45). Norman Thrower (1972: 123–24) also reminds us that it "was the 'lesser men' ... as well as women who took care of the day-to-day business of the map trade and occasionally made a breakthrough." The Farrers never set foot in America. Mary Ritzlin (1986: 2712–13) offers the following: Virginia, one of "the few women to be described as a compiler" of a map, was a member of an outstanding household led by her uncle Nicholas (1592–1637) and her father, John. The family, which included seventeen children, retired to the country to found the religious and literary community of Little Gidding. Both girls and boys were educated. We learn

CHAPTER 4

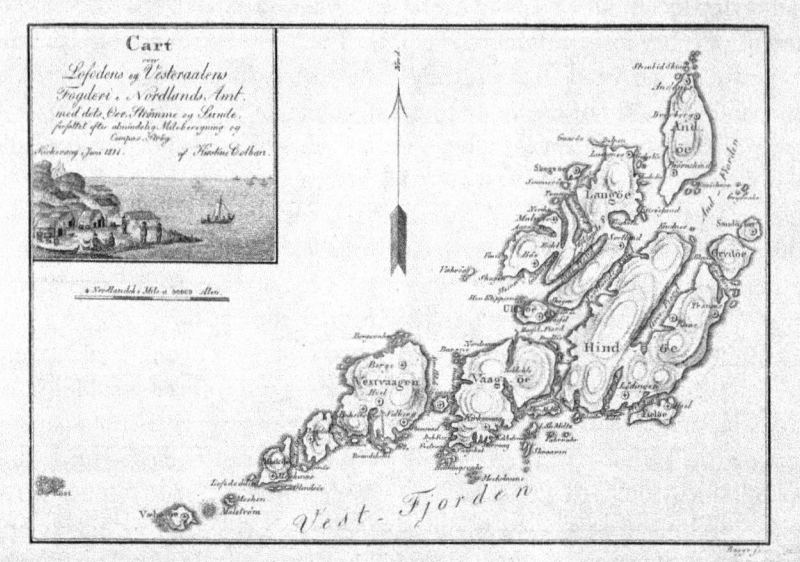

Figure 4.8 Kirstine Colban's first known map, of the Lofoten Islands. *Source*: http://ask.bibsys.no/ask/action/result?fid=forfatter&term=Colban,%20Kirstine1 kart kol. 23 x 34 cm på blad 32 x 46 cm (from Roald, 1988a).

that the "family was also taught the craft of book-binding and had their own small rollerpress."

Virginia and her father "felt a special relationship with her namesake land" and she assisted him "with his correspondence." She "was in direct contact with Lady Berkeley, the wife of Virginia's governor" and "found time to conduct experiments on raising silk worms, and to report her findings to those" in the United States (Ritzlin, 1986: 2712–13). Probably after her father's death, the maps began carrying Virginia's name instead of his (Tyner, 1997). She made many changes and corrections, adding decorations along the way (Ritzlin, 1986: 2713).

Elizabeth Bermingham was a cartographer, possibly from Charleston, South Carolina (Hudson and Ritzlin, 2000b: 9). In 1724, she produced an "extremely detailed chart using various sources" for her six-sheet manuscript, "The Atlantic Coast of North America from South Carolina...." Her sailing chart of North America, produced in 1727, represents the earliest sailing chart of this continent, showing degrees of latitude in its borders and four inset maps (Dymon and Kaye, 1999: 6).

THE EIGHTEENTH AND EARLY NINETEENTH CENTURIES

Figure 4.9 Perspective map of Kabelvåg by Colban. *Source and Permission*: Perspective Maps: Lofotmuseet, Lofoten Islands, Norway.

In Norway,[17] one finds the Lofoten Islands: some of the most remote and isolated islands, though they were surrounded by some of the world's richest fishing grounds. The Islands were also the home of Norway's first woman cartographer, Kirstine Colban, alias Stine Aas. It was Roald Aanrud, now eighty-five years old, who discovered quite by coincidence in the 1980s Kirstine's work as a cartographer. He had come across a volume by Pastor Erik Andreas Colban, entitled *Forsøg til en Beskrivelse over Lofodens og Vesteraalens Fogderie i Nordlands Amt* (1818), whose work covered a "range of local topographical descriptions" that had come "drifting in the stream of Norway and national self-awareness around 1800." Exuberant attachment to Norway's "heroic past and its peasants" (Anon., 1997) led many Norwegians, including Kirstine, the Pastor's daughter, to become interested in local topography.

Kirstine Colban, the second of three children, was born in Bø in Vesteralen in 1791. Roald Aanrud tells us that as a very rational priest, her father encouraged the fishermen's wives to cultivate the ground because he

CHAPTER 4

Figure 4.10 Church of Vågan by Colban. *Source and Permission*: Perspective Maps: Lofotmuseet, Lofoten Islands, Norway.

believed that "the cultivated land is better done by clever women than by lazy men." Kirstine's mother, Karen Nattanaelsen Angel (1762–1853), was powerful. She relieved or cured many illnesses, including the bubonic plague, by herbal medications, ointments, and poultices with ingredients from her own herb garden. Kirstine herself received a good education. She was competent in both Hebrew and Latin. She travelled extensively with her father. She was later the treasurer of a local governing council.

She was twenty-three years old when she finished her first known map: "*Cart over Lofodens og Vesteraalens Fogderi I Nordlands Amt, med dets Øer, Strømme og Sunde*" (see Figure 4.8). Kirstine measured the map by ordinary milestones and by compass. The map's real standard is about 1:1.000.000. The map is significant in that it shows the complex topology and geography of the Lofoten Islands with their countless channels, bays, and islands. It must not have been easy to traverse the rough area. Oluf Olufsen Bagge in Copenhagen finished the map in copper stitching. An antiquarian firm, Cappelens Antikvariat, describes this map as with a "small tear in title page, otherwise very nice."

She later took on the imaginative task of creating perspective maps that today evoke surprise and wonder for their artistic quality. In 1814, for example, she first created a delightful perspective map of the trading place of

Finnesset in Vågan. The French king Louis-Philippe stayed in the guesthouse there in 1795 on his way to North Cape, Norway. My personal favourite is the perspective map (a watercolour painting) of Kabelvåg, where she lived from 1797 to 1817 (see Figure 4.9).

No less fascinating is her famed perspective map of Church of Vågan (Kjerkvågen) (see Figure 4.10), a watercolour that juxtaposes the windswept landscape with a well-planned area around the church. The contrast between the natural and the human-made world is something that countless other painters have attempted to colour since Kirstine's time.

In 1817, at the age of twenty-six, Kirstine married widower John Henrich Aas, who was thirty-one years her senior. A judge in Senja and Tromsø, John was also an honorary treasurer for his royal services in northern countries during the Danish rule. Kirstine fitted well into cultured circles. Aanrud describes her as "versatile and gifted." After her husband's death—five years into their marriage—her talents turned to reformist causes, and she began referring to herself only as "Stine Aas."

Along with her craft in map-making, Stine Aas became one of northern Norway's most distinguished painters; the Lofotmuseet (Museum of Lofoten) displays her paintings. As if maps and art were not enough, she also wrote poetry. Her poetry was a testimony of her love for her parents and for northern Norway. She wrote some of her poetry as songs for children. Dwelling on her literary output would take us too far afield from discussing her as a painter of perspective maps.

Stine Aas, moreover, devoted much of her time to helping the poor and in 1837, at the age of forty-six, she headed Norway's first kindergarten, Trondhjems Asylum, inspired by the ideas of England's first socialist-industrialist, Robert Owen. Only one year later, she and her daughter moved to Christiania (Oslo), where she introduced another home for infants: the Grønland Asylum. Stine Aas's fame spread so far that Henry Wergeland, Norway's poet, orator, nationalist, intellectual, and social reformer, marked her fiftieth birthday by producing a poem for the occasion. Stine Aas died in 1863, at the age of seventy-two. Her cartographic and social reformist legacies continued through her descendants.

Her son-in-law, Ole Hartvig Nissen (1815–1874), became a politician and the foremost representative of the democratization of schools. His legendary school in Christiania in 1849 was the first to offer girls the same curriculum as boys.

Stine Aas's cartographic work, however, skipped a generation and came to rest on the shoulders of her grandson Per Schelderup Nissen

CHAPTER 4

(1844–1930), who, alongside his duties as a general during Norway's most formative period, did notable work as a cartographer and geographer and was Director of the Norwegian Geographic Survey from 1900 to 1906. He produced the *Economic Geographic Atlas of Norway* in 1921, a precursor to the contemporary National Atlas of Norway. Per's son, Nils Kristian Nissen (1879–1968), became Norway's foremost map historian.

Kirstine Colban's reaching out to areas that needed social attention foreshadows the work of many twentieth-century women pioneers in cartography who would become acutely interested in alleviating the burdens of marginal and indigenous groups.

Our attention now shifts to another part of the globe, namely Canada, where Elizabeth Simcoe (1762–1850), the spouse of the Lieutenant Governor, undertook expeditionary tours throughout the western part of the land that later was to become part of Canada, producing maps even on birchbark, guided by her uncanny observations for detail, her artistic talents, and her sense of adventure. We are also compelled to include the distinctive contributions of Shanawdithit (ca. 1800–1829), the last surviving member of the Beothuk tribe in Newfoundland, who, despite her tragic life, created a number of fine maps. The glow of these three women has steadily increased in lustre as the years go by. From the most privileged segment of society to the most abject, the contributions of these women inspire all who read their accounts.

THE ARTISTIC CARTOGRAPHY OF ELIZABETH SIMCOE (1762-1850)

The earliest of this small category of women (in Canada), and perhaps the best known, was Elizabeth Simcoe, whose diaries and maps of Canada evoked admiration in her native England.[18] She produced a manuscript map, the most remarkable of which was painted on birchbark.

Born in 1762, Elizabeth Posthuma Gwillim was from a wealthy family. Her mother was Elizabeth Spinkes; her father was Lieutenant-Colonel Thomas Gwillim. At sixteen years of age she married her thirty-year-old groom, John Graves Simcoe, who later became Lieutenant Governor of Upper Canada, dispatched to that area by his government in 1791.[19] The couple already had four children; eventually they would have eleven. Elizabeth decided to accompany her husband, and then lived in Canada for five years.

She possessed a sharp intellect as well as keen powers of observation, which stood her in good stead as a botanist and artist. She travelled extensively with her husband by whatever means available—open boat, sleigh, horseback, or carriage on rough roads. She saw the hazardous conditions as a challenge to overcome. One biographer avers that Simcoe "enjoyed everything" and that she was "eager to be pleased by new experiences" (Firth, 2000). The pages of her diary contained sketches and maps, in addition to the larger maps she produced with watercolour views. She had an artist's eye and showed interest in flora and fauna, food, and the medicines prepared from them (Firth, 2000). She was "fascinated" by native people, but her approach was anthropological or romanticized, not personal. Her cartographic work of the 1790s came to the attention of British and French officials. When she accompanied survey teams, she was able to offer opinions on the shortcomings or strengths of this or that survey.

Researchers have raised questions about the source of her cartographic interests and skills. No doubt her artistic bent in painting and drawing helped, but so did the geography lessons, we surmise, that she took from her governess. In some respects, her education in geography foreshadowed the results of a geography education that, as we have seen, would be fostered by women in the Western world. In several respects, Simcoe's manner of writing highlights the differences between women's and men's travel accounts, with or without the maps. In contrast to men, according to Mills, women employ less formal and less structured accounts, devoid of the "conquest narrative," sometimes contributing "to the imperial task of revealing the secrets of the colonized country" (Mills, 1994: 35, 43).

THE ICONIC YET TRAGIC WORK OF SHANAWDITHIT (CA. 1800-1829)

Far from the grandeur of material comfort of the Europeans, there appears a tragic but heroic figure of Shanawdithit in Newfoundland—the last remaining Beothuk who faced European settlers.[20] Through fear compounded by prejudice, a handful of settlers had sought the extermination of the Beothuk tribe. As the very last remnant of a tribe whose habits, culture, and language are now veiled in the mists of time, the young woman Shanawdithit stands as an icon of Newfoundland's tragic past (see Figure 4.11).

Born around 1800, as a child she witnessed the trials and sufferings of her whole family, most of whom were purposely eradicated by settlers who did not see the Beothuk as humans: they were to be hunted and killed. Any attempt by government to rehabilitate their fortunes was too late, although the few survivors were accepted into settled communities. Shanawdithit, one of the survivors, was allowed to return to her people. Unable to find members of her tribe, in 1823 she undertook the strenuous journey back to where the settlers lived. Through circumstances beyond anyone's ken, she entered the household of "successful and shrewd businessmen" (Whitby,

CHAPTER 4

Figure 4.11 Shanawdithit, the last survivor of the Beothuk, mapped the history of her people in Newfoundland. *Source*: Collection of the Newfoundland Museum, St. John's, Newfoundland.

2005: 112). Curiously, it was John Peyton Jr. at Sandy Point, an outport, who took her in. His father, John Peyton Sr., was an infamous "Indian killer" and was recommended by the government to leave the area to avert more persecution of the Beothuk (Whitby, 2005: 109). He had killed Shanawdithit's uncle and cousin.

In this ironic setting, Shanawdithit learned English customs, was forced to don English dress, and learned to speak some English. Her new life would be abhorrent to the Beothuk. Yet despite all her trials, she had an easygoing personality. Even when she was captured, according to Whitby (2005: 112), she seemed "irrepressible and overflowing with curiosity." Still, one medical doctor described her as "majestic, mild, and tractable," but also "proud and cautious" (Whitby, 2005: 112). There is some question of whether she was treated as a slave or left free to do as she pleased in the house. It is known that children were immensely attracted to her. The fine gifts that many visitors offered to the household showed she had become a celebrity, albeit a quiet one.

THE EIGHTEENTH AND EARLY NINETEENTH CENTURIES

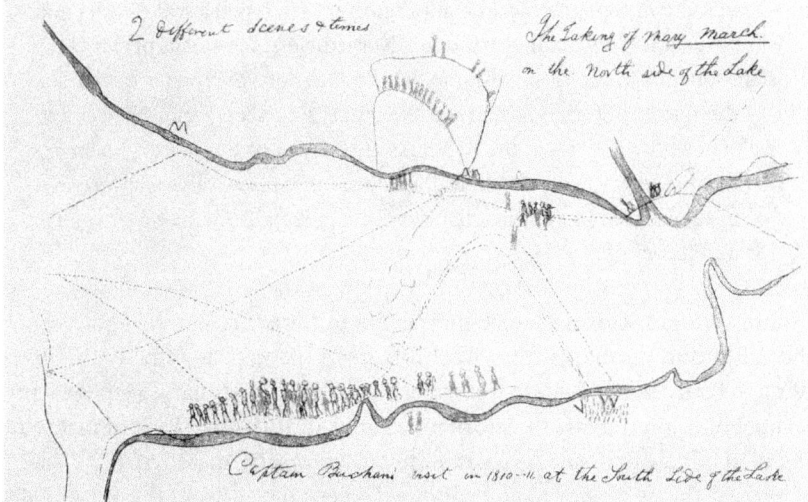

Figure 4.12 Red Indian Lake: Shanawdithit, the Beothuk map-maker, depicts the capture of her aunt Demasduit in 1819. Map and drawing by Shanawdithit, 1829. *Source*: See *Mercator's World*, Sept./Oct. 2000: 27–28; the original was published in J.P. Howley's *The Beothucks or Red Indians* (1915).

In 1828, explorer William Eppes Cormack managed to have Shanawdithit forcefully transferred to St. John's. Still, Cormack undertook to teach her more English than she had learned in the Peyton household. When he supplied her with paper and coloured paper, he discovered an artist. Even though she had been only ten or eleven years old at the time of the persecution of her people, Shanawdithit was able to recollect the dire events eighteen years later and portray them in a series of maps. Her maps compressed the historical events into one image, showing the movements of her people, the horrific actions of the marines against the Beothuk, and the slaying of a Beothuk woman.

Another map shows the capture of a young mother, Demasduit, Shanawdithit's aunt, whose small band of Beothuk had been chased by marines. Demasduit handed her child over to the band when she realized she might be captured and killed. When her husband, Nonosabasut, tried to rescue her, the furriers-cum-hunters killed him. Their child died within three days.

Shanawdithit managed to reveal much of the sufferings and recent history of the Beothuk through her five "action maps." According to Keith Winter (1975: 128–29), her drawings

CHAPTER 4

are accurate in topographical details, but they lack regular scale: rivers and lakes appear larger than they really are. Nevertheless, the details of shoreline, islands, bends in the river, falls, rapids, and junctions of rivers are accurate, and the relation of each of these to the other is correct. James Howley [an explorer who wrote about the Beothuks at the turn of the century] later remarked that "every fall, rapid and tributary or other remarkable feature is laid down, all of which I have no difficulty in recognizing from my own exploration and survey of 1875.

Still, William Cormack was insensitive to her situation: he showed her the skulls and remnants from Beothuk graves he had dug up. With dwindling resources, he had to return to England in January 1829, leaving Shanawdithit in the care of Attorney General Simms. On 6 June of that year, she died of tuberculosis. As part of the post-mortem, the physician Carson beheaded the corpse of Shanawdithit and sent the skull to the Royal College of Physicians in London. The skull was destroyed during a bomb raid on London in World War II. The sufferings of Shanawdithit evoke shame in later generations.

CONCLUSION

The premise of our argument is simple. The extent to which women participated in the map worlds of the eighteenth century and the early part of the nineteenth century does not reflect a linear, historical process. Rather, it is conditioned by the social organization of each prevailing map world. Such social organization was shaped by the advent of technological advances, social customs, scientific knowledge, and larger political forces.

Bringing the creation of maps inside the ateliers (as we have seen in the previous chapter) proved to be a boon for women wishing to participate in that process, but while the advent of outdoor surveying limited the women's participation, the discovery and colonization of new lands, especially in the nineteenth century, brought women into the vanguard of educational efforts in geography.

The eighteenth and nineteenth centuries in cartography opened up unimagined possibilities for women. In the incipient phase of this period (and throughout most of this time), the rise of national political entities drove the technology. Colonization drove education about maps, globes, and geography. Toward the end of the period, there were larger societal forces and institutional developments within cartography. The commencement of public education and the further consolidation of colonialism featured strongly in the ethos of cartography, along with the acquisition of new geographical knowledge. Schools for girls, in particular, women authors and teachers of geography, and embroidered maps became important apparatuses for developing a public interest in global geography.

However, in other areas, the participation of women was severely limited: the world of internationalizing cartography (and geography). If we bear in mind the imperial conquest and colonization in the nineteenth century, we realize that these processes produced the need to internationalize cartography as a handmaid of conquest (see, for example, Blunt and Rose, 1994: 10). If muted at best, the situation reflects the theme of many contemporary women cartographers who, as a later chapter shows, claim to be "good ghosts" in cartography. However, other chapters later in the book will also show that there are substantial implications as a result of the involvement of women in cartography.

CHAPTER 5

Cartography from the Margins: From the Early Twentieth Century to World War II

Aside from its exciting technological innovations, the nineteenth-century map world saw the early glimmerings of a new kind of cartography stimulated from the margins: namely, the rise of thematic, environmental, and Aboriginal maps in North America. Whether as engravers, geography teachers, colourists, or embroiderers, women offered their traditional talents to the field of cartography, which was finding its feet in a new map world as a consequence of new technologies and political aspirations echoing from efforts to colonize the world and from building the ideals of citizenship. A few women, however, stood outside this particular political ambit because of personal circumstances, opportunities, temperaments, and chances.

This chapter recounts the narratives of five women (as well as the group of women cartographers employed during World War II) who worked on the margins of cartography in the United States, Canada, and Britain. Florence Kelley (1859–1932) was distressed by the experiences of immigrants in Chicago and channelled that distress into creating thematic maps that eventually became important tools in bringing about a sea change in policies. Ellen Churchill Semple (1863–1932), deeply influenced by German thought, took to creating maps that were among the first to document and show how the environment and humans interact. Her approach constituted

CHAPTER 5

the first foray into a new field: human geography. Mina Hubbard (1870-1956) became an accidental cartographer in Labrador after accompanying her husband, who died on his mapping expedition. She completed his work with brilliance. Mary Adela Blagg of Britain offers us a vision of life fully dedicated to standardizing the names of features on the moon that have withstood the test of time for a whole generation, despite the fact she was not formally trained in any relevant field, including cartography. Phyllis Pearsall, another untrained in cartography, devoted her life to creating the highly popular *A–Z* city maps along traditional cartographic ways. Finally, we will look at "Millie the Mapper," the name for young women who joined the war effort as cartographers.

There were no doubt other women of this time who made notable achievements in this field, but whose record of accomplishments has not yet surfaced or is permanently lost.

THE THEMATIC MAPS OF FLORENCE KELLEY (1859-1932)

The nineteenth century witnessed the birth of a particular map genre that would later obtain a strong niche in cartography: the thematic map. Meteorology, hydrography, geology, and natural history—all new sciences in the century—were celebrated and became the object of thematic maps. But there was another side to the nineteenth century that aroused concern and worry: the social ravages brought on by industrialization. This problem was left to a number of social reformers and activists, mainly women, who seized upon the idea of using maps to create reforms to assuage these ravages. In their minds, thematic maps would be ideal conveyors of knowledge and major sources for raising the consciousness of politicians and social decision makers. Mapping epidemics (resulting in, for example, the famed cholera map by John Snow in 1854), population dynamics, and urban slums became *de rigueur* agents of change and reform.

Following Arthur H. Robinson's description (1982: ix), the thematic map portrays the "geographical character of a great variety of physical, social, and economic phenomena." The birth of the thematic map represents an "intellectual upheaval" (Robinson, 1982: x), because, as mentioned above, all kinds of new fields accompanied its birth. The thematic map "showed spatial distribution of one or several related characters or themes, usually of a physical, social, or economic nature" (Chancy, 1996: 75). The thematic map, according to Helen Wallis and M.H. Edney (2003: 1112), was invented in the latter part of the seventeenth century.[1]

Florence Kelley was one of the most dedicated social activists of the "Progressive Era" in the United States (1890s to the 1920s), which sought political reform and the end of corruption.[2] In her effort to end poverty, Kelley (along with others) produced an innovative series of Chicago maps that involved demographic and economic data presented in a fresh way. Justice Felix Frankfurter of the United States Supreme Court called her "A woman who had probably the largest single share in shaping the social history of the United States during the first thirty years of the [twentieth] century" (Blumberg, 1974: 606).

Born on 12 September 1859 in Philadelphia, she was the daughter of Caroline Bartram Bonsall and Democratic Congressman William D. ("Pig Iron") Kelley, who were Quakers. She was the third of eight children, five of whom died in childhood. Educated at home, she travelled widely with her father. She entered Cornell University in 1876, but ill health prevented her from graduating until 1882, when she earned a B.A. in literature. Refused because of her sex to study law at the University of Pennsylvania, she left for Switzerland for a short period of study of law and economics at the University of Zurich, which led her to read the works of Karl Marx and Friedrich Engels. Those readings created in her a desire to become "an ardent socialist" (Brown, 2009a: 1). In 1884, she married Lazare Wischnewetzky, a Polish-Russian physician. In 1886 she returned to the United States with her husband and their first child, a son (they later had two more children) and for five years was preoccupied with domestic duties. In 1894 or 1895, she finally obtained her degree at the Northwestern Law School.

Replenished with socialist ideals, Kelley translated Engels's *The Condition of the Working Class in London* (1887); she maintained contact with him for the remainder of his life. Upon her return to Chicago, she became a resident of the famed Hull House, the organization founded by Jane Addams. While broadly involved with helping impoverished immigrants and setting up educational, employment, and child services, Kelley became particularly noted for her interest in improving industrial working conditions and eradicating child labour.

When the United States Congress commissioned, in 1893, a wide survey of slums in major cities, it selected Kelley to lead the survey in Chicago. The most graphic way the conditions of slums could be portrayed, according to Kelley, involved maps. She then created a series of maps similar to Charles Booth's map (1840–1916) of poverty in London. Her maps represent early efforts to supplement social research with maps showing the spatial

CHAPTER 5

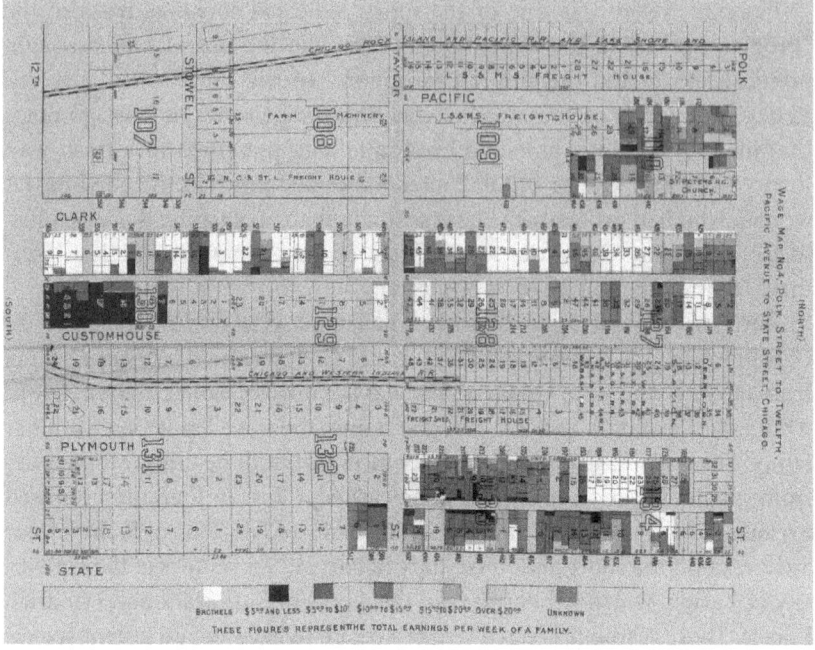

Figure 5.1 Thematic map created by Florence Kelley. *Source:* http://homicide.northwestern.edu/pubs/hullhouse/Maps/. Public domain.

patterns of demographic phenomena. The maps "showed each street in the district and each house is coloured to reflect the birthplace of the head of household and the family's wages" (Brown 2009a: 2). She resorted to creating cartograms—among the first times that cartograms were used—to show the proportional number of individuals in each nationality or wage group. Kelley published these maps in 1895 as *Hull-House Maps and Papers* (1970), providing much greater detail about Chicago demographics than the official government survey did.

Many organizations, such as Greenpeace, now use Kelley's maps as a model. One of her biographers captured the essence of her spirit: "She was a vigorous, dynamic person, whose method of approach was a head-on attack. Her voice was rich, clear, and commanding. There could be no doubt as to her attitude toward any problem, and she had only scorn for pretense and contempt for the socially selfish. Her mental processes were swift, her sympathy quick, and her courage invincible" (Breckenridge, 1944: 463). She died on 17 February 1932 in Germantown, Pennsylvania, at the age of seventy-three after a long illness.

THE INCIPIENT HUMAN GEOGRAPHY OF ELLEN CHURCHILL SEMPLE (1863-1932)

A pioneer in the study of human-environment interaction, Ellen Semple recorded her findings through maps.[3] Her concerns foreshadowed contemporary concerns with cultural and political ecology in the social sciences.

Born on 8 January 1863 in Louisville, Kentucky, she was the youngest of five siblings. Her mother, Emerine Price Semple, made a strong point of educating her children after Ellen's father, Alexander Semple, a merchant, died when Ellen was twelve years old. She attended Vassar College, graduating in 1882. For a few years she taught school in Louisville, but when she travelled with her mother to Europe in 1887, she heard about Friedrich Ratzel's theories of environmental determinism (which are currently discredited). Once back in the United States, she completed her master's degree and then returned to Germany to study with Ratzel at the University of Leipzig. She became immensely attracted to his idea that similar environments create similar societal patterns. The classes she audited had five hundred male students.[4]

Upon her return to the United States, she and one of her sisters opened Semple Collegial School for Girls in Louisville. She visited Germany again and in 1897 published her first scholarly article, "The Influence of the Appalachian Barrier upon Colonial History," in the *Journal of School Geography*. When doing fieldwork was still relatively unknown among geographers, she travelled on horseback through the backwoods of the Kentucky Appalachian area, often staying with local people. Her writing up this field trip led her to give lectures in Washington, D.C., and at the Royal Geographical Society in London, England. Her first book, *American History and Its Geographic Conditions* (Semple and Jobes, 1903) reflected her own interpretations of Ratzel's theory and correlated United States expansion and the geographical environment of North America. One biographer (Brooker-Gross, 2000:1) avers that this book was "widely" adopted as a textbook, while another (Cooksey, 2002:2) claims that the book was only "occasionally" adopted as a textbook.

Her innovative work was significant in that it moved geography away from its heavy reliance on physical geography and introduced the human aspect of geography, which included history and anthropology (Brown, 2009b: 1). While scholars now decry the idea of environmental determinism as advocated by Semple, she is recognized, as stated by Nina Brown, as a "pioneer in the study of human-environment interaction." She also foreshadowed contemporary interest in cultural and political ecology in the social sciences. Brooker-Gross (2000: 1) notes that the sweeping assertions in her work have been subject to much criticism, but that her "eloquence, enthusiasm, and thorough method ... led to her being highly regarded and influential."

After teaching a summer course at Oxford, and after teaching at the University of Chicago, Wellesley College, and the University of Colorado, she was hired in 1921 as the first female faculty member and as lecturer at Clark University in Worcester, Massachusetts, eventually becoming a professor of anthropogeography. Her appointment at Clark was her first permanent academic appointment, at the age of forty-eight. Clark University has consistently been among the "largest producers of professional geographers in the United States," more than even the University of Chicago (Monk, 1998: 1).

According to Gloria Cooksey (2002: 2), Semple's work was "fundamental" in establishing geography as a field of study in the twentieth century. Along the way, as a sessional lecturer at the University of Chicago, Semple taught Richard Harsthorne (1899–1992), who became one of the leading lights in geography (Martin, 1994: 481) and who eventually taught at the University of Wisconsin and the University of Minnesota, connecting with others who would shape the careers of women in academic cartography (see Chart 6.1 in Chapter 6). In recognition of her contributions, Semple received many honours, including being the first woman elected President of the Association of American Geographers (1921), of which she was one of the two women founding members (McManis, 1996: 1). She had already received the Cullum Medal of the American Geographical Society seven years earlier. She was awarded an honorary degree from the University of Kentucky in 1923 and the Helen Culver Gold Medal of the Geographic Society of Chicago in 1932—the first time the award did not go to an explorer. However, despite these achievements, the university paid her significantly less than her male colleagues.[5] Semple considered that a slight and retracted a planned donation to Clark University. Her tours took her around the world and included Japan, Mongolia, England, Greece, and other Mediterranean countries. One year before her death in 1932, she published *The Geography of the Mediterranean: Its Relation to Ancient History*.

Interestingly, United States women associated with maps were cosmopolitan in their orientation, while Canadian women linked with maps were highly contextualized in rural and Aboriginal settings.

Canada[6] represents an unusual case in the map world of the nineteenth century, perhaps inadvertently foreshadowing the idea of "power mapping" that would occur 100 to 140 years later. The uniqueness of Canada expressed itself in the fact that three of Canada's women cartographers were closely associated with Aboriginal peoples, and one was an Aboriginal herself. The women did not receive any training in map-making; they came upon map-making quite late in their respective lives; and they fell into map-making through coincidence or fortuitous circumstances.

Especially in Europe (and less frequently in North America), many women, in particular in the Victorian era, travelled widely, recording their experiences in diaries and sometimes creating maps along the way. For example, between 1789 and 1898, there is a known record of some thirty-six women explorers and travellers.[7] As Alison Blunt and Gillian Rose (1994: 9) point out, the interest in colonial women explorers has grown in literature, television, and films since the early 1980s. What is more, such interest has been guided more by accounts of women who have taken the spotlight, rather than by unassuming women, because being in the spotlight coincides with Western conceptions of what explorers and travellers ought to be. The difficulty resides in the fact that we are more likely to come across published records of "heroic" women than any other unadorned aspect of these women. Even the less-than-heroic women produced travel accounts, according to Sara Mills (1994: 35), that are "still implicitly producing knowledge" that shaped "the colonial presence."

THE INTREPID MINA HUBBARD, NÉE BENSON (1870–1956)

Mina Benson was born on 15 April 1870 in rural Ontario, Canada, to an immigrant family.[8] Her mother was born near Thorne, Yorkshire, and immigrated to Canada at the age of six. Her father immigrated to Canada from Ireland during the great Potato Famine of the 1840s. Mina's parents raised her and her six siblings with strict Methodist beliefs. Although her family's religion had mixed views on education, Mina's parents allowed her to attend school. She began elementary school, a typical single-room school.[9] From here she pursued high school, graduating at sixteen years of age.

After teaching for ten years as an elementary school teacher, Hubbard decided to move to New York, where she trained as a nurse. She attended the Brooklyn Training School for Nurses and worked at the General Memorial Hospital of New York. She eventually worked on Staten Island as a superintendent, where she met Leonidas Hubbard (1872–), a journalist. He came to the hospital with typhoid fever, and Mina nursed him back to health. The two married in January 1901 in a small New York church, with no family present. Settling in the Catskill foothills, they invited Leonidas's friend Dillon Wallace to live with them.

CHAPTER 5

Soon thereafter, in 1903, Dillon and Leonidas set off on an expedition to map the uncharted territory of the Naskaupi and George Rivers to Ungava Bay in Labrador. They were not successful in their endeavours, encountering numerous setbacks; Leonidas fell ill, and Dillon set out to seek help. Leonidas subsequently died, while Dillon survived. This sudden turn of events caused intense tension between Mina and Dillon, as Mina believed Dillon could have done more to save her husband's life. Mina, despite not being married to Leonidas for very long, was gravely affected by her husband's death. In effect, it inspired her to try to finish what he had started. It inspired Dillon to try as well.

In 1905, Mina and Dillon set off on separate missions to try to finish mapping the Naskaupi and George Rivers to Ungava Bay. One of the initial reasons for both the 1903 and 1905 expeditions was to meet the Naskaupi people, who were at the time said to be the most "primitive" Indians in North America. As a consequence, Mina's maps were seen as a major scientific contribution. They "stood as the accepted authority until the advent of aerial surveying" in the 1930s (Buchanan, Hart, and Greene, 2005: 369).

What might have been deterrents to some women, Mina saw as mere challenges. First of all, during the early 1900s, map-making and exploration itself were seen as something carried out by men exclusively. Second, she lacked any formal training in geography or other scientific disciplines. Finally, she was personally facing issues with gender and race in Labrador. She was also very frustrated with Dillon, but was inspired to finish her husband's work. The grief over her husband's death, it is said, became her religion.

On 20 June 1905, Mina set off on her expedition (see Figure 5.2). Both Mina and Dillon had asked George Ellison (a guide during the 1903 expedition) to join them on their separate trips; he choose to go with Mina and her group. The tensions between Dillon and Mina made their way into the press and people speculated over Mina's motives. Some claimed that she was not sincere in her efforts and that she was taking this trip to try and find out exactly what happened to her husband. Mina, however, stuck to her story that she simply wanted to finish her husband's journey because he could not.

In the end, Mina's expedition trumped Dillon's. After mapping parts of Labrador, she did not return to her regular occupation. She was no longer teaching or nursing. She began lecturing, writing articles, and working on formally documenting her time in Labrador. She spoke in a large variety of settings. In the first half of 1906 alone, Hubbard spoke as the "Ladies night" speaker at the Methodist Men's Social Union in Williamstown; at the

Figure 5.2
Mina Hubbard in Labrador.
Source: Roberta Buchanan and Bryan Greene and Anne Hart, *The Woman Who Mapped Labrador: The Life and Expedition Diary of Mina Hubbard* (Montreal: McGill-Queen's University Press), p. 142.

Present Day Club of Princeton, New Jersey; to the Adams School Teachers' Association; at a banquet of the Hamilton Club of Brooklyn, New York; and on a lecturing trip in Canada. In May of that year she returned to New York and continued lecturing. In 1907 she moved to England, still lecturing about her work. In the meanwhile, she named a number of geographical features in Labrador after some of her young nieces, such as Lake Agnes, Marie Lake, Maid Marion Falls.

During her lecture tours she continued working on her book of exploration, entitled *A Woman's Way Through Unknown Labrador*, completing it in 1908. Published by the American Geographical Society and adopted by the Canadian Geological Survey, Hubbard saw her book as a "purely pioneer work" (Hart, 2005: 369). The British (and later the American and Canadian) press awarded the book "approving if patronizing reviews" (Hart, 2005: 365).

Hubbard married Harold Ellis, the son of "one of the wealthiest industrialist families in England" (Hart, 2005: 375), in September 1908, and at the age of thirty-nine gave birth to her first child; two others would follow. She took great interest in the enfranchisement movement and followed the meetings with avid interest. However, during World War I she received the news that her husband, who had left to fight in the war, had taken up living with another woman. The divorce settlement took place in 1925.

In old age, Mina displayed signs of confusion and memory loss (Hart, 2005: 431). She died on 4 May 1956 at the Coulsdon South Station. Rather

CHAPTER 5

than using the passageway to get to the train, she "opened a gate bearing a sign, PLEASE DO NOT CROSS THE LINE, and walked straight into the path of an oncoming train." She was struck by the train and died in the ambulance on the way to the hospital. The only visible injury was a bruise on her head. (Hart, 2005: 432).

Thanks to various towns, the Ontario Heritage Foundation, and the District Historical Society, a plaque was erected as a testimony to her achievements:

> Mina Benson
> (1870–1956)
>
> An early twentieth century Canadian female explorer of Labrador and writer, Mina Benson was born on this property, lot 28, concession 7, township of Hamilton ... from June 27 to August 27, 1905, Benson completed the ill-fated 1903 Labrador expedition of her late husband, Leonidas Hubbard Jr., a distance of 576 miles from the Northwest River to Ungava Bay. Her maps, accepted by the American Geographical Society, were some of the first to record the Naskaupi River and George River systems. In 1908 she published an account of her trip, A Woman's Way Through Unknown Labrador. Later she remarried, had three children and moved to England where she died. (Hart, 2005: 428)

MOONSTRUCK MARY ADELA BLAGG (1858–1944)

Mary Adela Blagg typically belongs to a generation of amateurs who pursued science (astronomy, in this case) to the fullest extent possible and to whom scientists owe much of their knowledge.[10] She was born on 17 May 1858 in Cheadle, Staffordshire, to Frances Caroline Foottit and Charles John Blagg,[11] a lawyer. She had five sisters and four brothers. Mary remained her whole life in Cheadle, aside from staying at a private boarding school in London, where she studied algebra and German. She taught herself mathematics using one of her brothers' textbooks. When she was in her mid-forties, she decided to take a university course in astronomy at Wellington College. Her brilliance shone through so clearly that one of her professors, J.A. Hardcastle, invited her to solve a major problem in selonography: namely, the lack of standardization of nomenclature for features on the moon. Such an invitation came on the heels of a 1905 decision by the International Association of Academies to set up a committee to delve into solving this particular problem. The committee appointed Blagg to collate the names of all known lunar features. Her work took her to 1913, when she published her *Collated List* (see Figure 5.3). Reviewer A.H. Joy (1916: 88) describes this work as "a long step in the problem of lunar nomenclature and will be the basis from which a uniform system may be devised."

>by Mary A. Blagg (1935)
>"Do you lack basic drafting skills? So did Mary Blagg back in the 1930s, but she penciled the official lunar map for the International Astronomical Union that was the last graphical word on lunar nomenclature until the mid-1960s when the System of Lunar Craters officially replaced it. Miss Blagg, as she was always referred to, was an indefatigable lunar nomenclature researcher. She had worked diligently to identify and resolve differences in names and letters given on different maps by nineteenth century lunar authorities, and the publication in 1935 of the *Named Lunar Formations* catalog and maps memorialized the sanity she brought to a confused situation." Words by Chuck Wood.

Blagg served as a volunteer for Professor H.H. Turner, director of the University of Oxford Observatory, being associated with his work for more than twenty years of untiring effort. According to Roger Hutchings (2004), Blagg "sorted out a mass of original variable star observations by Joseph Baxendall, editing the whole by 1918 for discussion by Turner and subsequent publication." She and Turner jointly published nearly a dozen papers in *Monthly Notices*.

Despite her desire to continue living in Cheadle (and lead a quiet and isolated life), Blagg performed volunteer work that included caring for Belgian refugee children during World War I.

In 1920, the newly created International Astronomical Union appointed her to the Lunar Commission, under whose aegis she continued her work on standardization. When she published her *Named Lunar Formations* in 1935 with Karl Muller of Vienna, it became the standard reference work for thirty years, naming some six thousand features of the Moon.

Blagg had to work her way through numerous inaccurate, misleading, and conflicting nomenclatures found in the works of a large number of crucial scientists, such as W.G. Lohrmann's measurements of 1824–36. Thanks to all their efforts, measurements, and drawings, Blagg managed, in 1922, to draw maps to be included in an atlas depicting the moon's ten outer portions and the four inner quadrants.

Although reserved, Blagg spoke straightforwardly and did not refrain from criticizing earlier astronomers; the scientific community accepted her criticism, which rested on the solid foundations of having done "masses of tedious work for others." Her objective was to disentangle earlier errors and

CHAPTER 5

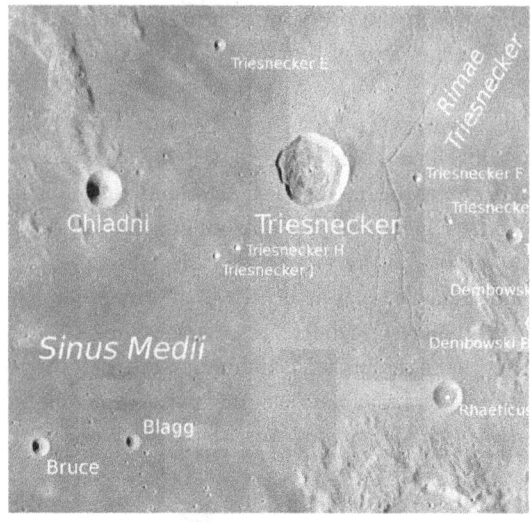

Figure 5.3
The Official Lunar Map for the International Astronomical Union.
Source: http://lpod.wiki spaces.com/July+25, +2009.

to create order out of that chaos. Not unlike amateurs in other fields, such as botany and geology, to name two, Blagg published her work in scientific journals. She published six papers in the *Journal of the British Astronomical Association* and, in 1916, became one of the first five women to be elected a fellow of the Royal Astronomical Society. Blagg never married. She died on 14 April 1944 at the age of eighty-six after suffering from heart disease for many years. The International Astronomical Union named a lunar crater for her after her death. There is a report of an obituary that describes her as being of "modest and retiring disposition, in fact very much of a recluse," who rarely attended meetings.

THE PERIPATETIC PHYLLIS PEARSALL (1906–1996)

Phyllis Pearsall of the United Kingdom created *A–Z Maps*, which is now well known by millions of tourists and residents alike.[12] She travelled more than 3,000 miles in London alone, sometimes eighteen hours a day, eventually covering 23,000 landmarks (Knowles, 2003: 16). Her first map (of London) was printed in 1936 and was an instant success (Straight, 1997: 4). The Design Museum in London describes it as "one of the most ingenious examples of early twentieth century information design" (Design Museum, 2006). Writers on urban planning sometimes refer to Pearsall as the only connection between early forms of feminism and entrepreneurial businesswomen (see, for example, Greed, 1994: 109). Unlike in North America, where academic or professional societies recognized pioneers in cartography, the United Kingdom was wont to recognize individuals under the aegis of public institutions. Pearsall became a Member of the British Empire and, in 2005, Southwark Council affixed a plaque to the house where she was born: Court Lane Gardens, Dulwich.

Phyllis Isobel née Gross was born on 25 September 1909 in Dulwich, in southeast London. Her Irish-Italian mother, Isabella Crowley, was an artist (Design Museum, 2006). Many saw her as "elegant," "brilliant," "genuine," and a "remarkable business woman" (Pearsall, 1990: 25). Pearsall's father, Alexander Gross (1879–1958), was a "truculent Hungarian immigrant" (Grace's Guide, 2008: 1) who never gave his wife credit for her contributions to the mapping company, Geographia, he had founded in 1908 (Pearsall, 1990: 25), and who, after going bankrupt, left for America. Although she attributed to her father a beneficial influence on her life, that is where his influence stopped: "This is not to denigrate an impossible father (loved in retrospect) in order to magnify myself, but to demonstrate that a business woman true to her nature can be successful in her own right.... I'm no Warrior Queen" (Pearsall, 1990: "Foreword").

Her mother's lover, Alfred Orr, the royal portrait painter, threw Pearsall out of the house after she turned fourteen (Heald, 2006: 2). Pearsall was pulled out of Roedean, an expensive girl's boarding school (Greed, 1994: 109), and left for Paris to teach English at a girls' school. She learned French and attended philosophy classes at the Sorbonne. Having no money, she slept on the streets (Pearsall, 1990: 21) under newspapers; she warmed herself on library radiators (Pearsall, 1990: 81). She took up portrait painting and etching in addition to writing articles, all for rather meagre wages. She returned to England in 1926, where she later married Richard Pearsall, a friend of her brother Anthony's; they were married for eight years and separated in 1935.

A new phase of life began when her father asked her to publish a world map produced in the United States. In her own words, she became a map publisher "[i]nadvertently, and without ambition or envisaged aim" (Pearsall, 1990: "Preface"). Aside from learning the technical jargon of cartography, she had to take it upon herself to sell the map directly to customers. In one of those trips[13] in 1935 to the Belgravia area in London, she lost her way and came up with the idea of creating an up-to-date map of London. Working from her bedsitter on Horseferry Road near Victoria Station, she began cataloguing every street in London (Design Museum, 2006). She founded the "Geographers A–Z Map Company." When she produced her first *A–Z Map* of London in 1936, she had walked some 23,000 streets in London,

CHAPTER 5

"collecting street names, house numbers, along main roads, bus and tram routes, stations, buildings, museums, palaces, etc." (F. Bond, 1996: 1). From the earliest days of her mapping of London, Pearsall faced even the simplest challenges: "Slapdash or non-existent records or petty bureaucrats' refusal of entry sent me checking on the ground—chaotic after checking maps. Often in the maze of many a turning off many a side street I found myself back where I started, or, completely lost, had to ask the way" (Pearsall, 1990: 27).

What bedevilled her were illogical house numbers and designations, misspelled street names, and the like. The border of the next postal district was enough to start renumbering the houses and to change the name of the contiguous street.

Eventually she also created a complete index of street names—an essential feature of any usefully good city map. To achieve her exploration of the streets, she would get up at 5:00 a.m. and walk for eighteen hours a day. James Duncan, a draftsman, was her only colleague.

For Pearsall, there were many struggles along the way. For example, just when she had completed her map records for London, the City decided to change the names of two thousand roads to avoid duplication of names. A flimsy shoebox filled with those changes, moreover, had spilled out of her window, and despite her efforts to recover most of the cards in flowing traffic, some of them (including "Trafalgar Square") "disappeared on the top of a passing bus" (Knowles, 2003: 18). She also faced the conservative map trade, who frowned on outsiders. The first person to buy her map was a Mrs. Naylor, a widowed newsagent in Clapham (Pearsall, 1990: 47). Pearsall sold one copy. When she apprised someone of her maps, he snorted, "Tell me the name of the Company you represent, and I'll know what maps not to buy!" (Pearsall, 1990: 27).

As publishers rejected her map, she decided to publish ten thousand copies herself (Heald, 1996: 2). It took persistence to join the crowd of salesmen each day at the head office of bookselling chain W.H. Smith and Son, and to be ignored day after day, until someone in the crowd took pity on her as the sole woman and pushed her to the front of the line (Knowles, 2003: 18). Eventually, her efforts met with success, as orders started to flow in. She rose as early as 4 a.m. to fill these orders, and on the same day would push a wheelbarrow filled with her maps (Pearsall, 1990: 52).

While interest in her distinctive maps grew (each member of the Metropolitan London Police was issued a uniform, whistle, truncheon, and an A–Z Map; "the A–Z," says one officer, "was by far the most used" (Peter Bolt in Heald, 1996: 4). During World War II, maps such as hers were curtailed by the government and she worked for the Home Intelligence Division of the Ministry of Information. Once the war was over, she returned to her mapmaking enterprise. However, calamity soon followed. A 1948 plane crash[14] left her with a badly fractured skull and spine. She forced herself back to work, but went blind for several months. Two years later, she had a stroke (Knowles, 2003: 18). Her frail health required long periods of recuperation. Staff had to carry her upstairs to her office (Bond, 1996: 2). There was also betrayal among her staff, as the then manager of her firm led the business into a downturn (2). Nevertheless, she stood her ground and did not let these trials get in her way. As one of her managers described this challenging period, "From Pearsall's sick bed she began to have suspicions that all was not well. Yet again her astute foresight and awareness of danger saved the company, but the drama took its toll ..." (2). A designer, Ian Griffin, would later say that she was "both determined and inspirational, knew exactly what she wanted to do and carried everyone with her" (Heald, 2006: 2).

When she came upon a company with staff who were inconsolable because they were about to lose their jobs, she hit on the idea that the staff in her company would henceforth hold a hundred percent of the shares. She gave up "any legal right for a dividend, a pension, or even employment" (Knowles, 2003: 19). She despised bureaucracy, but placed "great faith in spontaneous decision making" and banned Board meetings (Design Museum, 2006: 2). In fragile health, she died in Shoreham-by-Sea on 28 August 1996, just before she turned 90. Today, her company employs 129 people and publishes over 350 titles (Heald, 2006: 1).

Aside from her *A–Z Maps* legacy, Pearsall demonstrated what turned out to be remarkable foresight about the needs of businesses well into the future when she stated her business objectives: "a commitment to natural and sustainable growth ... in the hope of bringing together a work team that would appreciate and thrive (both in their work and in their private lives) in an atmosphere of stability, mutual trust, honesty, and high endeavour" (Design Museum, 2006: 3).

"MAPPING MAIDS"/MILLIE THE MAPPER: 1941–1945

World War II fuelled an unexpected demand for women to enter occupations traditionally left to men. Geology and cartography were some of the fields marked for an infusion of women. In some respects, we can trace the rise of women pioneers in cartography to this infusion. For example, the "Mapping Maids"—a term Avril Maddrell used to refer to women the United States government had recruited to replace men cartographers who

CHAPTER 5

had gone off to war in 1942—entailed educating and training women in several fields, paving the way for other women to enter the field and exercise their due influence.

Many technological strands define this prelude; it might be difficult to pinpoint either its starting date or end date. Some would claim that the use of kites and balloons in the mid-nineteenth century signifies a start.[15] Others would assert that the invention and use of planes, especially in World War I, indicates the start of the prelude (Hodgkiss, 1981: 55) of sizable mapmaking companies, such as Rand McNally (est. 1868) and Hammond (Tyner, 1997: 50), attracting women to their employment ranks. However, the rise of academic cartography in the twentieth century allows us to learn more about the participation and contributions of women.

Of particular note was the rather unexpected rise of young women to cartographic service in the United States during World War II. In December 1941, President Roosevelt signed for additional monies for the Army to secure the services of one thousand to two thousand additional employees to produce maps (Tyner, 1999: 23).[16] It was already recognized that the United States lacked suitable maps for most areas of the world, including in those regions where war was breaking out. After the United States declared war, it had to rely on local "second-hand" maps provided by civilians and refugees from Europe (Mandell, 1996: 45).

By 1942, ninety-nine courses related to topographic mapping were approved by fifty-seven institutions in thirty states. The courses included topographic map drawing, the use of surveying instruments, surveying field procedures, plane-table topography, and photogrammetry. The expected pay would be between US$1,440 to $1,620 per year; the women would work up to seventy hours a week on secret maps, many of which would never see the light of day (Straight, 1993: 2). An overwhelming number of women (usually referred to as "girls" in those days) attended these courses. In other courses, such as photogrammetry, about one-fourth of the students were women (Tyner, 1999: 24). For the most part, the women enrolled in courses in departments of geography, rather than civil engineering or surveying.

By the end of World War II, "thousands of women," according to Judith Tyner (1999: 24), "had been involved in cartographic activities at all levels, but in particular drafting, geographic research, libraries, and training in cartography and in map reading." However, the civil service and other governmental agencies hired "Millie the Mapper" at "sub-professional" levels. The pay was good. Agencies that involved cartography seemed particularly open to hiring women. The women who were not working at the sub-professional levels were hired as professional women geographers for research at the Office of Strategic Services. In this category, there were some two hundred women. The number of professional women *cartographers*, however, remained small: namely, two at the United States Coast and Geodetic Survey and a small number at the United States Geological Survey (USGS). The Tennessee Valley Authority, the major employer of women, hired ten women as photogrammetrists (Tyner, 1999: 25). The Navy also recruited some women with special training in cartography. The Military Map-making Corps employed two hundred women (Straight, 1993: 1). The students had recently graduated from universities on the East coast and in the Midwest. The "3M Girls" (Military Map-making Girls)—some two hundred worked for the Military Mapmaking Corps—received a government-sponsored sixty-hour training course (Straight, 1993: 1). Eventually, the women made some forty thousand maps covering 40,000 square miles of the world, resulting in more than five hundred million military maps compiled from these originals (Straight, 1993: 2).

At the level of state agencies, women increasingly came to occupy relevant positions. The reaction by men was best summed up by a statement found in the 1944 report of the American Congress on Surveying and Mapping, which stated, "Miss Elisabeth M. Herlihy, the first woman to appear on a Congress program [and Chair of the Massachusetts State Planning Board], glamorized surveying and mapping. She covered a broad field of state planning and pointed out specific cases with respect to the importance of surveying and mapping information. She won the attention and hearty applause of all those present, and even the 'hard boiled' engineers and surveyors were made to realize that woman's place transcends the boundaries of the home" (Tyner, 1999: 26).

A truly large question arose: What would happen to "Millie the Mapper" after the war? The consensus was in: women would succeed in such work. But there was also the belief that after the war, it would be difficult for young women to complete an engineering degree between the time they graduated from high school and the beginnings of married life (Tyner 1999: 26). Some, including the Women's Bureau of the United States Department of Labor, felt that instruction in cartography, which included mathematics and civil engineering, would proceed well to the advantage of women. Map librarianships would also open up, according to the Women's Bureau (Tyner, 1999: 26), mainly due to the War Department's earlier search for older maps, including those of cities, ports, and the like. The twentieth century stands associated with the most iconic works in cartography.

CHAPTER 5

There was, however, a more dramatic side to map-making awaiting women. World War II, as we shall see in the next chapter, led the military to draw many women in Great Britain and the United States into cartography. Academic cartography emerged as a new field for women.

CHAPTER 6

Mid- to Late-Twentieth-Century Pioneers and Advancers in North America

In comparison to other continents, North America has been a particularly fertile field for women in cartography. Marie Tharp constitutes the most significant offshoot of the infusion of women into fields traditionally held by men. Tharp was an oceanographic cartographer whose maps of the world's ocean floors became iconic of the twentieth century, only to be outdone by pictures of Earth from the surface of the moon. Another veritable pioneer was Mary G. Clawson, an important contributor to digital mapping. She represented the applied arm in cartography, not only configuring the future role of global positioning systems (GPS) in cars, twenty years ahead of its time, but also becoming the highest-ranking cartographer in the United States military. Women in academic cartography represent the other strain in cartography where nearly a dozen women in North America (and eight in Europe, Eastern Europe, and Asia)[1] left their mark on the field. This chapter traces the interrelationship between the "Mapping Maids," Marie Tharp, and the academic cartographers.

The following fifteen vignettes of North American women (and eight women elsewhere, in Chapter 7) underscore the larger discourse about women in science that Joyce Tang (2006) so succinctly presented in *Scientific Pioneers: Women Succeeding in Science*. Based on the works of Dean

CHAPTER 6

Keith Simonton and his many treatises on the genius, Tang identifies three types of distinctions among scientists who advance the field:

1. *Pioneers or precursors* [who] venture into unknown or unexplored domains,
2. *Advancers* [who] are the majority of scientists, [and] who try to build on an existing body of knowledge, ... [and]
3. *Revolutionaries* [who] rearrange the pieces to generate more explanatory power over previous configurations. (Tang, 2006: 2)

Appendix A explains how I selected the twenty-three women cartographers as pioneers and advancers in the field. Appendix C offers a thumbnail sketch of all twenty-eight pioneers, including the five presented in the previous chapter. Each vignette at least incorporates what Tang (2006) thinks are the essential characteristics of pioneers and advancers, namely a creative thinking style (10–13); personal qualities (such as intellect, drive, ambition, determination, risk-taking) (13–14); family and home background (15–17); and "good timing" (17–18). The reader should bear in mind that given the divergent and scattered nature of information often available about the women, it is not always possible to present the same outline of information. Some areas show large gaps; other areas are well filled with information.

In concrete terms, there was a rise of professionally recognized cartographers, although a few stayed on at the military mapping agency, "but most married [and] developed other careers, such as teaching" (Straight, 1993: 2). The ACSM (American Congress on Surveying and Mapping) had no women members at its founding in 1941, but by April 1942, four of the four hundred members were women. By 1945, women's membership had grown to twenty-four. In 1952, the American Society of Professional Geographers had 252 members with a special interest in cartography, forty of whom were women (Tyner, 1999: 27). From available data, it seems that many of the women professional cartographers worked in Washington, D.C., as teachers, map librarians, editors, and researchers. Significantly, Arthur Robinson headed the cartographic division of the Office of Strategic Services. Robinson would later prove to be a major nurturer of women in academic cartography. It is also noteworthy that while it was the engineering departments that traditionally inculcated the skills in cartography, the education and training of the "Mapping Maids" paved the way for departments of geography to assume cartographic training.

MARIE THARP (1920-2006)

Marie Tharp (see Figure 6.1) made the most notable contributions to seafloor cartography; her *Ocean Floor* panorama (created with Bruce C. Heezen, an oceanographer at Columbia University) "helped to spark one of the greatest scientific revolutions in history": namely, the theory of continental drift (Lawrence, 1999: 37) (see Figure 6.2). Their map threw into relief the "largest uncharted landscape of the world's ocean floor" (Fox, 2006), becoming the "first comprehensive map of the entire ocean bottom, illuminating a hidden world of rifts and valleys, volcanic ranges stretching for thousands of miles and mountain peaks taller than Everest" (Fox, 2006). It was called "one of the most remarkable achievements in modern cartography" (Wilford, 2000: 327). Doel, Levin and Marker (2006: 620) confessed that "[n]o single narrative adequately captures the impact" of that map that Tharp and Heezen compiled. The map touched commercial and military interests, geology, global geophysics, and plate tectonics, and eventually defined concerns that led to the United Nations Law of the Sea Conference in the 1970s and 1980s. Doel, Levin, and Marker (2006: 621) put its significance at the same level as the Martin Waldseemüllers 1507 map of "America."

It was Tharp herself who pointed out that her map compares in some ways to the "Copernican revolution" and had "shaken the foundations of geology!" (Lawrence, 1999: 42–43). Until the eighth decade of her life, Tharp "rarely got scientific recognition," but in 1997, the Library of Congress named her as one of four individuals "who have made major contributions to the field of cartography" (Lawrence, 1999: 42). In 1978, Tharp and the late Heezen received the Hubbard Medal of the National Geographic Society (Anon., 2004: 2). In 2001, Columbia University recognized her work with the Lamont-Doherty Heritage Award (which now also has a fellowship in her name to promote women in science).

Marie Tharp was born in 1920 in Ypsilanti, Michigan.[2] Her mother, Bertha Louise (née Newton), taught German and Latin as a schoolteacher, but left teaching upon her marriage. Her mother was forty and her father was fifty when she was born. Her father, William Edgar Tharp, was a soil surveyor and the family moved often. He told his daughter "to choose a job simply because she liked doing it." She had a half-brother, seventeen years her senior. Tharp often accompanied her father on his surveying expeditions. Her education covered not only a wide geographical area (she attended two dozen public schools, and by the time she was fifteen had

CHAPTER 6

Figure 6.1 Marie Tharp (1920–2006). *Source*: http://www.columbia.edu/cu/news/06/08/tharp.html. http://www.marietharp.com/. Photo by Bruce Gilbert. *Permission*: The Earth Institute, Columbia University.

moved thirty times), but she almost flunked out of fifth grade; still, she took in a diverse body of knowledge.

Tharp's first choice of university was St. John's College in Annapolis; it did not admit women, however. She completed an undergraduate degree in English and music and four minors in 1943 at Ohio University, where she created her own version of the University of Chicago's literature classics course. World War II opened up employment possibilities and she followed up with an M.Sc. in Geology (University of Michigan)—a program designed to attract women. Soon she worked as a junior geologist for Stanolind Oil and Gas Co. in Tulsa while she earned a bachelor's degree in Mathematics (University of Tulsa) in 1948. Because women were not allowed to search for oil, Tharp found herself organizing maps for all-male crews. Still, she was searching for something more challenging. She moved to New York, where she found a job as a research assistant at the Lamont Geological Observatory at Columbia University, becoming one of the first women hired by the Laboratory, arriving there in 1948.

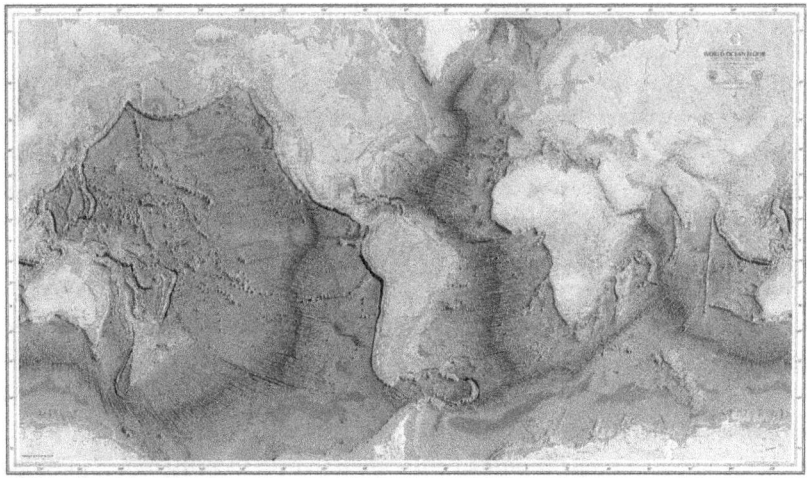

Figure 6.2 *World Ocean Floor Panorama*, Bruce C. Heezen and Marie Tharp, 1977. *Source*: Marie Tharp Maps, New York.

Initially asked by Maurice Ewing, a prominent geophysicist, to assist graduate students with drafting and computing, she soon was working closely with Bruce Heezen, whose interests included the Atlantic ocean floor. As Navy convention (until the mid-1960s) did not permit women to go out on its vessels to explore the sea, Tharpe stayed in the laboratory, assembling the data collected by Heezen, who was permitted to go out on these vessels. Pulling together the data from government, but also from commercial and scientific vessels, was not simply a sinecure, as the data from the echo soundings were erratic, unsystematic, and given to huge fluctuations. To her own surprise, she discovered a rift in the middle of the Atlantic Ocean whose only explanation was continental drift, a theory advanced by Alfred Wegener in 1912. Mid-oceanic earthquakes, she later found, supported her theory. She and Heezen announced the global rift system in 1956 at a meeting of the American Geophysical Union in Toronto. (The following year, *Sputnik* traversed space.) Today, their dramatic *World Ocean Floor Panorama* evokes as much admiration as it did when it first came out in 1977 after thirty years of collaboration (Barton, 2002: 216). Their enthusiastic collaboration as a scientific couple benefited each of them—there was no more than a platonic relationship. He pursued data out at sea; she collated and interpreted his data on shore. They co-authored maps, books, and articles. Initially, Heezen was skeptical of her interpretations and brushed them off as "girl talk." When Tharp first showed him what she had found, "he groaned and said, 'It cannot be. It looks too much like continental drift'" (Tharp, 1999: 3). Tharp persisted in her "heresy" and conviction that the rift represented the separation of Europe and North America (Lawrence, 1999: 40); after about two years, Heezen was forced to recognize the strength of her interpretations, which flew in the face of fixist theories. Tharp (1999: 6) admitted that she "was so busy making maps" that she "let them [scientists] argue."

In the face of a personal vendetta and challenges, a particular quarter of Columbia University first tried to fire Heezen (an unsuccessful gamble, as he was tenured), then banned him from sailing the expedition ship needed for measuring the oceans' depths, and then prevented him from using the Lamont Laboratory. That quarter did fire Tharp, who continued to bring her maps home, where she could work on them. Heezen died of a heart attack in 1977, just a few months before their *World Ocean Floor Panorama* was published (Tharp, 1999: 8). Tharp died in 2006. In reflecting on her career, she said, "I worked in the background for most of my career as a scientist, but I have absolutely no resentment. I thought I was lucky to have a job that was so interesting. Establishing the rift valley and the mid-ocean ridge that went all the way around the world for 40,000 miles—that was something important. You could only do that once. You can't find anything bigger than that, at least on this planet" (Tharp, 1999: 8).

MARY G. CLAWSON (B. 1948)

Mary G. Clawson is considered a pioneer in the use of digital geospatial data to solve complex problems, including the development of advanced weapons systems, and a leader in transforming the research, identification, evaluation, analysis, and acquisition of geospatial source information in support of critical national security issues.[3] Her professional experience is firmly rooted in the geospatial sciences—cartography, geography, digital geospatial information, and remote sensing.

She was born in Philadelphia on 20 October 1948 to Mildred Florence Hamlin (1915–2003), a homemaker, and Frank Rosario DiPaula (1901–1972), a medical doctor. A brilliant chess and bridge national champion, he always encouraged Mary to do whatever she wanted. Mary attended schools in Louisana, California, Alabama, New York, and Maryland. Among women cartographers who were influential in her career were Alberta Wood and Barbara Petchenik.

Her love for cartography and mapping dates back to fifth grade. At Towson University in Maryland she studied mathematics and geography for a career in teaching. She married David Clawson, her college sweetheart, in 1969. While he was drafted into the Army soon after, she finished her degree. She was enjoying cartography more and more and worked part-time in the Cartography Lab. However, it was her experience in student-teaching eighth graders, in 1970, that convinced her that she really did not want to teach, but rather wanted to be a cartographer.

After graduating in 1970 with a bachelor of science in geography, she took a job as a draftsperson working for the Baltimore Regional Planning Council. Her job was to build a crude digital outline map of the Baltimore region. Using a large grid to determine points that were then key-punched, she got a taste of automated cartography in its infancy. When her husband returned from Vietnam in 1971, they moved to Kansas.

With no work in sight as a cartographer, she worked with maps as a volunteer at the local Soil Conversation Office. Afterwards, she worked in the soils lab at the University of Kansas and then, under the direction of Dr. Bidwell, she compiled, designed, drafted, and completed all of the graphic arts colour separations for the map. In 1972, the first *Soils of Kansas* was published and catalogued in the Library of Congress. It was her first published map—and she now regarded herself as a cartographer.

In 1974, she started her work for the Military Charts Branch at the Defense Mapping Agency (DMA) Hydrographic Center. She had never seen a nautical chart before. She said, "I was hired to make something I'd never seen or used…. Wow! Did I have a lot to learn!" (Mary Clawson, email to author, 25 March 2012). She thinks of herself as a pioneer in the use of digital mapping data and in digital maps for the Department of Defense. Her next projects involved a map series showing the key cities across Europe, compiled and drafted using automated cartography techniques.

While working full-time at DMA from 1975 to 1980, she attended University of Maryland night classes to obtain her master's in cartography. Based on her earlier curiosity about nautical chart symbology, she researched the "Evolution of Symbols on Nautical Charts Prior to 1800." She spent Saturday mornings at the Library of Congress reviewing every available nautical chart, from the earliest portolan charts drawn on animal skins in the 1400s up to nautical atlases printed from copper plates and hand coloured. Starting in 1975 and continuing throughout her career, participation in conferences was integral to her growth as a professional. Her main professional association was the American Congress on Surveying and Mapping (ACSM); later she was an elected official of ACSM. She was also the Director of Auto-Carto 5.

CHAPTER 6

Starting in 1980, Mary began working with a weapons system developer to use digital mapping data—probably one of the earliest uses of digital map data in a deployed military system. In 1981, she left DMA to take a position with the IIT Research Institute (IITRI) in Annapolis, where she quickly learned that the imagined uses for digital maps far outpaced the capabilities of computers in those days, and in terms of accurate positioning, global positioning systems, or GPS, was just being launched. Between 1981 and 1984, she helped set the agenda for the use of digital maps in advanced weapons to be developed over the next two decades. She was part of the team that conducted the Army's first requirements assessment for digital geospatial information.

One of the most interesting research projects she conducted while at IITRI was for Ford Motor Company: namely, to research how people navigate in automobiles and how this information would translate into design considerations for in-car navigation systems. Working with Dr. Barbara Petchenik, the pre-eminent cartographer, and Dr. Arnold Greenland, a professional statistician, she conducted a literature review of human navigation and subsequently designed a large telephone survey to learn more about how people navigate while driving. Three key needs emerged from this research: detailed digital maps; an accurate positioning system; and navigation guidance provided as voice commands from the system. Time has proven that their recommendations were sound and are now reality. It took more than twenty years for everything to come together.

Based on her work at IITRI documenting digital map data requirements for weapons systems, she was recruited by Dr. Don Durham to work for the Navy in December 1984. She became the lead for all geospatial issues for the Department of the Navy and Chief of Naval Operations. Over the next eleven years, Mary worked with weapons system developers and Navy program offices to ensure that required digital geospatial information was available in time to meet operational deployment. Her in-depth knowledge of all digital geospatial products and their appropriate uses, as well as her thorough knowledge of weapons systems and their characteristics, enabled her to serve effectively as the conduit between developers, program managers, and her counterparts at DMA. As the Navy expert on digital geospatial information, she participated extensively in the design of avionics, navigation, mission planning, and command and control systems. She also often coordinated with colleagues from the Marine Corps, Air Force and Army, and prepared the Navy position on mapping, charting, and geodesy (MC&G) issues for the Chief of Naval Operations and the Secretary of the Navy.

In 1996 she returned to National Geospatial-Intelligence Agency (NGA), formerly DMA, where she took on several challenges related to understanding and integrating the core geospatial information and imagery intelligence segments into one element—Geospatial-Intelligence (GEOINT).

She was promoted to the Defense Intelligence Senior Level (DISL) in 2005 and served as the Senior Advisor for Regional Analysis and Open Source. She occupied a number of important positions at the NGA, including Imagery Analysis-Geospatial Information Integration Manager for the Strategic Planning Division (1998–2001); deputy chief, Imagery Analysis Corporate Planning and Operations Division (2001); business executive and executive officer, Office of Asia Pacific (2001–2005); DISL for Regional Analysis (2005–2006); and deputy director, Office of Global Navigation (2006–2008). She retired from the NGA in 2008 with more than thirty years of federal service.

Rather vast are her accomplishments in the leadership of her professional organizations: Fellow in the American Congress on Surveying and Mapping (1998); Bureau Member and Congress Director, International Federation of Surveyors (FIG) (1998–2002—the first woman Congress Director); and numerous leadership positions in the American Congress on Surveying and Mapping, 1980–2002.

Her awards are too numerous to cite individually, but they include the Meritorious Civilian Service Medal (2008) and the National Intelligence Meritorious Unit Citation (2001). Earlier, she was honoured with NGA Special Acts Awards, the Joint DMA and NIMA Meritorious Unit Award, Navy Performance Awards, the Defense Mapping Agency Outstanding Performance Award, Defense Mapping Agency Suggestion Awards, American Congress on Surveying and Mapping Presidential Citations, and the American Society for Photogrammetry and Remote Sensing Presidential Citations.

In reflecting on her work, Mary said,

> I was frequently the only woman in the room. But that never mattered, because my expertise was a key contribution to many successful firsts in the realm of applying digital map data to support critical systems for the US military and intelligence community. In the professional associations, it wasn't that I was a woman that put me in leadership positions; it was that I was recognized for being able to do the job. There are now many more women in leadership positions, including Ms. Letitia A. Long who was appointed Director of the National Geospatial-Intelligence Agency on August 9, 2010, and the first woman to head an Intelligence Agency.

CHAPTER 6

ACADEMIC CARTOGRAPHERS

After World War II, cartography moved from engineering departments to geography departments, initially only for training purposes, but by the 1950s, according to Judith Tyner (1999: 27), academic cartography became an established feature in these departments, too. There was a real sentiment at that time that cartography should leave civil engineering and establish its own home (McMaster and McMaster, 2002: 314).[4] From 1946 to 1986, major graduate programs in academic cartography flourished at the universities of Wisconsin, Kansas, and Washington, although with the development of geovisualization, GI science, GIS systems, and remote sensing, its future might be uncertain (McMaster and McMaster, 2002: 305). One of the outcomes of that process was Pat Gilmartin, a historical cartographer and graduate of George Jenks at the University of Kansas. Her particular interests include cognition of space and maps (Gilmartin, 1985), namely "the cued spatial response approach to macro-scale cognitive maps." She is now retired from the University of South Carolina.

In a number of respects, if it were not for the enthusiastic response of university geography departments to the American appeal for hiring cartographers who could assist in the war effort, the entry of women into cartography might have been delayed still further. As Tyner (1999: 27) shows, academic cartography quite naturally fell into geography departments, which increasingly enrolled young women. It was acceptable now to write a thesis on technical cartographic subjects. Of the forty-two dissertations in the 1970s and 1980, eleven were authored by women. These freshly minted Ph.D.s became active in professional organizations, contributing their share to women's participation in cartography.

The advances for women in academic cartography, however, are not as straightforward in, for example, one European country. The woman president of a Royal Geological Society had this to say in an interview with the author:

> I'm convinced that the Geological Society is both a professional as well as a scientific professional and scientific charter. And for decades it had been dominated by the academics, which meant they were losing members and they were simply not responding properly to the demands of the professional members. So we shifted that a lot.... So what we see now is that academia is turning its back on us. We shifted to a proper 50–50 balance. Academia is turning, all the professors are now patting each other's [back], scratching each other's back in the National Science Foundation Committee where they are

all exclusively male ... and because this has now become their domain. They simply don't want to share the domain with professionals and so we try to have to get some of them on a different footing. (Interview with "Leslie": 7)[5]

Academic cartography arrived, attracting many women to the fold. One group of women centred on the University of Wisconsin (under the tutelage of Arthur H. Robinson) include Mei-Ling Hsu (the first woman to write a Ph.D. dissertation in the field, in 1966), Barbara Bartz Petchenik, and Judy Olson. One would find another group on the west coast: Judith Tyner, Patricia Caldwell, and Evelyn Pruitt. Norman J.W. Thrower and Clifford Zierer can be counted as two of their mentors. The University of Kansas and Washington University were the home of a third group, involving Barbara Shortridge, Barbara Buttenfield, and Patricia Gilmartin. Their mentors included George Jenks and John Sherman.

Thus, a small band of men cartographers was able to change the gendered face of cartography. Academic cartography entered a very productive period. Several women wrote master's theses in the 1960s, but by the 1970s and early 1980s, as mentioned above, women had written eleven of the forty-two dissertations in the field (nearly 26 percent). A number of these women became actively involved in professional organizations as presidents, or became executives in private industry (Tyner, 1999: 27). Table 6.1 illustrates the percentage of women members in various professional organizations.

All nine women academic cartographers are Ph.D. graduates from four universities (Wisconsin, UCLA, Washington, Kansas); at least three were tutored directly by Arthur Robinson at the University of Wisconsin (and a further four indirectly, through other students he has taught). Another two were tutored by Norman Thrower, and two by George Jenks. Especially in light of the fact that there were seventeen other venues[6] that developed significant programs in academic cartography, what was it about those four universities that rendered them so hospitable to women cartographers? In particular, what was it about the environment created by Robinson that nurtured women? It would be most useful to explore the status and social structure in these universities that permitted such an abundance of women academic cartographers. In the end, seven worked in universities or community colleges, while one worked for the Department of the Navy, and another worked in a private firm.

Arthur Robinson (b. 1915) of the University of Wisconsin had been a graduate student at Ohio State University when Richard Hartshorne hired

Table 6.1. Membership in selected cartographic organizations, 1987

Organization	Total Members (in N)	Percentage of women
American Cartographic Association	813	16%
Cartography Specialty Group of the Association of American Geographers	>500	31%
North American Cartographic Information Society (1995)	385	33%

Source: Tyner, 1999: 27–28.

him to come to the newly established Geography Division in the branch of Research and Analysis of the Office of Strategic Services.[7] One of Robinson's many accomplishments was his *Robinson Projection* of the world, which has become the standard projection for all the world maps of the National Geographic Society (http://www.amergeog.org/ags_in_memoriam.htm). Karen Cook, however, points to his greatest achievement: his talent of inspiring graduate students (Cook, 2005: 196), particularly women students. Wisconsin University produced, according to McMaster and McMaster (2002: 312), "several hundred students with Master's degrees in cartography and well over twenty students with doctoral degrees in geography, but specializing in cartography." Arthur Robinson, the "Dean of American Cartography," had many students, including Mei-Ling Hsu, Judy Olson (past President of the Association of American Geographers [AAG]), and Barbara Bartz Petchenik. Another of Robinson's students, Norman J.W. Thrower, had female students who included Judith Tyner and Patricia Caldwell.

John Sherman (1916–1996),[8] born in Canada, was part of the Department of Geography at the University of Washington for almost half a century, from 1939 to 1986. He was the founder of the modern cartography program at that university, the first full-time cartographer at any American university for over three decades, and one of the leading cartographers in the country (http://faculty.washington.edu/krumme/faculty/sherman.html). Judy Olson described him as a fine, humble, and thoughtful man who was one of the "triumvirate" (Robinson, Jenks, and Sherman) of his generation. He is best remembered for his works on maps for the blind, but he was a general cartographer, not limited to a specific area (Judy Olson, email to author, 1 October 2002).

George Jenks (1916–1996), a professor of geography at the University of Kansas from 1949 until his retirement in 1986, was born in upstate New York.[9] A large number of contemporary women cartographers were his students, including Barbara ("Babs") Buttenfield, Jill Marino, Susan Waldorf, Barbara Shortridge, and Patricia Gilmartin. One woman cartographer referred to him as "definitely a character. He never hesitated to ask pointed questions or stir things up (intellectually) when people gave talks or reported on their work. And everyone loved it and few hesitated to argue with him (and all laughed a lot)" (Judy Olson, email to author, 1 October 2002). Olson also recalls hearing him give a paper at a cartography meeting: "[a]nother prominent geographer got up afterward and said something like, 'Now, George, you know it's impossible to do that.' His response was, 'Yes, but we're going to do it anyway.'" (Judy Olson, email to author, 1 October 2002).

Another leader in promoting the status of women in cartography was Richard Dahlberg (1928–1996),[10] "a rather unassuming gentleman who over the years had a far greater influence on the field than many might have realized"; he was extremely influential in Illinois, as well as at Northern Illinois University in the maps, GIS, and remote sensing areas (Judy Olson, email to author, 1 October 2002; Carter, 1997).

EVELYN PRUITT (1918–2000)

Evelyn Pruitt apparently coined the term "remote sensing" (*Latitudes*, 2002: 1).[11] She worked for twenty-five years as an applied geographer for the United States Department of the Navy (*Washington Post*, 28 January 2000). Her work greatly advanced the study of coastal environments and the use of remote sensing in geographical studies from the 1940s into the 1970s.

Born on 25 April 1918, Evelyn Lord Pruitt displayed a free spirit and a love for the sea and its coasts. Her mother Ethel Lord, of San Francisco, married Conrad Douglas Pruitt, who was from Mississippi. This would partly explain Evelyn's later attachment to Louisiana State University, but she also had a soft spot for the university because of its work on coastal areas, and a number of geographers there were her friends. Her maternal great-grandfather, William A. Lord, captained ships that went all around the world, including around Cape Horn, one of the most treacherous bodies of water. For the Pruitts, life was hard; the family apparently lived in an "unfinished" house—a house with plumbing fixtures but no wall board

(Walker, 2006: 432). Evelyn's father exercised an important influence on her life; she inherited from him the traits of a happy, free spirit.

Despite (or because of) their own lack of education, the parents wanted their two daughters to receive a university education. Evelyn fell into geography quite by accident. Because her sister, who was already in university, did not drive, Evelyn had to drive her to a university-held field course every Saturday: "My first introduction to geography, which had been nothing but a dull thing in school, ... was wonderful! It was outdoor mapping, you had plane tables and you paced things off, and you went interesting places, and you talked to people ... that's the best part of geography, anyway, the field.... So, I entered UCLA as a freshman in geography" (Walker, 2006: 432).

While initially interested in meteorology and geology, Pruitt soon learned that these fields were not welcoming to women, so she stayed with geography. She received a B.A. in geography from the University of California–Los Angeles in 1940 and then, in 1943, an M.A. in that field from the same university. (During this time she took a geology class taught by William Putnam, who later received a grant from the Office of Naval Research when Pruitt was heading it.) Before she finished her thesis, the United States Coast and Geodetic Survey, Aeronautical Chart Branch, took her on board for wartime employment at the age of twenty-four (Walker, 2006: 433). She was the first woman to be ranked in the "Professional" category. In 1947, she took an interest in attending the summer school in geography at McGill University in Montreal, and assumed that her position at the Branch would be left open for her return. At McGill she attended a course on the Arctic; after returning to Washington and learning that she had been dismissed from the Branch, she turned to the Head of the Human Ecology Branch at the Office of Naval Research, who had an intrinsic interest in the Arctic. Soon the Branch, impressed with her work, hired her full-time (Walker, 2006: 433).

Occupying such a key position in the newly emerging Office of Naval Research, she made it possible to strengthen the work of her old and new friends and colleagues that allowed them to expand their laboratories and research programs, such as M.C. Shelesnyak's Arctic research program at Point Barrow, Alaska. Taking the broad view, she encouraged and funded research on "coastal Louisiana, land ownership in Kansas, sequence occupance [sic] in Costa Rica, Saharan transportation and ... Antarctic coasts" (Walker, 2006: 434). At the same time, she took the lead in developing cultural geography, which, however, fell into disfavour.

Pruitt thus had an active engagement with researchers and scholars outside her immediate realm of work. She showed a great deal of enthusiasm for her work, and had a good sense of humour. This engagement sometimes led to surprising developments and inventions. The invention of the triangulated irregular networks (TIN), for example, arose out of a lunch she had with T.K. Peucker (now spelled Poiker) back in early 1972. Poiker was searching for a way of integrating triangulation of regular and irregular surfaces, which would later become an essential part of GIS (Mark, 1997: 3). Pruitt understood the complexity of the issues, expressed interest in having this innovative research funded, and made it possible for Poiker to conduct that research. She got along well with her male colleagues—she was not, according to Janice Monk (email to author, 13 September 2011), pompous, but did not back off when she thought she was right.

Monk (2004: 3) underscores the point that Pruitt "demonstrated the importance of effectively representing the discipline within emerging governmental bodies that funded research." Significantly, her approach led to the awarding of grants to enable research in "coastal studies, remote sensing, and spatial analysis." The establishment of the Coastal Studies Institute at Louisiana State University constitutes a clear example of such work. Pruitt, who was at that time the new director of the Coastal Geography Programs of the Office of Naval Research, was convinced that such an institute would benefit a region visited regularly by hurricanes. The students who worked with the Institute became endearingly known as "marsh rats" (Coastal Studies Institute, 2007: 1).

Walker reports that by the time Pruitt retired from the Office of Naval Research in 1973, "she was being credited with being a major driving force behind the conversion of coastal research from that done by individual scientists to that involving large, well-trained, talented teams; from short-term studies to long-term systematic research projects; and from one of descriptive reconnaissance to highly refined ... investigations." (Walker, 2006: 435).[12]

Her Geography Branch funded many pioneers in the development of dynamic/quantitative geomorphology, a veritable revolution (1954–1965) that reached out to the University of Washington, the University of Chicago, the American Geographical Society, the University of Michigan, the University of Wisconsin, and Northwestern University (Walker, 2006: 436).

Another area significantly touched by the Geography Branch was photo interpretation. This area was especially challenging because new technologies inundated the traditional approach in aerial photography. It was during an afternoon session with her colleagues that Pruitt came up with the term "remote sensing" (Walker, 2006: 436). An ensuing major conference on remote sensing pointed to the need to organize vast amounts of data; the conference also pointed to one of the major weaknesses of American geographers: namely, the lack of experience in areas outside the United States. Here again Pruitt stepped in and resolved that dilemma: her office funded ninety-seven individuals (out of 432 applications) between 1955 and 1966.

Her stint as geographer extended to other areas of research, including the editorship of *The Professional Geographer*. She also became more widely known after she co-authored, with Louisiana State University professors Fred Kniffen and Richard Russell, the major textbook in geography, *Culture Worlds* (*Latitudes*, 2002: 1). Richard Russell had served on the Advisory Committee of the Office of Naval Research. Pruitt served on the Association of American Geographers as a councillor and as chair of both the Publications Committee and the Marine Geography Committee. As an indicator of her forward thinking, she chaired the first conference of the Coastal Society, held in 1975 in Arlington, Virginia. In 2010, the organization held its twenty-second conference.

In 1973, Pruitt was the highest-ranking woman scientist in the United States Navy, which awarded her its Superior Civilian Service Award that year. Throughout her career, resentment and prejudice greeted her at most steps. Apparently, she considered these potential obstacles as challenges. As women began to enter the profession in the 1970s, she gave them support while serving on the Committee on the Status of Women in Geography (Walker, 2006: 438). Her desire to have more women educated in geography is reflected in her later largesse, notably the Pruitt National Fellowship for Dissertation Research through the Society for Woman Geographers and a large endowment to Louisiana State University. She was known to be a frugal person; this personal attribute allowed her to acquire a "sizeable estate" (Walker, 2006: 438).

At the age of fifty-seven, Pruitt's life began to be filled with recognition. In 1972, she received the Citation for Meritorious Contributions to Geography from the Association of American Geographers, and in 1981 received the Outstanding Achievement Award from the Society of Woman Geographers. In the meanwhile, she served as the first president of the Coastal Society in 1975. In 1983, when she was sixty-eight, Louisiana State University conferred upon her an honorary degree for her contribution to "research in coastal environments," for "promoting the field of remote sensing," and for "helping LSU gain national and international prominence in geography" (*Latitudes*, 2005: 1). In 1984, Pruitt was the recipient of the James R. Anderson Medal of Honor in Applied Geography. The Applied Geography Specialty Group reserves the award for one who has rendered "the most distinguished service to the profession of geography" (Applied Geography Specialty Group, 1984).

Evelyn Pruitt died on 19 January 2000 of pneumonia in Arlington, Virginia, at the age of 81 (http://www.thecoastalsociety.org/pdf/bulletin/TCS_24_3e.pdf). The Louisiana State University Department of Geography and Anthropology established the Evelyn L. Pruitt Lecture Series in 2003 following an endowment of more than $900,000 from her estate to assist women graduate students in the Department (*Latitudes*, 2005: 6).

MEI-LING HSU (1932-2009)

Mei-Ling Hsu was a highly regarded scholar and the first woman to obtain a Ph.D. in (academic) cartography (McMaster and McMaster, 2002: 318).[13] Her particular contributions relate to symbolization in maps, bringing recognition of Chinese cartography to the Western world, and map projections.

Born in 1932 in mainland China, as a young girl she escaped with her family to Taiwan (Judy Olson, email to author, 22 June 2011). She received her B.A. in 1954 from Taiwan Normal University. Given the lack of opportunities in Taiwan, her family encouraged her to return to the United States, where she attended South Illinois University in 1956 for her master's degree. She proceeded through the program very quickly. However, after spending an unhappy year at University of Illinois at Urbana, she chose the University of Wisconsin in 1959, completing her Ph.D. in cartography in 1966, with agricultural geography as a secondary area of research. While the earliest cartographic Ph.D. dissertation belongs to Arthur H. Robinson (*Foundations of Cartographic Methodology*, Ohio State) in 1947, and women academic cartographers wrote their master's theses starting in the early 1960s, by 1972, Mei-Ling Hsu "wrote the first true cartographic dissertation" (Tyner, 1999: 27) on isarithm mapping, under Robinson's supervision.[14] Usually she was the only woman in her program. Hsu's approach to cartography had delighted Robinson because it was theoretical, statistical, and quantitative—quite rare in those days.

Hsu had no trouble finding a position, especially with statistics combined with cartography. She said in an interview with Jan Monk that "if I were only to know how to draw pretty maps, I might have not had much opportunity" (Monk, 1990: 6). A number of faculty members at the University of Minnesota passed through Madison, and once apprised of her talents, were ready to offer her a position in the Department of Geography at their university—starting as research associate setting up a cartography lab in 1965. She had not yet finished writing her dissertation, but Robinson encouraged her to go. She quickly realized that being head of a laboratory would not do her career any good. In the end, the department accepted her as a full-time teacher, the only woman in the department. With a few exceptions, she received encouragement from her department. She eventually served as chair of the department from 1994 to 1997. As Zhou (2010: 3) noted, Hsu was "unquestionably the first female Chinese geographer in any subfields of geography in the United States." She was a role model for many young scholars pursuing work in cartography or China or both. She became highly respected in the field of cartographic symbolization, Chinese cartography, and map projection (see, for example, Hsu, 1972).

Among her significant early contributions one can include her population map of Taiwan. She spent six months (in 1966–67) in Taiwan to accomplish that task. She managed to find a way of resolving some inherent problems in tabulating demographic statistics so that the population could be shown with sufficient accuracy (Thrower, 1969: 612). In 1970, she co-wrote (with Arthur H. Robinson) *The Fidelity of Isopleth Maps: An Experimental Study*. She was an active member of the China Geography Speciality Group of the Association of American Geographers, where she also served as its chair from 1995 to 1996. She also produced, in 1969, a detailed map with extensive notes on population growth and urbanization, "Taiwan Population Distribution, 1965," Map Supplement No. 11 in the *Annals of the AAG*, 59: 3 (1969), and cartographic designs for a number of influential college texts.

She built up the University of Minnesota's connection to China and was a member of the first University of Minnesota delegation to China in 1979. She later served at the first director of the University's China Center, from 1980 to 1984 (China Center, 2009: 13). Hsu seemed to have been particularly diligent in helping Chinese students dispel homesickness, as Zhou (2010: 4) recalls:

> In some of the early photos of the Geography Department at Minnesota [University], I remember seeing Mei-Ling stood out as a young Chinese woman among all her white male colleagues. The fact that she thrived in this field during the early era testifies [to] her hard work and courage. In her

later years, she has shown great interest[s] in the development of China. In the early 1990s, she organized a semester-long lecture series on China for the departmental coffee hour, at a time [when] such an extensive discussion on China in geography was very rare.

Aside from being the first director of the university's China Center, she was instrumental in building the cartography and GIS programs in its Department of Geography. After leaving the China Center, she remained at the university and was most recently an adjunct professor emerita of the East Asian Studies Program within the Department of Geography at the University of Minnesota (OIP, 2009: 3).

Judy Olson, among others, recalls Mei-Ling as "very welcoming and engaging" (email to author, 22 June 2011) and

> was also a feisty sort of person (not particularly noticeable on that first visit), which I suspect was a reflection of the sort of determination she had to have had to do the things she did. She did not direct that feistiness "at you" (though she certainly said what she thought), but would "use it" to express herself about conditions, actions of others, etc. "Foolish" was a favorite descriptor [of hers]). "What do you think of blah blah?" ... "Foolish." (Not "foolish," I think, "just the declarative huff 'foolish.'" And I'm smiling as I write that.... She could also be highly supportive and encouraging.

When, in an interview, Janice Monk asked her what explained her success, Hsu said that she was "a little stubborn," and that she hated telling herself, "You failed." She described herself as having an "inner persistence," more than being very ambitious (Monk, 1990: 20).

She never married, and died on 23 May 2009 after a few years of illness.

JUDITH TYNER (B. 1938)

Judith Tyner is an academic cartographer who has written extensively on map design, history of cartography, women in cartography, map samplers, textile maps, persuasive cartography and lunar cartography.[15] Tyner was among the first to research and write about a number of these issues, especially the idea that maps are subjective expressions of their maker.

She was born in Lima, Ohio, to Francis X. Zink, a tool and die maker and precision machinist, and Virginia R. Wood, a homemaker. The family moved to California in 1948. Tyner is married to Gerald Tyner, also a geographer, who is retired from the California State University System. They have lived in Kentucky, upstate New York, and Germany. While a few of the women in the vignettes in *Map Worlds* are married, Judith appears to be the only one with children. She and Gerald have two sons: David A. Tyner, an artist, and James A. Tyner, a professor of geography at Kent State University in Ohio.

Tyner completed all of her university degrees in geography at the University of California–Los Angeles (B.A. in 1961, M.A. in 1963, and Ph.D. in 1974). She had started her education as a mathematics major and was generally the only woman in her class. She switched to geography, which she described as much more welcoming to a junior. When she was working on her M.A., a six-week summer field class was required, but she was the only woman eligible. The professor and the department "graciously" waived the requirement. In her dissertation she used the term "persuasive cartography" to underscore the social-constructivist and subjective nature of maps. Persuasive cartography also refers to the social and political functions of maps, in that one can also explore the assumptions and ideologies that enter maps even before they are made—a significant insight. Later, other map thinkers, such as Mark Monmonier in *Maps, Distortion, and Meaning* (1977) and J. Brian Harley in his latest *The New Nature of Maps* (2001), would come onto the scene. Today, many cartographers are aware that maps can or should be deconstructed.

Tyner had three mentors "without whom [her] career … would not have been possible" (Tyner, 2010: ix): Richard Dahlberg; Gerard Foster, who taught her how to teach cartography; and Norman J.W. Thrower.

Some two dozen publications represent Tyner's contributions to cartography; they often open up new historical vistas or reveal a keen understanding of interactions between maps and society. Her three books delve into cartography in general (1973), thematic cartography (1992), and map design (2010). This latest book (*Principles of Map Design*) is a remarkable contribution in that the digital manipulation of maps has become so pervasive that many practitioners do not know how to apply an informed and aesthetically pleasing map design. Her articles also provoke intense interest in particular topics that can pique one's curiosity: lunar cartography (1969), persuasion and cultural interactions (1982, 1987), embroidered maps (1994, 2001), and silk maps (1996). Tyner endowed cartography with other interesting essays, such as historical pieces on women in cartography (e.g., Tyner, 1997). She also served as map editor for *California Patterns on the Land* (Tyner, 1976). There are seven articles forthcoming in two volumes of the multi-volume work history of cartography, as well as two books:

Geography through the Needle's Eye, and *The World of Maps: Maps and Map Reading for the 21st Century.*

Tyner served as chair of the Department of Geography at California State University–Long Beach for six years and founded the Cartography/GIS Certificate Program, heading it for twenty years. She introduced classes in remote sensing, advanced cartography, seminar in cartography and GIS, computer cartography, and history of cartography. She taught at California State University–Long Beach for over thirty-five years, retiring in 2005 as professor emerita in geography.

Tyner's own reflections on the topic of who was the first woman cartographer or who has primacy maintained that this issue is fraught with difficulties (maphist@geo.uu.nl, 14 April 2008). Each name would provoke a discussion and disagreements:

> It is always risky making statements about primacy. We don't know who the first female cartographer was.... It also comes down to what you mean by cartographer. There is evidence of a woman embroidering a map on silk in the 3rd century A.D. to make it more permanent. Was her work mapmaking? You could get quite an argument on this. I have done work on globes embroidered by schoolgirls in Pennsylvania that pre-date James Wilson's globes by several years, but many would dismiss these as school projects. (maphist@geo.uu.nl, 14 April 2008)

BARBARA BARTZ PETCHENIK (1939–1992)

Barbara Bartz Petchenik is particularly noted for developing cartographic and spatial comprehension by children and for her diligent production of maps for *World Book* and the Newberry Library in Chicago.[16] The biannual Barbara Petchenik Children's Map Competition under the auspices of the International Cartographic Association is the internationally acknowledged legacy of her outstanding contributions to cartography.

Born on 17 August 1939 in How, Oconto County, a rural northern Wisconsin settlement, to a well-established family, Petchenik grew up in comfortable, though not wealthy, circumstances. She had four sisters and one brother. Her mother was Margaret Clara Schreiber and her father was Roland August Herman Bartz. The family provided her with "a secure and nurturing environment" (Auringer Wood, 1993: 1). When she was in high school, the school recognized "the Bartz girls" (her four sisters, three cousins and herself) as "achievers and leaders, both academically and socially";

one account tells us that she was "more the former than the latter." She did not see herself as a "prom queen" (Auringer Wood, 1993: 1).

Apparently, two books exercised a major influence on her life: Karl Menninger's *The Human Mind* (1930) and Ralph Lapp's *Atoms and People* (1956). Both books fascinated her: she believed she could be a psychiatrist or a nuclear physicist. Her mother hoped she would become a nurse (Auringer Wood, 1993: 1–2).

In 1961, Petchenik obtained a B.Sc. in chemistry (with a minor in English) from the University of Wisconsin–Milwaukee. She then worked for a year at that university as a teaching assistant in geography and as map librarian. She started a graduate program there with a National Defense Education Act Fellowship in cartography. In 1964, she completed her master's degree. Although she had initially intended to do the Ph.D. in physical geography, in the hopes of teaching at the university level, she admitted that "a whole lot of things happened that changed my course" (Auringer Wood, 1993: 2). Things just "happened" to her. In 1969, she completed her Ph.D. in cartography and educational psychology, incorporating her work on maps for children.

Wishing to live in Chicago, she chose, in 1964, to work with the Field Enterprise Education Corporation as cartographic editor. This work encompassed much research to produce maps for nine- to fourteen-year-olds for the *World Book Encyclopedia*. This very start of her work life remained identified with some of her major contributions to cartography. She spent one year in San Francisco as a cartographic consultant with *World Book*.

In 1970, Petchenik moved to a five-year position at the Newberry Library in Chicago as cartographic editor of the *Atlas of Early American History*, which she planned, designed, and produced, working with its editor-in-chief, Lester Cappon (Cappon, Petchenik, and Long, 1976). She made a number of innovations to this atlas, including a series of thirty-eight chronological military campaign maps of the American Revolutionary War. According to Robinson (1994: 174), her work on the *Atlas* "established her reputation as an expert, versatile cartographer."

From 1975 until her death in 1992, she served as the senior sales representative of cartographic services for R.R. Donnelly & Sons Company. Through her efforts, this company sponsored an award for a map designed by a student on behalf of the American Congress on Surveying and Mapping.

Her interest in education, and especially her desire to link maps to children, stayed with her all her life. To deepen her own knowledge about what children would like to see in maps, "she interviewed over a thousand kids from all over the world about their cartographic knowledge. It turns out that kids prefer clear and uncluttered maps without too many extraneous elements. No surprise there!" (http://www.thedailychannel.com/journals/, 24 July 2005). In 2003, Herzig and Jarausch (2003) provided an in-depth analysis of her legacy through the Barbara Petchenik International Children's Map Competition, which Fraser Taylor established in 1993 after her death. This competition has remained one of the most popular and enduring aspects of the biannual conferences of the International Cartographic Association.

Although her career unfolded mainly in private companies, Petchenik's influence extended across numerous publications and journals, including academic ones. She integrated scholarly research and active participation in commercial cartography (Mapping Science Committee, 1990). Over her lifetime, she contributed significantly and produced some sixty articles, papers, reviews, and other materials to seventeen journals and conference proceedings (Auringer Wood, 1992) on map design, education, cognitive psychology, and computer-assisted vehicle navigation.

Petchenik was a student and colleague of Arthur H. Robinson and a mentor of many students herself (Robinson, 1994: 174). A third book by Arthur Robinson, *The Nature of Maps* (1976), was co-authored with Petchenik and fostered the intellectual development of the field of cartography in the 1970s; it constituted the philosophical approach to the understanding of maps (http://www.amergeog.org/ags_in_memoriam.htm). She was the primary author of the 1990 National Research Council's "Spatial Data Needs: The Future of the National Mapping Program."

Petchenik was very active in professional organizations, including the American Cartographic Association of the American Congress on Surveying and Mapping, the Association of American Geographers, the Society of Automotive Engineers, and the International Cartographic Association (ICA). As managing editor of a special issue of *The American Cartographer*, she was devoted to the transition from analogue to digital representations of space around 1970. In 1991, delegates to the ICA General Assembly elected Barbara Bartz Petchenik as its first female vice-president to the executive in its thirty-two-year history (Robinson, 1994: 174).

Henry Steward, a cartographer at Clark University, offered the following about Petchenik:

> I agreed with David Woodward when he suggested that she might be the brightest mind in cartography. Alas, cut down too soon. A few memories

of her recall her breadth: her insightful comments about my *Cartographica* Monograph on Generalization; hearing her talk at Yale in a Tufte-linked Graphic Design Class; her belief that cartographers trained in academia should get out and make maps and not spend time talking about them; and a memorable conversation in Chicago drinking beer and talking about the novels of Trollope. To adjust that famous quote: "If she had lived we might have learned something." (Henry Steward, email to author, 10 September 2010)

Her 1972 marriage to Kenneth H. Petchenik meant she had to stay in Chicago—not an unwelcome prospect at all. Kenneth had a publishing career and three young sons by his previous marriage, which made it impractical for her to move (Auringer Wood, 1993: 3). They had a home in Baileys Harbor, Wisconsin, in Door County, where they were owners of Pine Street Press, specializing in books relating to Door County. This experience made her realize some of the problems that young professional women can face early in their career.

Described as someone with "great intelligence and good humour" (Auringer Wood, 1993: 4), she was also part of an informal group called TWIC–Tall Women in Cartography, along with Patricia Caldwell and Alberta Auringer Wood. She died on 7 June 1992. She was fifty-two.

PATRICIA GILMARTIN (B. 1941)
Born in Atlanta, Georgia, in 1941, Patricia Gilmartin grew up in Atlanta and in West Palm Beach, Florida.[17] When she was old enough to get out on her own, she moved back to Atlanta for college, work, and college (in that order). She dropped out of college the first time after only a year, then worked for several years in low-level jobs in retail and banking (long enough to learn that she needed more education "if I was ever going to get anywhere" (Gilmartin, 2011: 1); and finally went back to school to complete her education. At a young age, she had already said she wanted to be an airline pilot—an "unrealistic" goal for women in those days. Her mother encouraged her to become a dental hygienist. She rejected that line of work and had no interest in the usual other kinds of work open to women: secretary, elementary school teacher, or nurse. She even refused to take typing classes. Most of her family had not finished high school, let alone pursued university studies. In all, it took her twenty years of following "a circuitous route from college freshman to cartography professor" (Anon., 1991: 3). That route, she believes, was an advantage, rather than a disadvantage.

Returning to school at age twenty-nine, she worked nights as a restaurant server while earning her B.A. in psychology. She took an undergraduate course in thematic cartography; her first job after graduation was doing cartography at "a small firm that specialized in cadastral mapping" (Anon. 1991: 3). Finding the job too structured, she resigned, wanting more flexibility and independence. She received her master's degree from Georgia State University. Patricia obtained her Ph.D. in geography (specifically, cartography, or map-making) from the University of Kansas in 1980, where George Jenks was her supervisor (McMaster and McMaster, 2002: 317).

Gilmartin's first university position was teaching geography at the University of Victoria in British Columbia, Canada. She then taught at the University of South Carolina. Eleven years into her teaching career, she was the only female faculty member in a department with sixteen professors (Anon., 1991: 6). With hardly any other women in her field, her conference costs were considerably higher, because there was no one to share hotel costs with.

She is now a distinguished professor emerita. She was a member of the Association of American Geographers and chaired its Publications Committee.

Her cartographic interests took her to cognitive mapping (see, for example, Gilmartin, 1981, 1985), the use of maps in education, and women explorers and travellers. She also directed her interests in maps toward her love for arts and design. She adhered to the idea that cartography is sometimes defined as the art and science of map-making. While she always had a "feel" for art but never any formal training, cartography enabled her to combine her attraction to artistic endeavours with a scientific application "in a discipline that provided a job and paycheck." She found this a "good meshing of my needs and interests, although I was mostly unaware for a long time of how neatly things had fallen in place for me." At some point a few years ago, she avers, "I had the crazy notion that I wanted to try doing some clay sculpture." So she took a class with Britta Cruz at the local Parks and Recreation studio in Columbia, South Carolina. She was immediately in love with sculpting. As she explains, "It felt so right, so natural." Within a year she had retired from her position at the university and began her sculpting full-time. As she says, "It's a new, wonderful, fulfilling life for me ... not a second childhood but a second adulthood" (Gilmartin, 2011: 1–2).

Her husband, Will Graf, is a geographer at the University of South Carolina.

CHAPTER 6

BARBARA G. SHORTRIDGE (B. 1943)

Barbara Shortridge developed thematic maps dealing with food, restaurants, and consumption patterns.[18] She teaches in the Department of Geography at the University of Kansas.

In 1964, as an undergraduate at the University of Wisconsin (junior and senior years) she worked in Cartographic Services in the Department of Geography on a part-time basis. For two summers (after her junior and senior years) she was an intern in the map division at Rand McNally in Chicago. Her job was to take towns off a master base map that had been "drowned" by new reservoirs being constructed. She held a National Defense Education Act Title IV Fellowship for three years at the University of Wisconsin–Madison. Her undergraduate supervisor was Arthur H. Robinson; she obtained her B.A. in geography in 1965. Within three years, she obtained her M.A. at the University of Kansas, with a thesis entitled "Some Aspects of Error on Dot Maps"—her graduate adviser was George F. Jenks. She obtained her Ph.D. in 1977 in geography with the same supervisor, with a thesis devoted to "Map Lettering as a Quantitative Symbol."

From 1968 on, she was committed to working at the University of Kansas, moving from being a graduate research assistant to teaching associate to assistant professor or lecturer. She achieved these latter positions some thirty years after her start as a graduate research assistant. She interrupted this long stretch of devoted service with an appointment as visiting assistant professor at Dartmouth College.

Shortridge is an active member of the Association for the Study of Food and Society and of the Association of American Geographers, in addition to the American Folklore Society (Foodways). She authored an early atlas on women, the *Atlas of American Women* (1987), in addition to her other notable work, *The Taste of American Place: A Reader on Regional and Ethnic Foods* (1998), which she co-edited with her husband, James. One reviewer, ironically named Pillsbury (1998), did not find that edited books, including this one, "stretch[ed] either the methodological or theoretical frontiers." However, this particular compilation did "make an effective presentation of much of the literature" (Pillsbury, 1998: 605). The Shortridges, moreover, "have provided an effective instrument to help human geography instructors take one more stride toward bringing the outside world into the classroom." H. Mark Livengood (1998: 202) calls the book a "cogent collection of [nineteen] essays" and comments that this "engaging framework geared toward students" moves "the study of food and eating out of the margins." Her research filled a particular niche in the scholarly world, as witnessed by her contributions to a number of encyclopedias (*The New Encyclopedia of Southern Culture, The American Midwest, The Greenwood Encyclopedia of American Regional Cultures*, and *The Encyclopedia of the Great Plains*). She was the book review editor for *The American Cartographer* from 1982 to 1988.

If one were to list the topics of her research and writing, one would be confronted by an unusual list: farming, rice, Mexican restaurants, Appalachian regional foods, ethnic heritage food, representative foods of Minnesota, consumption of fresh produce, stimulus processing models, the discrimination of town size on maps, and map reader discrimination of lettering size.

Among the awards she received, one must include the National Council for Geographic Education's *Journal of Geography* Award for the best content article, "Apple Stack Cake for Dessert: Appalachian Regional Foods," from October 2005.

JUDY OLSON (B. 1944)

Judy Olson is one of the first women to have developed and contributed to academic cartography, cartographic research, and GIScience in the United States.[19] She was an active participant in the NASA/ASEE programs and a consultant to the United States Census Bureau, to the Department of Transportation, and to private industry. She was born and raised in central Wisconsin in a family of Norwegian heritage. Her father, Leonard Olson, was a mechanic and foreman; her mother, Hilma, was a nursing assistant at the Grand Army Home for Veterans.

Olson obtained a B.S. in geography in 1966 at Wisconsin State University–Stevens Point, which she followed with a master's in geography (1968) at the University of Wisconsin–Madison. She soon received a Ph.D. in geography (1970) at the same university. Joel Morrison, who was finishing his degree and teaching at the time she entered graduate school, and other graduate students, including Karen Severud (later Pearson, Cook), Barbara Bartz (later Petchenik), David Woodward,[20] George McCleary, Michael Conzen, and others were her peers and mentors who exercised an important influence in graduate school. Also, she was a research assistant under Arthur H. Robinson at the University of Wisconsin. Of those featured in this chapter, Robinson directly trained three and indirectly trained five others. In a tribute to Robinson (1915–2004), Olson stated that she owed him much due to his support of her scholarly career, especially when she was a young assistant professor and he asked her to become associate editor of the new journal, *The American Cartographer*: "I was tremendously honored that he asked me to join him as associate editor and remember well that he specified the title as 'associate' and not 'assistant.' Consistent with the title, he gave me complete independence in handling my share of the manuscripts, in effect grooming me to take over in a few years. He was a textbook model of someone starting a venture that would continue with a life of its own" (Olson, 2005: 133). Olson also counts Barbara Petchenik, who had come back to finish her degree, as an important influence.

Olson commenced her work life[21] as assistant professor at the University of Georgia (1970–74), which was followed by her appointment as associate professor at Boston University (1974), and then full professor at Michigan State University (1983).

Her work included quantitative mapping and the psychology of maps. One of her major contributions lies in the field of the use of colour in mapping and designing maps for people with defective colour vision (Olson and Brewer, 1997). Some nine million people in the United States are affected adversely by maps that are not appropriately coloured. She would be quick to point out that it was Daniel Heist and other seminar students who deserve credit for getting the research under way, and her former graduate student, Cynthia Brewer (see also Roth, 2010), who became the real expert in colour mapping and whose work fleshed out the findings and transformed experimental results into usable tools. However, Olson's publications also extensively touch upon map complexity, two-variable maps, map projections, and multimedia in geography.

In roughly mid-career, when multimedia was getting wide attention by younger cartographers and geographers and receiving mixed reactions at best by the older generation, Olson was among the early advocates for using multimedia in cartography and geography. In her 1996 Presidential Address to the Association of American Geographers, she persuasively argued that veterans in the field had a distinct role to play, and that by ignoring the new media, they would be a disservice to the field (Olson, 1997: 571). Despite some of their potential negative effects, Olson provided substantive positive examples where multimedia could help the discipline to progress in new ways. Fourteen years later, scholars would still refer to Olson's pronouncements about cartography and GIS (for example, Crampton, 2010: 11). Others, such as Fitzsimons and Turner (2006), would also find Olson's work seminal or highly illustrative, such as her observations about job descriptions in cartography relative to such fields as remote sensing and GIS (Olson, 2003). The whole issue of technology has made Olson step back from her cartographic work, and she now looks at it more philosophically. The introduction and the wider use of GIS have pulled Olson into newer levels of cartography, and by the mid-1990s had changed her identity as well. In 1999, she said:

> I always find it a little difficult to say that I am doing anything different than what I used to because many of the things that are identified now with the GIS ... I was teaching in my "computer mapping" class or "automation in cartography" ... That's an evolutionary change in identity rather than something that was ... sudden. If it [were] kind of suddenly that I found out that, oh yes, that's, that is what GIS is supposed to be about, well, I had been calling it cartography and so had every one else. (Interview, 19 August 1999)

Olson recalls a comment by a colleague that reflects her dilemma of identity: "A colleague who is traditionally a geodesist said at a meeting once—jokingly, but seriously, you know—"I don't know what I am anymore." And everybody said, "Yeah, yeah, that's the way it is. Yeah, we don't know what we are anymore" (Interview, 19 August 1999).

While uncertainties had shaped cartography, especially as it related to GIS and other technological advancements, Olson believed that creativity was the mainstay of any cartographic endeavour, whether classical or "new":

> But also some combination of creativity [contributes to one's reputation], and I don't mean in drawing maps, I mean in just, thinking, reliability in ... getting things done, participating, and engaging, I think they are all important [as well]. I mean, some people can get along without some of those traits. They can be hermits and do very well, but then they have to be extremely strong in some other trait. There are some people who are so creative that they don't have to sell their ideas, their ideas sell themselves. (Interview, 19 August 1999)

Judy Olson has a particular penchant for getting her ideas across to any audience. Her presentation at the 2011 ICA conference in Paris on map typologies was one of the most succinct and clearest talks I attended at that conference (Olson, 2011).

Making her professional colleagues realize that they had a seed of creativity was one of Olson's talents, especially when she was editor of *American Cartographer*. A staff member of a mapping agency had called in with complaints about the way his agency had made certain decisions. When she realized what a wealth of ideas he had, she suggested he write about those decisions. And when he resisted because he "was not a writer," she offered to send him questions to which he could respond. He also told her he did not have anyone to type up the answers, but she encouraged him to send her his handwritten replies because she had received handwritten letters that were perfectly legible. Within a few weeks, she had fifty pages of handwritten answers that turned out to be quite fascinating, and eminently suitable for publication (Interview, August 1999).

Among her significant contributions were her editorships of a variety of cartography-related publications: namely, associate editor of *The American Cartographer* for three years from its inception in 1974, and then its editor for six years (1977–82). She also served as cartography associate editor of the *Annals of the Association of American Geographers* (1988–91) and as a member of the editorial boards of the *Professional Geographer* (1972–77), the *Annals of the Association of American Geographers* (1982–84 and 1991–93), and *Cartographic Perspectives* (1998–2000).

Olson's interest in promoting cartographic knowledge found expression in her dedicated service to the American Cartographic Association (ACA, later known as the Cartography and Geographic Information Society), and the Association of American Geographers. For the latter organization, she held the posts of vice-president, president, and past president between 1994 and 1997. This was a very active period for her, because she also served as a member of the steering committee for the University Consortium for Geographic Information and Analysis, organized as the University Colloquium for Geographic Information Science at the time of its inception in 1994.

The range of Olson's activities clearly extended beyond the confines of the United States. She served several years as member or chair of the United States National Committee for the International Cartographic Association. In the latter organization, she served as vice-president (1992–99), and also chaired the ICA commission on Map and Spatial Data Use from 1987 to 1991. She is formally a member of cartographic associations in Canada, Britain, and in her own state.

In 1990, Olson was bestowed with "AAG Honors" (of the Association of American Geographers) (AAG, 2010); in 2001, with the ICA Honorary Fellowship award; and in 2005, with the Award of Distinction of the Canadian Cartographic Association. The Board of the Cartography and Geographic Information Society awarded her, in 2008, the prestigious Earle J. Fennell Award for her long-standing contributions to the field (Anon., 2008: 11).

PATRICIA S. CALDWELL (B. 1945)

Patricia Caldwell owned a cartographic consulting and production firm for twenty-three years, lectured worldwide to help audiences understand rapidly evolving mapping-science technology, and held a variety of leadership roles in cartographic professional societies.[22]

She obtained a B.A. in geography from the University of Washington in 1967, and a master's in geography from the University of California–Los Angeles (UCLA) in 1972. She received her Ph.D. from UCLA in 1979, under the supervisorship of Norman J.W. Thrower (like Judith Tyner).

Caldwell started as a staff cartographer in the Department of Oceanography at the University of Washington, followed by a stint of service (1967–70) at the Central Intelligence Agency, producing maps for cabinet-level officials (Straight, 1990: 5). She taught cartography both at Santa Monica City College and UCLA before founding her own company, Caldwell & Associates, a cartographic production and consulting firm. From 1979 to 2002, Caldwell assisted clients that included Prentice-Hall Inc., University of California Press, McGraw-Hill, Sequoia and Kings Canyon National Park, UN Environmental Program, ESRI, and *Consumer Reports*. During this period she also taught GIS at Sonoma State University (1987–91). Thereafter, she joined the Los Rios Community College District as a project manager for the Business and Economic Development Center (2002–2007), followed by a position as a statewide coordinator of a project encouraging cooperation between the state's 115 community colleges and the Workforce Investment Boards.

Her initial cartographic focus was design, with an early interest in the role of typography on maps. She later explored the design of maps for low-resolution images for television, which morphed into the methods of developing maps for computer screens. Her innovative work involved matching the interests of educational systems with the capacities and potential of graphic digital systems. She is one of the very few women cartographers who have stayed out of academic life to pursue their cartographic contributions in a large variety of private and governmental sectors. Caldwell's eagerness to promote the wider acceptance of newly arising technologies is well known. For example, when Gunter Greulich advanced the idea, in 1990, that the scientists involved with developing GPS should be nominated for a Nobel Prize, it was Caldwell (then president-elect) who wholeheartedly supported the idea, despite the surprising opposition by the ACSM itself (Greulich, 2006: 24). Soon after Caldwell became president of the ACSM, she appointed an ad hoc committee to pursue the idea of a Nobel

Peace Prize for these scientists. Despite her best efforts (and those of the ad hoc committee), the idea stalled. The committee realized that no single person had been the inventor of GPS, but that it was a military effort by two institutions, namely the United States Navy and the United States Air Force, under research programs called "Timation" and "621-B," respectively (Greulich, 2006: 25).

Caldwell's numerous publications include analyses of television news maps; the effects of the medium on the map; election-night maps and graphics; the responsible use of computer graphics in broadcasting (Straight, 1990: 5); and an analysis of road atlases for *Consumer Reports* magazine. She has lectured on mapping-related topics to large professional and lay audiences in Qatar, Scotland, Finland, Canada, and throughout the United States.

Caldwell served on the ACSM Board of Direction, the Northern California ACSM Section, and as president of the American Cartographic Association (ACA). In 1990, she was elected president of the ACSM, the second woman in its history, succeeding Alberta Auringer Wood, who presided in 1987. Caldwell also served as secretary of the International Geographic Information Foundation, and was a member of the United States Delegation to the International Geographical Union. She also served as vice-president of the California Map Society and the Society of Woman Geographers.

BARBARA ("BABS") BUTTENFIELD (B. 1952)

Barbara Buttenfield is a recognized expert in cartographic generalization, multi-scale databases, representations of uncertainty, and cartographic information design.[23] Her research interests ask how cartographic knowledge is constructed, and how it changes in content, appearance, and underlying geometry with resolution and scale changes.

Her mother, Gretchen Steinmeyer, worked as a unit clerk in a women's hospital. She was adventurous, travelling to New Caledonia in 1943 as an Army Red Cross nurse; she worked in the South Pacific until the end of World War II. Buttenfield's father was Charles Campbell Buttenfield, a chemical metallurgist for U.S. Steel. He loved classical music, baroque art, and formal gardens.

Buttenfield graduated in 1969 from Ellis School, one of Pittsburgh's age three to grade twelve independent girls' schools (*Ellis Magazine*, 2009: 20). She obtained her B.A. in geography at Clark University in 1974, followed by a master's in geography at the University of Kansas in 1979 on grid representations of sketch map distortion. The thesis analyzed spatial distortion in cognitive sketch maps (remembered geography) as a traditional map projection problem. Buttenfield's work was supervised by John Sherman, a "fine, humble, and thoughtful man," who is best remembered for work on maps for the blind (Judy Olson, email to author, 1 October 2002). However, it was George Jenks who was Buttenfield's master's degree supervisor. Jenks's influence led Buttenfield to a long-standing interest in statistical analysis and quantification in cartography. In 1984, she obtained a Ph.D. at the University of Washington, again in geography; her thesis title is "Line Structure in Graphic and Geographic Space." The research topic for her dissertation marks the initial work on cartographic generalization. John Sherman enlisted the assistance of Tom Poiker (formerly Peucker, professor emeritus from Simon Fraser University) to collaborate on the dissertation research.

Buttenfield arrived at the University of Colorado–Boulder in January 1996 "to catalyze the GIScience curriculum" (Buttenfield, 2011: 1). She still directs the Meridian Lab, "a small research lab in the Geography Department where three faculty and roughly a dozen graduate students work on research using GIScience, as well as on research to advance knowledge about GIScience" (Buttenfield, 2011: 1). She is also a research faculty affiliate with USGS Center for Excellence in GIScience (CEGIS) (http://www.colorado.edu/geography/babs/webpage/). She designs software for multi-scale cartography for *The (United States) National Map*.

Before her arrival at the University of Colorado–Boulder in 1996, Buttenfield conducted research or taught at a number of American universities: the University of Kansas (1974–78), University of Washington (1980–82), University of California–Santa Barbara (1982–84), University of Wisconsin–Madison (1984–87), and SUNY–Buffalo (1987–95). During this time, she also cultivated connections with the University of Maine–Orono (1995–98) and the University of Vienna, Austria (1997).

Because her skills as a cartographer were in high demand, she did stints of research at the United States Defense Mapping Agency (1978), the Bureau of Census in Seattle (1981), the National Center for Geographic Information and Analysis (NCGIA) (1988–96), and the United States Geological Survey (1993–94). Her career is demonstrably quite different from many other women pioneers in cartography. She was an incumbent in successive university positions and engaged herself in numerous governmental agencies. Her curriculum vitae runs forty-two pages, longer than those of other women cartographers. Extraordinarily well connected, she is often guest speaker at the annual meetings of professional societies and give public lectures in universities. Until 2009, she served on the Board of the North

American Cartographic Information Society (NACIS). Moreover, she is not a stranger to high schools. In 1998, for example, Buttenfield served as academic advisor to the project on GIS and the Internet at the Blackhawk Middle School (Smith, 1998).

A committed researcher, Buttenfield would on occasion complain about the increased workload of online fiscal ethics tests and changes in purchasing and travel procedures (Dodge, 2006: 2). She is sometimes impatient with bureaucracy and is at the forefront of trying to install more substantive dialogue between faculty and the university administration (Buttenfield, 2007).

Buttenfield's work has landed her some half-dozen awards. In 1997, she was appointed as a Lifetime Fellow, American Congress on Surveying and Mapping, and in 1999 was recognized as having written the "Best Research Paper of the Year" by the Geoscience Information Society (affiliate of the Geological Society of America). In 2001, she became the Inaugural Recipient of the National Award for GIScience Educator of the Year, University Consortium for GIScience (UCGIS), and in 2007, she became the LEAP Research Fellow (as part of the NSF ADVANCE Program "Leadership Education for Advancement and Promotion" at Colorado University–Boulder). In 2010, she was elected to membership in the Society of Woman Geographers, the same year that one of her papers was heralded as one of the "Top Four Short Papers" at the GIScience conference, Zurich, Switzerland (Buttenfield, Stanislawski, and Brewer, 2010).

THE CANADIAN CONTINGENT

The Canadian contingent of women cartographers is distinctive from those in the United States.[24] For example, the Canadian women did not follow the path of academic cartography as much as their United States counterparts did. Some six of the fifteen known women cartographers in Canada were (or are) associated with universities. Jacqueline Anderson worked at Concordia University (Montreal) and is noted for her active engagement in the ICA Commission on Children and Cartography, as well as her devoted work in the international Barbara Petchenik Children's Map Competition. The late Barbara Farrell taught cartography for many years at Carleton University, Ottawa. Grace Welch, former assistant chief librarian (Access) for the University of Ottawa library system, according to an award she received from the Association of Canadian Map Libraries and Archives in 2003, "provided persistence, strength and a great skill set to the negotiation of educational licenses to Canadian colleges and universities for spatial data from Natural Resources Canada, a step in the process of providing free access to all Canadians; she was a proponent of the educational license partnership with DMTI Spatial; and was a vocal advocate in the recent lobbying of Natural Resources Canada for the continuation of the availability of print topographic maps."[25] Sue Nichols is one of the most internationally recognized geodesists, working at the University of New Brunswick.

Several other Canadian women hold up the firmament of women's fully fledged participation of women in cartography. Diane Richardson, an expert in map generalization, worked for the Government of Canada. Diane Mann made significant contributions to mapping at the Department of Natural Resources. Claire Gossan, now retired, also made numerous contributions in that department, as does Donna Williams today. Alice Wilson (1881–1964) was the first female geologist in Canada and the first woman to become a member of the Royal Society of Canada; she worked for the Geological Survey of Canada. Moira Dunbar (1918–1999) was a pioneer of Arctic sea-ice research and the first woman to conduct scientific observations on Canadian icebreakers.[26] Finally, Diana Rowley became an early Arctic explorer in the 1930s and 1940s.

The lack of space prevents us from delving into the specific contributions of most of these women. However, Chapters 8 and 9 contain anonymous interviews with at least three Canadians, in addition to the vignettes of Helen Kerfoot, Alberta Auringer Wood, Eva Siekierska, and Janet Mersey found below. Helen Kerfoot and Eva Siekierska were both born outside of Canada and were thus immigrants. More significantly, both women are deeply committed to maps that reflected indigenous or local culture. One devotes her life to creating maps that are meaningful to Canada's Arctic peoples; the other stimulates, on an international scale, the adoption of toponyms that reflect local designations.

HELEN MARGARET KERFOOT (B. 1938)

Helen Kerfoot is a Canadian toponymist who has become an international advocate and expert on the standardization of geographical names.[27] She has played an even more important role for cartographers worldwide, however, as chairperson of the United Nations Group of Experts on Geographical Names (UNGEGN) standing advisory committee of the Economic and Social Council of the United Nations.

Born in 1938 in the suburbs of London, England, Kerfoot spent her childhood years living on the North Downs, enjoying fossil collecting in the chalk escarpment, playing in the fields and hedgerows, and seeing foxes and badgers—at least, when the bombing of the war allowed these activities.

CHAPTER 6

Maps and geography became a fascination; during school years, seeing different parts of the United Kingdom through courses at outdoor centres in different parts of the country became a bonus.

Kerfoot obtained a B.Sc. in geology, geography, and anthropology at King's College, University of London, and enjoyed the fieldwork involved and the leadership of staff of the Joint School of Geography of King's-LSE (London School of Economics). She remembers many interesting lectures and discussions with professors John Pugh and Keith Clayton, among others. Life in London was full of interest, and held Kerfoot for another year after university (working for Unilever) before she left for Canada in 1961.

In Vancouver, British Columbia, she was employed in the business world and later took up teaching at a local girls' school. She spent the summers as a field assistant in the Mackenzie Delta and western Arctic, and after moving to St. Catharines, Ontario, and then Ottawa, was a geomorphologist and field party leader in the northern Yukon and Northwest Territories for the Geological Survey of Canada, where she became interested in toponymy. In 1975 she joined the Secretariat of the Canadian Permanent Committee on Geographical Names at the Ministry of Energy, Mines and Resources (now Natural Resources Canada) in Ottawa, where she was appointed researcher for Northwest Territories toponymic data, under the leadership of Alan Rayburn. In 1987, upon Rayburn's retirement, she became head of the Secretariat and also of the Geographical Names Section. As a manager, she was responsible for developing the geographical names website, which included the national toponymic database—the first in the world to be available on the Internet—and for various publications on Canadian toponymy (including a bibliography on Canadian indigenous geographical names). In 1998, she retired from Natural Resources Canada, but remained with the department as an emeritus scientist.

Kerfoot was delegated by the Canadian government to the United Nations Group of Experts on Geographical Names (UNGEGN) in 1987, was elected as vice-chair in 1994 and elected as chairperson of UNGEGN from 2002 to 2007, and for a second term from 2007 to 2012. During this time, according to Rystedt,

> she has succeeded in turning UNGEGN into a more professional body and in bringing its work in line with current Spatial Data Infrastructure initiatives. She is one of the few experts with hands-on experience in practically all fields of toponymic standardization. Her drive to attend all the meetings of UNGEGN working groups, her participation in toponymy courses worldwide and

in scientific seminars and technical meetings, as well as her endeavours to make all UNGEGN "jurisprudence" on geographical names accessible through its website, have benefited the entire spatial information community, as geographical names standardization is a most important aspect in the exchange and linking of geospatial data. (Rystedt, 2007: 6)

Kerfoot has been at the forefront in encouraging countries to establish national authorities to standardize their geographical names for international use. She has been particularly pro-active in supporting efforts in Africa to create a more solid foundation for countries to become involved with the initiatives and standardization aims of UNGEGN. Not only have her goals been to see the wide dissemination of endonyms, she has worked to heighten respect for indigenous toponyms and promoted the value of geographical names as a vital part of our language, our heritage, and our identity. Accurate information associated with toponyms as part of the social infrastructure can facilitate social and economic development. Her article (Kerfoot, 2004) and a report about her work (DESA, 2007: 4–6) elucidate the problematique of toponymy in Canada and elsewhere.

In recent years, Kerfoot has presented numerous lectures on the work of UNGEGN and the technical, economic, social, and cultural benefits of standardized names, including keynote addresses to the ICA and the International Congress of Onomastic Sciences (ICOS). For her services and contribution to cartography and geographic information, Helen Kerfoot was awarded an Honorary Fellowship of the International Cartographic Association. Kerfoot has been a governor of the Royal Canadian Geographical Society, president of the Canadian Society for the Study of Names, chair of the Ontario Geographic Names Board, and a contributor to a variety of journals, both in Canada and internationally.

Kerfoot has received several awards for her work, including the 2010 Camsell Medal for outstanding service to the Royal Canadian Geographical Society.

ALBERTA AURINGER WOOD (B. 1942)

A leading map librarian who is conversant with many other important aspects of cartography, Alberta Auringer Wood has long been recognized on a national and international scale for her salubrious care of atlases, map collections, and catalogues.[28]

She was born in Detroit, Michigan, on 19 February 1942 to Gjertine Therese Silnes Auringer (1911–1995), a medical secretary, and Frank John

CHAPTER 6

Auringer Jr. (1908–1998), who had an academic degree in biology and advanced degree work in marine biology, including at Woods Hole, a highly regarded marine research institute in Massachusetts.

She obtained all of her degrees from the University of Michigan: an A.B. with honours in geography and mathematics (1964), an A.M. in Library Science (1965), and an A.M. in geography, specializing in the history of cartography (1973). Among her mentors she counts Richard W. Stephenson and Andrew M. Modelski, both retired librarians from the Geography and Map Division, Library of Congress; professors Waldo Tobler (now at the University of California, Santa Barbara) and George Kish (deceased in 1989), formerly at the University of Michigan; and Richard H. Ellis, university librarian (now retired) at Memorial University of Newfoundland. Among women cartographers she numbers as special friends and influential colleagues Jackie Anderson, Barbara Bond, Trish Caldwell, Mary Clawson, Judy Olson, and Barbara Bartz Petchenik (deceased in 1992).

Auringer Wood whetted her appetite for geography and maps through books, living in the country, and having wanderlust. Her publications reflect her vast interests in topics and regions, whether it is historical Detroit, new and old atlases, curiosities of the publishing world, acquisition philosophy and cataloguing priorities for university map libraries, map scales, topics relating to cartography and to the publications of the American Congress on Surveying and Mapping, and collections in public and university libraries. Her reviews of new books constitute an important addition to scholarly knowledge. She co-edited several works (such as in 1979, 1980, 1981). She penned a number of citations honouring the achievements of cartographers.

Increasingly, cartography-related organizations have sought her talents in writing précis and proceedings, whether for the Association of Canadian Map Libraries (later the Association of Canadian Map Libraries and Archives), the Map Curators' Group of the British Cartographic Society, the North American Cartographic Information Society, the Special Libraries Association (Geography and Map Division), the American Congress on Surveying and Mapping (ACSM), the International Cartographic Conference of the International Cartographic Association (which she served as vice-president), and the Congress of Cartographic Information Specialists Associations, the Canadian Committee on Bibliographic Control of Cartographic Materials, the ICA Executive Committee Working Group on Archiving, the Canadian National Committee for the International Cartographic Association, and the American Society for Photogrammetry and

Remote Sensing. She normally served in an executive capacity in virtually all of these organizations. She also contributed substantially to the Barbara Petchenik Children's Map Competition (with Jacqueline Anderson).

It is evident that Alberta Auringer Wood is a staunch supporter and nurturer of the work of her colleagues, either while serving in various capacities or informally. She hopes that her "participation increased the likelihood of other women following to do similar activities" (Alberta Auringer Wood, email to author, 23 February 2012).

Women are still dominant in the field of librarianship, though not necessarily in cartography. The wide range of her involvements proved to be a profound source of comfort to the few challenges she faced. She said that "aside from my going to Washington, D.C., to work at the Library of Congress, my work locations were mostly dictated by where my husband wished to go or was located. Instead of giving up work upon having a child, I continued working, despite the problems with child care" (Alberta Auringer Wood, email to author, 23 February 2012).

From an early age, she had an interest in maps, including collecting maps of travels as a child or requesting road maps of various states. In short, she has been interested in maps per se since she was quite young.

Although reserved about saying anything about her creativity—she claims she does not have any in the artistic sense—she has "sometimes adopted a topic that then caught on with a lot of others. This was the case with the article on acquisition policy and cataloguing priorities, which I did in 1972. I like to think that I just do things that need to be done" (Alberta Auringer Wood, email to author, 23 February 2012).

Married twice, both times to geographers, she has a daughter, Jennie, from her first marriage. Clifford H. Wood and Alberta Auringer Wood have been married since 1975. Clifford is a cartographer who has practised and taught in the field, with primary interest in design aspects.

Having contributed substantially to so many different areas in her field, she received much in the way of formal recognition: the American Congress on Surveying and Mapping Presidential Citations (1983, 1984, 1985, 1986, 1989, and 1991), the Special Libraries Association Geography and Map Division Honors Award (1987), the American Society for Photogrammetry and Remote Sensing Presidential Citation (1988), the Association of Canadian Map Libraries and Archives Honours Award (1999), the Earth Sciences Sector Merit Award, Natural Resources Canada (1999), and Honorary Research Librarian (2005–) at Memorial University of Newfoundland.

CHAPTER 6

EVA SIEKIERSKA (B. 1945)

Eva Siekierska has made a number of contributions to cartography, namely, the development of electronic atlases, tactile maps, cartographic visualizations, and multilingual topographic maps of Canada's Arctic and Sub-Arctic.[29] She has played a leading role in the promotion of women in cartography.

Born in 1945 in a labour camp in Leipzig, Germany, Siekierska came from a socially active and highly educated family. Her mother, Klementyna-Kama (née Kahan), was an economist and had a stronger influence; she had a university education, which was unusual at the beginning of the twentieth century. Eva's father, Stanislaw Siekierski, was a lawyer-economist. He was an avid painter, which stimulated Siekierska's interest in the visual arts.

Siekierska's first advanced degree was an M.Sc. in cartography from Warsaw University in 1971. She designed a comprehensive wall map of the North Sea as she was planning to do a Ph.D. in oceanography. Indeed, what became the identifying moment for Siekierska—that is, what made her turn to cartography—was a combination of nature, art, and innovative applications.

In 1980, she obtained a doctorate in the Natural Sciences from ETH-Z, the Swiss Institute of Technology in Zurich, in the Department of Cartography and of Computer Science. The topic of her thesis was "Raster Mode Processing of Topographic Data on Small Computers." The most interesting part for her was developing a Digital Elevation Model of the Grindelwald Region, with the famous Eiger Nord Wand. It gave her a good reason to spend time hiking in the mountains. She immigrated to Canada and worked for six months at the University of Waterloo and later found a permanent position with the federal government in Ottawa.

The people who most deeply affected her career and professional outlooks date back to the time of her master's degree at Warsaw University under the supervision of Professor Pietkiewicz, nestor of Polish cartography. Another important influence was professor Lech Ratajski, author of theories on cartographic generalization thresholds and on cartographic communication. Siekierska's stint as a master's student seemed to have been an important anvil upon which her creative cartographic interests took shape.

However, the strongest impact on Siekierska, and the one whom she admires the most, was Professor Fritz Mueller—a very warm and "humane" person who coached her in the adaptation to foreign culture. Her current research interest in multilingual mapping of the Arctic has been inspired by him and his wife, Barbara. Siekierska owes a lot to Professor Fraser Taylor, who facilitated the founding of the Commission on Gender and Cartography at the ICA and who encouraged her to chair the Commission, and to Dr. Theodor Blachut, an accomplished photogrammetrist and very active member in the Canadian Polish Heritage society.

The federal government restricted much of her scientific cartographic work, which is viewed as a luxury. Siekierska helped to develop the concept of an "electronic atlas." She implemented the concept, in prototypes, of graphic station Mark 1 and Mark 2 in 1981–1984 (Siekierska, 1984). Electronic atlases were not developed at the time (Siekierska and Taylor, 1991), although there were GIS-type systems available, and electronic charts have been in the early implementation stage in the Canadian Hydrographic Service. The electronic atlases were a combination of the two. Another type of an atlas was a GEOSCOPE—Interactive Global Change Encyclopedia, published in 1993. The development involved several Canadian federal government departments and the Canadian Space Agency. Siekierska co-chaired the ICA Commission on National Atlases from 1987 to 1991, and National and Regional Atlases from 1991 to 1995.

"Automated cartography" was a non-conventional occupation for women; she had to face these issues as the sole woman in many cases. She did not get credit for the work she did, but "needed to work extra hard for the work to be 'noticed.'" On another level, Siekierska felt she was forced to choose between having a family and pursuing a Ph.D.; she choose the latter.

Siekierska was one of the participants in formulating the proposal for the Treasury Board of Canada to conduct a five-year program for "People with Disabilities On Line." Several interactive tactile, audio-tactile, and audio-tactile-haptic maps (haptic maps are based on force feedback devices) were developed (Siekierska and McCurdy, 2008). The first edition of the Tactile Atlas of Canada appeared in 1999, and the last in 2003 (http://www.datadot.gc.ca/dataset).[30]

Siekierska began to explore visualization based on maps. The first was the Historical Evolution of Canada, produced for the 125th anniversary of Canada (Siekierska and Armenakis, 1999). Another was the visualizations developed to commemorate the creation of Nunavut, Canada's latest (and largest) territory, in 1999. It contains an audio map with Aboriginal names for settlements (http://maps.nrcan.gc.ca/visual/index_e.php). In 2003, she was able to create a dynamic time-based visualization to show the evolution of Iqaluit, an early four-lingual website to explain the history of that community (http://maps.nrcan.gc.ca/iqaluit/index_e.php). She served as

Canadian representative on the ICA Commission on Visualization and Virtual Environments (1999–2007) and on Maps on the Internet (1999–2003).

Most recently, she has taken on the challenging task of being project leader of "Customized Multilingual Topographic Maps of Arctic and Sub-Arctic," envisaged as a tool for community engagement and development of partnerships, as well as for promotion of Inuit and First Nations cultures. The multilingual topographic maps of Nunavut carry all four official languages of that territory. Siekierska sees the need to develop a new process for inclusion of traditional knowledge and collection of user volunteer information.

In Canada, Siekierska began to contribute extensively to improve the status of women in cartography; in Poland, the majority of cartographers were women, but this was before the computer era. Her first effort led her to conduct an international survey on barriers and incentives to more active participation of women in the International Cartographic Association (Siekierska, O'Neil, and Williams, 1991). She chaired the ICA Presidential Task Force on Women in Cartography (1989–1991), which she immediately followed as chair of the ICA Working Group on Gender and Cartography (1991–1995), and then as chair of the ICA Commission on Gender and Cartography (1995–1999). Siekierska sees her cartographic work as part of the larger loom to contribute substantially to the betterment of society.

A number of relevant recognitions came her way. In 1983, the Government of Canada selected her as Exemplary Scientist for the NRCan career development video, and in 1993, she was awarded the 125th Anniversary of Canada Medal. For the contributions to the community of the blind and visually impaired, she and her team received an APEX award for Leadership in Service Innovation—"for its creative blend of partnerships, technology and new approach to client services to benefit Canadians with special needs."

JANET MERSEY (B. 1953)

Janet Mersey's primary research interests focus on the application of geomatics technologies (geographic information systems, remote sensing, and spatial statistics) to problems related to land management and resource assessment, especially in environmentally sensitive areas.[31] Recent studies have involved soil erosion mapping and modelling biodiversity at the watershed scale. Her other research initiatives explore issues in cartographic design, data symbolization, and the history of thematic mapping.

Janet Mersey was born and grew up in Moncton, New Brunswick, the youngest of three children. She completed an honours B.A. in mathematics and economics at Mount Allison University in Sackville, New Brunswick (1975), which was followed by a second B.A. from Carleton University, Ottawa (1977). A post-graduate program in cartography attracted her to the University of Wisconsin–Madison, where she earned an M.Sc. in 1980 and a Ph.D. in 1984, under the tutelage of Arthur Robinson and David Woodward, respectively.

Now an associate professor in the Department of Geography at the University of Guelph, Ontario, Mersey is currently (2012) acting associate dean (academic), College of Social and Applied Human Sciences. She has also served as associate chair of the Department of Geography at the University of Guelph (2006–2011). Mersey's responsibilities in the Department of Geography include teaching undergraduate courses in geographic information systems and in remote sensing and cartography, and managing the department's GIS laboratories.

Her published work involved the use of geographic models, such as a soil-erosion potential model, to plan effective conservation strategies in ecologically sensitive areas, and the importance of community-based participation in the success of such projects. Mersey's interest in the use of the Internet in GIS education is also reflected in her work, and she developed a series of GIS/RS modules for an online course in GIS and landscape restoration at the University of Wisconsin.

Mersey has contributed to teaching, research, and service in the area of cartography. First, in the area of teaching, she has taught in the constantly changing field of geographic information systems, cartography, and remote sensing for over twenty-five years. Although this means constantly renewing course materials, incorporating the use of new software programs and digital data, and upgrading her own skills, she finds that many students have a natural affinity for maps and satellite images and are intrigued by computer-based approaches to their exploration and analysis. Cultivating students' interest and enthusiasm for the mapping disciplines and providing them with a sound understanding of the concepts, methods, and applicability of spatial techniques are among her primary goals as an educator. She wants students to leave her courses equipped with the theoretical and technical expertise required to be competent and creative problem solvers who are capable of communicating (both graphically and in writing) the results of their analyses.

The experiential learning that goes on in the laboratory sessions is a fundamental component of these geomatics courses. Apart from the obvious goal of providing training in the spatial methodologies involved in the

CHAPTER 6

subject, she also emphasizes the importance of critically questioning computer-generated output.

Second, in the field of research, her recent projects involved monitoring landscape diversity and change in Mexico, land use change and forest fragmentation in Ontario, land use conflict in China, soil erosion mapping in Mexico and biodiversity modelling in Ontario, and investigating the spread of the emerald ash borer in Ontario. Other research initiatives focus more specifically on map communication, graphic design, and cartographic symbolization. She also has a long-standing interest in the design of thematic maps, and has recently written a chapter for a history of cartography text on the development and use of choropleth maps since 1700 (Mersey, 1990).

Third, in the area of service to cartography, Janet Mersey continues to serve as Canada's principal national delegate to the International Cartographic Association (ICA). This involves chairing the Canadian National Committee for Cartography and holding annual meetings throughout Canada. On two occasions (2007 and 2011), she was responsible for preparing and submitting to the ICA Canada's five-year report on cartography in Canada. These comprehensive documents, published as special issues of the journal *Geomatica*, included reports from over fifty geomatics specialists in Canada in government, academia, and private industry. She was responsible for soliciting, editing, and collating these reports, and for coordinating the review of the research papers included in the document. Other duties of this post included coordinating Canada's National Map Exhibit at the ICA conferences in La Coruna, Spain (2005), Moscow (2007), Santiago, Chile (2009), and Paris (2011), and overseeing Canada's entries to the Children's National Map Competition at these conferences.

During the time that Janet Mersey was technical chair of the Canadian Institute of Geomatics, the Institute highlighted her various important involvements with the Canadian Cartographic Association, including as President (1995–1996) and her services on the Editorial Advisory Board of the journals *Cartographica* and *Cartographic Perspectives*. She has also served on the Board of Directors of the North American Cartographic Association. Her professional work earned her the Earth Science Sector Award of Merit from Natural Resources Canada in 1999. She also received the "Top Ten Site" Citation Award for an online GIS bibliography linked to the GIS resources page of the Department of Geography webpage, 2002.

CONCLUSION

In reviewing these vignettes, one would be impressed by their diversity of chosen topics and approaches to cartography. The empirical (thematic) basis of cartography was related to oceanography, children, foods and restaurants, consumption patterns, and seacoasts. Some of the women in the chosen vignettes also delved into representation on maps, involving symbolization, digitalization, and visualization. Some have focused on what map-makers (intentionally or unintentionally) conveyed through maps ("persuasive cartography" and toponymic issues) and have researched the comprehension of maps by readers, such as spatial comprehension by children, and the psychology of maps. It is noteworthy that any awards that were forthcoming often arrived late in the life of the women cartographers.[32] On a more theoretical level, some have investigated how cartographic knowledge is constructed, while others have concentrated their attention on the physical aspects of making maps, such as mapping-science technology, GIscience, remote sensing, and creating tactile maps. It is clear that the boundaries among these areas are not only quite permeable, but that a number of women cartographers have tackled a variety of issues throughout their cartographic career.

The next chapter (7) takes us to pioneering women cartographers in the rest of the world, notably Britain, Austria, Russia, China, and Latin America. Do their contributions to cartography vary from the North American ones, and, if so, in what way? What other significant features can be attributed to them?

CHAPTER 7

Late-Twentieth-Century Pioneers and Advancers in Europe, Asia, and Latin America

In North America and elsewhere, many different participants inhabited this map world. It would be impossible to find a more diverse group of cartographers than those represented by the ten women in this chapter. Helen Wallis of the British Library became the leading map librarian in the world, while Eila Campbell—a close friend of hers—put all of her energies into historical cartography through her eminent editorship of *Imago Mundi*. Barbara Bond demonstrated her mettle in continuing the worldwide stature and work of the British Admiralty charts. Ingrid Kretschmer, one of the few notable women cartographers outside of Britain, took special interest in Austrian folk culture and wrote the world's first encyclopedia of cartography. Kira Shingareva of the former Soviet Union (now Russia) pioneered the creation of lunar maps, especially of the far side of the moon.[1] Liqiu Meng took the unusual step of moving from China to Germany, where she established an international reputation in geo-data generalization. Interestingly, the two remaining women cartographers, Carmen Reyes (Mexico) and Regina Araujo de Almeida (Brazil), both paved the way for making maps accessible to minorities, indigenous people, and children, sometimes producing tactile maps. Map curators, such as Helen Wallis, would serve as presidents of the British Cartographic Society and lead national and international efforts to raise the "scholarly nature and academic acceptance of cartography" (Ravenhill, 1996: 300). The linkages between those who profess historical cartography, those who are incumbents of map librarianship, and those who are engaged in practising cartography were both implicit and explicit. Members of each group fed into each other's interests and knowledge as a matter of course, but also stepped in to take an active lead in developing areas outside their original interests. Starting in 1972, five European women—Helen Wallis, Eila Campbell, Olga Kudronovska, Leena Miekkavaara, and Ingrid Kretschmer—as well as Dorothy F. Prescott of Australia consecutively led the effort to put the history of cartography on the map of the International Cartographic Association (ICA), but their underlying bond was their love and enthusiasm for maps.

EUROPE

Carol Beaver (1993: 5) has found six primary areas of work where women are currently active as map-makers: in cartography, remote sensing, surveying, GIS, geodesy, and geomatics. Women cartographers in the United Kingdom felt discouraged from applying to governmental mapping services; one of the consequences of this discouragement was that women became freelance map-makers, usually as the heads of their own private cartographic firms. The rise in freelance cartographic firms thus offered a new niche for women in cartography.

There was, however, an indirect effect of the vast production of maps by military establishments, for within twenty years after World War II, Helen Wallis suggests, "large consignments of cartographic materials" passed on to libraries for preservation (Wallis, 1989). As a later chapter indicates, map libraries around the world were chiefly sustained by women. With the increased interest in preserving maps in libraries and archives came a deepened interest in the history of cartography—an area that has flourished from the founding of *Imago Mundi* in 1935 as the academic journal for the history of cartography and was heightened after 1967, with the establishment of biennial international conferences (Wallis, 1989).

EILA CAMPBELL (1915-1994)

A close friend of Helen Wallis's, Campbell made significant contributions to cartography in at least four areas.[2] As an eminent geographer at London University, she was among the first to document the participation of women in geography, and through her wide activities paved the way for women in geography. She substantially bridged the gap between geography, cartography, and history (Wallis, 1994: 361), not only through her own scholarship, but also through her early involvement with, and later meticulous editorship of, *Imago Mundi*, the primary journal of historical cartography. The innovative part she played in analyzing the eleventh-century *Domesday Book* serves as her fourth major contribution.

There is virtually nothing recorded about Campbell's parents, with the exception of one reference to her mother in an acceptance speech for an award in 1979, in which she thanked her mother "who for nearly twenty-five years acted as my 'honorary' wife doing as many of the chores that the wives of explorers and scholars have to—keeping house, answering the telephone and feeding the cat" (Royal Geographical Society, 1979: 527).

Born on 31 December 1915, Campbell was educated at Bournemouth High School for Girls; later she entered Brighton Diocesan Training College for Teachers. Having moved from Hampshire, she became a Londoner, with Birkbeck College, London University, at the centre of her life. According to Michael Hebbert of the University of Manchester, Campbell "disliked the silence of the countryside," and the background music of London became part of her life (Hebbert, 2000: 1). She taught school from 1934 to 1945. She then graduated with honours in geography from Birkbeck College. By 1941, she was already starting to move to an academic life—she had become part-time assistant in geography at the college (while still managing to fit in her schoolteaching). In 1945, she worked as assistant lecturer in geography, eventually earning her master's. For fifteen years (from 1948 to 1963), she was a lecturer at Birkbeck College, then became a reader from 1963 to 1970. From 1970 (when she was fifty-five) to 1981, she was professor of geography at London University and head of the department at Birkbeck. When she retired, in 1981, she was honoured with the rank of professor emerita. In the end, she served her college for forty years.[3]

Campbell exemplified an unsurpassed loyalty to her students. One of her colleagues, John Wells, averred that she always had time for students. He recalls, "I can remember her walking down the corridor on many occasions carrying the globe without which she would never teach that aspect of geography" (Birkbeck, 2007: 1). One cartographer who never had much direct contact with her work "did hear a lot about her care and encouragement of her students, including, to my surprise, in the field of map projections" (Henry Steward, email to author, 10 September 2010).

Campbell received numerous awards, including the Murchison Award for geographical sciences of the Royal Geographical Society, with the Institute of British Geographers (1979), where she had also served as the council of the Institute. On many occasions, Campbell was able to turn a moment of praise for her work into something that was potentially self-deprecating, such as when she accepted the Murchison Award: "The awards are delightful because one never expects or even hopes to receive one. They fall into one's lap rather like rays of sunshine—perhaps some of you may feel rather that they fall like rain on the just and on the unjust—on the deserving and the not deserving. But hope not" (Royal Geographical Society, 1979: 527).

The International Map Collectors' Society honoured her with the Tooley Award in 1989 for her contributions to the history of cartography (Wallis, 1994: 361). She was also elected member of the Society of Woman Geographers, Washington, D.C. Like Wallis, Campbell served many societies faithfully: the Library and Maps Committee of the Royal Geographical Society (twenty-one years); as secretary-editor of the Hakluyt Society (twenty years), becoming its first honorary secretary (1962–1983) and then its vice-president in 1983; and was able to connect the work of that society with the American organization, the Society for the History of Discoveries, as chairman of the Royal Geographical Society's Sub-committee for Cartography (six years); the British National Committee for Geography, as president of Section E of the British Association for the Advancement of Science (1975); as President of the Society for University Cartographers (now the Society of Cartographers), from 1974 to 1979; served as editorial advisor to *The Map Collector* since its launch in 1977; on the Council of the Society for Nautical Research, which she joined about 1952, later becoming its trustee in 1989, but also as vice-president from 1993 until her death a year later; regularly attended meetings of the International Cartographic Association, occasionally as head of the British Delegation; also followed the congresses of the International Geographical Union, and took part in its mid-term regional meetings; and the Federation of University Women (Wallis, 1994: 361).

Even in 1979, the status of women in geography was still on the margins in the membership roles of the Royal Geographical Society. In that year, they accepted 226 new Fellowships, both full and associate; 23 percent were women (Royal Geographical Society, 1979: 527)—a figure comparable to geographical and cartographic societies in North America. An interview with Anne Buttimer, a professor at the School of Geography, Planning and Environmental Policy, University College, Dublin, reveals the extent to which Campbell was the first to document the participation of women in geography. Buttimer also reveals that "her male colleagues have not been very kind to her" (Buttimer in Maddrell, 2009a: 757).

Buttimer concluded that Campbell "may have felt very much alone" (Maddrell, 2009a: 757) and that during the war women had to do all the teaching of geography, but when the men returned from war to resume their professional careers in geography, they did not shed their traditional image of women and found it hard to work with them.

Campbell often recounted her experiences as head of the British delegation of the International Cartographic Conference in 1980 in Tokyo, when she asked the Japanese officials whether they found it strange for a woman to head the delegation. Wallis (1995: 10) reported that they had replied, "Not at all. For the purpose of this conference you are elected an honorary gentleman."

Campbell worked tirelessly for two scholars, Leo Bagrow and Clifford Darby. It is clear that the reason "Ella [sic] did not publish more in her name was because she devoted so much of her time to work that has appeared under the names and to the credit of others" (Barber, 1995: 11). The circumstances under which she worked with Leo Bagrow (d. 1957) did not particularly mark a high point where women were taken seriously. Bagrow would badger her almost daily in his letters and mount intrigues behind her back. But when Bagrow asked Campbell to be a referee when he applied for a visa to visit Britain, she agreed and wrote to Britain's Home Office that Bagrow was "a scholar and a very eccentric elderly man.... I am prepared to support his application but know nothing of his activities apart from his editing of *Imago Mundi*." Bagrow apparently did not receive his visa (Barber, 1995: 11).

As a young scholar she was nurtured by Eva Taylor (1879–1966) (Wallis, 1994: 361), the English geographer and historian and the first woman to hold an academic chair in geography; Sidney Woolridge (1900–1963), a geologist, geomorphologist, and geographer; and Edward Lynam (1885–1950), author of *British Maps and Map-makers*, who introduced her deeply to historical cartography (Royal Geographical Society, 1979: 527). These three academic and professional influences on her life contributed to her desire to bridge the gaps among geography, cartography, and history. Her special interest in fieldwork brought home the idea of the inseparability of physical and human geography; the unitary character of geography stayed with her for the rest of her life as a scholar (Mead, 1995b: 8). Apparently, Campbell became disillusioned and disenchanted with later developments in her field (8).

All of these interests flowed quite naturally into the considerable and meticulous editing work she did for *Imago Mundi*. Along with Helen Wallis and two men, Campbell organized the first of the International Conferences on the History of Cartography, in 1967 (Wallis, 1994: 361)—a biannual conference closely associated with the profound scholarly style of *Imago*

CHAPTER 7

Mundi (*Imago Mundi*, 1988: 56). Wallis described Campbell's work for the journal as her "greatest achievement" (Wallis, 1994: 361). She was first corresponding editor until 1974, and then took over as editor and chairman of the Board of Directors. In all, she worked for *Imago Mundi* for forty-eight years. The journal would have come to an abrupt end if it were not for Campbell's decisive action. With the departure of Cor Koeman of Utrecht University[4] and Nico Israel, a publisher in Amsterdam no longer wished to undertake this task. In short, *Imago Mundi,* under the new editorship of Campbell, appealed for subscriptions to cover the cost of printing. The appeal fell £1,000 short of the goal. The publication of the next issue of *Imago Mundi* might have fallen by the wayside if an anonymous guarantor had not come through with the difference. It took five years to learn that this guarantor was Eila Campbell herself. For eighteen years, Campbell produced *Imago Mundi,* until 1993, the year before her death. As Harry Margary (1995: 8) wrote, "neither illness nor age had diminished her enthusiasm."[5]

Monique Pelletier, director, Maps and Plans Department, Bibliothèque nationale, Paris, expressed the lengths that Campbell would go to welcome newcomers to the field, including Pelletier, back in 1981 (1995: 9). "Never a passive attender," says Pelletier, "Eila was always on the alert, ready to be attracted by a broad variety of topics.... [S]he constantly acted as a teacher concerned abut the transmission of knowledge and the way the history of cartography was so often ignored in academic programmes" (1995: 9). When Campbell proffered a course at the Institut Cartogràfic de Catalunya (Barcelona), she spoke on thematic mapping, another one of her many interests (1995: 9).

Early in her career, Campbell worked with a colleague, professor H.C. Darby (later Sir Clifford Darby), as an editor and contributor to the first volume of his well-known *Domesday Geography of England,* 1954. Now a multi-volume descriptive and analytic rendering of the original, *The Domesday Book* (1086 edition) was a significant survey in England and Wales under the aegis of William the Conqueror (William I of England). According to Peter Barber, map librarian at the British Library, Campbell played a "major role and innovative part in the analysis" of the *Domesday Book,* published in 1962 (Barber, 1995: 11).

One hears of Campbell's personal attributes. Such tributes as "generous," "indefatigable," "warmth of heart," and "loyal" were invariably associated with her name. It was well known, though, that Campbell "wore her undoubted scholarship lightly" and had a "down-to-earth approach and sense of adventure." One writer stated, "She had a particular skill of relating to all types of people whatever their station in life.... She often said one could discover more about university departments from porters, cleaners and secretaries than from professors, senior lecturers and other academic staff" (Birkbeck, 2007: 2).

Eila Campbell died on 12 July 1994. After her death, the history of cartography disappeared from the curriculum of London University. Academic geographers were already adopting quantitative techniques, leaving historical cartography in the dust (Wallis, 1994: 361).

HELEN WALLIS (1924-1995)

Helen Margaret Wallis was the leading figure in map librarianship in the United Kingdom and "the best-known contemporary British historian of cartography" (Ravenhill, 1996: 299), even globally (T. Campbell, 2004: 1).[6] She was the first woman to hold the post of superintendent of the Map Room at the British Museum and then the British Library.

Wallis was one of a set of twins born on 17 August 1924 at Barnet, Hertfordshire, north of London. Her parents were known for their "integrity and high principles, though wise and tolerant," too (Ravenhill, 1996: 299). While there is very little information about her mother, Mary McCulloch Jones (1884–1957), who was a teacher, we know enough about her father, Leonard Francis Wallis (1880–1965), to know he was headmaster of Willesden Grammar School (Mead, 1995a: 241).

Her father sent her initially to the small Colet Girls' School, but at the age of ten she went to St. Paul's, winning scholarships there, including one for classics and another for piano. At twenty-one, after spending the last two years of the war teaching, she secured entry in St. Hugh's College, Oxford. It was at Oxford that she acquired a deep taste for geography under the tutelage of J.N.L. Baker, historian of geographical exploration, a founding member of the Institute of British Geographers, and its guide during "the difficult postwar period" (Ravenhill, 1996: 299). Through Baker, Wallis gained an enduring interest in cartography. After her undergraduate education, she chose to research the exploration of the South Sea, 1519–1644 (Wallis, 1954), and graduated from St. Hugh's with a D.Phil. in 1954. She stayed devoted to St. Hugh's and served on its association of senior members.

To enable her to gain access to source materials, the university brought in R.A. (Peter) Skelton as an additional supervisor; he "had already achieved international fame" (Ravenhill, 1996: 300) in the history of cartography, exploration, and discovery. She had spent three years as member of staff at the

Museum by this time. It helped that Skelton occupied the position of superintendent of the Map Room at the British Museum. Skelton was so "impressed" with this "discerning scholar" that in 1951, when she was twenty-seven years old, he appointed her as assistant keeper of the Map Room. Relations with Skelton, however, became strained. Tony Campbell, her successor as map librarian, wrote: "Wallis worked hard to overcome a natural diffidence [with Skelton], even to the extent of radically changing her appearance, but she had to face further setback when the time came to choose Skelton's successor. Although she was the obvious candidate, his unfavourable assessment of her caused a morale-sagging delay before she was appointed in 1967 as the first woman to hold that post" (T. Campbell, 2004: 1).

She was thus forty-three when she succeeded Skelton. In 1973, she became the first map librarian in the British Library in its new form. The Map Library became an important international node for geographers, cartographers, map librarians, and "learned booksellers." Under her tutelage, the Library acquired significant collections of pre-twentieth-century maps rather than contemporary maps and atlases. She worked assiduously to prove wrong her predecessor's judgment of her (T. Campbell, 2004: 1). Upon her retirement in 1986, her bibliography contained 250 items.[7]

Wallis never considered her steady stream of publications in isolation from her involvement in cartographic, geographic, and curator societies. Perhaps her earliest major contribution, some twenty-five years after being elected Fellow of the Royal Geographical Society, was her "courteous, but penetrating" exposure of the fraudulent "Vinland" map in 1974 (Ravenhill, 1996: 300). As the first chair of the working group on historical cartography of the International Cartographic Association, she wrote *Map-making to 1900* (1976), for which she compiled and edited submissions by twenty international scholars (Ravenhill, 1996: 300), to be followed in 1987 by the much larger *Cartographical Innovations: An International Handbook of Mapping Terms to 1900*. Interestingly, she co-edited this latter work with A.H. Robinson—a name we came across in the previous chapter, and someone who clearly stands associated with nurturing women cartographers on both sides of the Atlantic Ocean. Another co-authored work with Robinson was "Humboldt's Map of Isothermal Lines: a Milestone in Thematic Cartography," 1967. Her scholarship also rescued cartography from the margins of geography (Ravenhill, 1996: 300).

As the curator of one of the world's finest map collections, she stood at its scholarly helm with "exceptional brio and versatility" (T. Campbell, 2004: 1). She extended knowledge of the collection worldwide through her writings. The same year she joined the British Museum, she published her first paper—a study of British globes, and discovered England's first globe, by Emery Molyneux, at Petworth House (T. Campbell, 2004: 3). She also acquired "the only known copy of the woodcut wall map of the world by Giacomo Gastaldi (1561)" and completed the Museum's purchase of its largest map of the Royal United Services Institution (T. Campbell, 2004: 2). While her research on claims of a sixteenth-century Portuguese discovery of Australia was one of her more typical works, Wallis's reputation as a scholar involved the production of the facsimile of Jean Rotz's *Boke of Idrography* (1542), which the Queen presented to President Reagan on his visit to Britain. During the last months of her life she managed, despite her illness, to "complete the first national inventory of cartographic resources," *Historian's Guide to Early Maps of the British Isles* (1994), for the Royal Historical Society. Even so, her "most notable publication, and her most controversial scholarly position" (T. Campbell, 1995: 187), was her argument that the first Europeans to cover Australia's coast were the Portuguese, in 1530.

In rendering the Map Room more visible to the outside world, Wallis developed "new standards" of scholarship and scheduled topics outside the ken of dusty collections, such as the voyages of Cook, the American War of Independence, the voyages of Francis Drake, and so on. As Campbell said, Wallis's approach was interdisciplinary, long before it became fashionable (T. Campbell, 2004: 2). Her exhibitions and scholarship brought her to many places in the world; in North America, she received an honorary degree from Davidson College, North Carolina (1985) for her contributions to American history. Wallis's interests extended to the Jesuit mapping of China and thematic mapping (Barber, 2004: 1). Her work, as one commentary indicated, showed historians how to work with maps, and cartographers how to work with texts (Barber, 2004: 2).

Wallis's contribution to cartography was solidly vested in her participation in a score of cartography-, library-, and geography-related societies. She participated for almost fifty years in the Hakluyt Society as someone interested in the history of discoveries and edited for them *Carteret's Voyage Round the World, 1766–1769* (1965). The Society for Nautical Research received almost thirty years of her incessant service, especially as its chair (T. Campbell, 2004: 2). Other societies included the Society of Antiquaries, the Internationale Coronelli Gesellschaft, the British Cartographic Society, the Library Association, the Foundation for Science and Technology, the British Committee for Map Information and Catalogue Systems, the International Cartographic Association, the Société de Géographie (Paris),

CHAPTER 7

the Society of Woman Geographers, the National Maritime Museum, the Archival Research Task Force, *The Map Collector, The Canadian Cartographer,* the biennial international conferences on the history of cartography, the Royal Historical Society, and the Royal Geographical Society. In all of these matters, Wallis stepped into a man's world. In 1986, she became president of the International Map Collectors' Society and was a founder of the Geography and Map Libraries Section of the International Federation of Library Associations. As Tony Campbell notes (2004: 2): "She was the first woman to hold several of the positions ... and these distinctions awarded merit, not assertiveness. She found it amusing to be classed as an honorary man in Japan."

Her intersecting scholarship and librarianship, and her work on a large number of professional and scholarly societies, brought in many awards, including the Order of the British Empire (1986).[8] Wallis enjoyed travelling and took it upon herself to give lectures in Singapore, Australia, Canada, the United States, Hawai'i, and Japan. In her honour, the British Library established The Helen Wallis Fellowship.

According to Tony Campbell, Wallis had a "bubbling and urgent enthusiasm" for her work and interests (T. Campbell, 2004: 3). He also penned her personality as follows: "Her breathless lecturing style, with commentary and slides seemingly racing one another, was engagingly inimitable. She was equally encouraging to her own staff and to younger scholars. She was natural and friendly to all she met and found it impossible to refuse commissions. Unmarried, she devoted most of her free time to academic and society matters, as well as choral singing" (T. Campbell, 2004: 3).

Peter Barber, the then head of Map Collections at the British Library, simply stated that she "loved people and maps—in that order" (Barber 2005: 195), and "It was this same human warmth and capacity for friendship, allied to a belief in the virtues of structured co-operation, that made Helen such a firm advocate of national and international collaboration" (Barber, 2005: 195).

Barbara Backus McCorkle, a keynote speaker at the 2008 annual meeting of the Society for the History of Discoveries, relates the following about Wallis: "Yet Helen was still completely unpretentious. She trended [sic] to speak rapidly, but behind that ordinary appearance was an extraordinarily sharp mind" (McCorkle, 2008: 1).

Norman J.W. Thrower mused about Wallis as follows: "[She] was an accomplished pianist and a devotee of the theater. She was optimistic, friendly, and democratic.... Helen Wallis was a lady of great faith, and when this reviewer telephoned her only a few days before her death ... she remarked that she was ready to go, having had a good life" (Thrower, 1999: 15).

Henry Steward of Clark University confessed that he met her for the first time when he was a student. He had walked in off the street into the British Museum Map Room, where "she handled some inane query I had, not letting me know that she was the boss" (email to author, 10 September 2010).

Although she had friends across the globe, the one friend that she was close to was Eila Campbell, founder and editor of *Imago Mundi*.

She died of cancer in London on 7 February 1995 at the age of 71. Still, she had maintained a cheerfulness throughout it all; and even in the video interview taken toward the end of her life she recounted her life "fluently and amusingly" (T. Campbell, 2004: 3). At least a dozen obituaries appeared, some of them in Britain's most prestigious newspapers.

KIRA B. SHINGAREVA (B. 1938)

Kira Shingareva, professor at Moscow State University for geodesy and cartography, is principal scientist at the Planetary Cartography Laboratory and the Laboratory of Comparative Planetology at the Institute of Space Researches at the Academy of Science.[9] She is one of the most eminent cartographers of extra-terrestrial bodies and was among the first people to succeed in mapping the "dark" (reverse) side of the moon. She heads the Commission on Planetary Cartography of the International Cartographica Association.

Shingareva was born in 1938 in Russia. Her mother died when she was five years old. Her father was a chemical engineer. It was he who suggested, at a critical point in her studies, that she study mathematics in the university's astronomical curriculum. She admits that "she is forever grateful to him for that, loving him dearly."

She studied at Dresden University, where she graduated from Technical University in 1961 (at the age of twenty-three), obtained a Ph.D. in 1974, and a Doctor of Science in 1992. Before then, she had gone to the University of Moscow. She had wanted to become a mathematician and to study the theory of mathematics. During the exams, she did not have enough points to be allowed to continue with mathematics (she missed it by one point). As a result, she went to another university, which included mathematics in the astronomical curriculum.

After returning to Moscow from Dresden in 1962, Shingareva connected with a friend who was heading the moon project; he asked her to work for him at the Laboratory of Comparative Planetology at the Institute of Space

CHAPTER 7

Researches under the aegis of the Academy of Science. In October 1959, the Soviet *Luna 3* had already succeeded in photographing the moon's far side. Three years after her arrival at the Institute, then at the university, she participated, in 1965, in the National Space program, mapping the moon, Mars, Phobos, and Venus. As a twenty-seven-year-old, it excited her very much to work on the project. Her main task was to select the landing sites for the moon probes. On 3 February 1966, *Luna 9* was able to land safely on the Moon (the first ever to do so) and take surface close-up images in the Oceanus Procellarum; *Luna 13* was able to follow up on these images on 24 December that same year (Williams, 2005: 2, 3).

A turning point early in her career was the 1967 Congress of the International Astronomical Union (IAU), where she presented, for the first time, the nomenclature of the reverse side of the moon. She was then only twenty-nine years old. The Soviet presentation of *Atlas Obratnoi Storony Luny, Ghast 2, 1967* (*Atlas of the Far Side of the Moon, Part 2*) at the Union failed on several accounts. Shingareva claims, "the images were of bad quality and there were mistakes." Ewen A. Whitaker (1999: 176), who was closely involved with the proceedings, noted that the map and a list of new names seemed like a *fait accompli*. Moreover, some 45 percent of the names were Russian. In any case, when the USSR delegation presented its nomenclature of the moon, it faced opposition from the United States National Committee on Lunar Mapping and Nomenclature. The US Committee suggested that only numbers be assigned to the 450 features on the reverse side of the moon, and that "we should be very conservative in assigning names" and "use names of permanent renown" (Commission de la Lune, 1967: 104).

According to a participant in the triannual meetings of the IAU congresses in the 1960s, the controversy started a year earlier, in 1966, when Dr. A. Mikhailov of the USSR Academy of Science sent a letter to Dr. D. Menzel, President of the Lunar Nomenclature Commission. Dr. Mikhailov suggested that "names of poets, painters, composers, etc. be used to identify the newly imaged craters on the Zond 3 photos" (Letter from Ewen A. Whitaker to W.C. van den Hoonaard, 28 March 2011). Later that year, the USSR published a list of 153 new names, of which some sixty-six were Russian, bypassing the rules of the IAU Lunar Nomenclature Committee.

When Shingareva presented her map, it became evident that the standards that applied to the near side of the moon could not apply to the far side. The near side showed the south pole on top of the map; the far side would show it at the bottom of the map. And where would "east" and "west" be (Whitaker, 1999: 173)? The United States scientists already had much information from their own lunar orbital photographic missions (1966–1967) involving 600,000 high-resolution images (Lunar and Planetary Institute, 2010), but the Soviets wanted her to select craters and name them.[10] The scientists from Europe agreed with the approach taken by the Russian delegation.

After Shingareva had returned to Moscow, a United States colleague sent her a map with a small crater named "Kira" in recognition of her remarkable achievements. She has that map on her wall still. All of her grandchildren know about the Kira crater.[11] She fondly recounts the story of a 102-year-old Russian lunar scientist, naming something after him and believing that he was dead. Soon she received a letter from him, saying, "I'm very much alive!" It is the International Commission of Nomenclature of the IAU that then ruled that one could now name craters after people who are over one hundred years old. Shingareva was busy for ten years at the USSR Academy of Sciences, participating in the Moon Exploration Project until 1977.[12]

With more than 150 publications to her name, including "Atlas of Terrestrial Planets and their Moons" and "Space Activity in Russia: Background, Current State, Perspectives" (Shingareva, Karachevtseva, and Cherepanova, 2003), the International Cartographic Association appointed her as cochairman of the ICA Planetary Cartography Working Group (1995–1999), chair of the ICA Planetary Cartography Commission (1999–2003), and, according to the *Proceedings of the International Cartographic Conference*, "managed such projects as a series of multilingual maps of planets and their moons, glossary on planetary cartography, and specialized map-oriented DB on planetary cartography in the frames of commission activity" (Shingareva, Karachevtseva, and Cherepanova, 2007). On the initiative of the Moscow State University for Geodesy and Cartography (MIIGAiK), several groups in Europe involving Shingareva are working on a Multilingual Planetary Map Series (Hargitai, 2004: 150).

More recently, Shingareva has been trying to bring her graduate students to more earthbound projects, such as bringing her experience to bear, in 2006, on finding solutions related to the Moscow Megacity Road and Transport Complex (Sinitsyna and Shingareva, 2006).

Shingareva is well recognized and was elected Honorary Fellow of the International Cartographic Association (*ICA Newsletter*, Dec. 2007: 5).

INGRID KRETSCHMER (1939–2011)

As an eminent professor of cartography at the University of Vienna, Ingrid Kretschmer brought maps of Austria to the attention of international cartography.[13] Having published more than 250 scholarly papers, she has also published at least six books on Austrian maps and cartography. Her best-known work is the two-volume *Lexikon zur Geschichte der Kartographie* [*Encyclopedia of the History of Cartography*] (Kretschmer, Doerflinger, and Wawrik, 1986), the first of its kind. Her active association with international cartographic organizations constitutes another of her major contributions to the discipline.

She was born in 1939 in the industrial town of Linz, the first of three children; her parents were Hildegard and Ernst Kretschmer. Her father was an electrical engineer. When the war came to an end, she completed her primary school and gymnasium. She began to earn some money by working at the editorial office of the *Austrian Folklore Atlas*—an interest she took with her into her later graduate work. At the age of twenty, in 1959, Ingrid Kretschmer entered the University of Vienna to begin her studies in geography and anthropology. Six years later, she completed a Ph.D. with a thesis on the thematic map and its relationship to folklore. In 1966, she occupied the newly created Chair of Geography and Cartography at the University of Vienna. By 1975, she already had more than thirty scholarly publications to her name and had completed her *Habilitation* on the subject of geography and cartography. Fritz Kelnhofer, renowned for his topographical maps and related publications, and Wolfgang Pillewizer (1911–1999), who was responsible for setting up the cartographic institute at the University of Vienna,[14] consecutively paved her way in university.

Her contributions to cartography gained significance with her editing and designing the *Austrian Folklore Atlas*, 1966–1981. In 1975, she had published with Erik Arnberger *Die Kartographie und ihre Randgebiete* [*Mapping and Its Outlying Areas*], a two-volume work, along with *Wesen und Aufgaben der Kartographie: Topographische Karten* [*The Nature and Tasks of Cartography-Topographic Maps*]. A prodigious scholar, she wrote four hundred pages of this work. As if this reference work were not sufficient (and it is still regarded as an important one), she edited, with two of her colleagues, the acclaimed *Lexikon zur Geschichte der Kartographie* [*Encyclopedia of the History of Cartography*] (Kretschmer, Doerflinger, and Wawrik, 1986), the world's first encyclopedia on cartography. Cor Koeman, a noted authority on cartography, remarked that "it is quite possible that this book is going to be the second great classic of the twentieth century in the German language," although he also levelled criticisms at its relativism, various important omissions, and inadequate treatment of some topics (Koeman, 1980: 386). Interestingly, while some of her work focused on what might be considered peripheral topics (such as folklore) and marginal regions, she devoted much of her attention to bringing Austrian cartography to the attention of the international community. In addition, her extensive work on theoretical and methodological aspects of cartography led to new developments in the field.

After 1988, when Kretschmer was forty-nine, the university promoted her to university professor. She continued unabated with her committed scholarship. She was member of the Board of Directors of *Imago Mundi* and published *Atlantes Austriaci,* a Catalogue of the Austrian Atlases from 1561 onwards, in two parts.

Kretschmer's influence extended through her memberships as chair of the Cartographic Commission of the Austrian Geographical Society (1997–2006), while also later serving as president of that Society. From 2003 to 2010, she headed the editorial board of *Kartographischen Nachrichten,* a major cartographical newsletter. In light of her involvement with the International Cartographic Association and the Coronelli World League of Friends of the Globe, she wrote at least twenty-four technical reports, usually unpaid and as a volunteer. Many have noted her enthusiastic manner of working with others.

One sees a number of significant awards that emerged from her scholarly contributions. The German Cartographical Society invited her, in 1996, as an Honorary Member in light of her contributions to German-Austrian cooperation in cartography; it was this same society that in 2004 awarded her the Mercator Medal, the highest scientific award of the Society.

Throughout all of these cartographic contexts and events, Kretschmer's life was bound up with the numerous other scholars who have held sway over the field, including Arthur H. Robinson, whom she met in September 1995 at the International Conference on the History of Cartography. Ulrich Freitag (2004: 4) avers that Kretschmer's work touched and affected folklore, ethnography (of the European kind), mathematics and geodesy, economics, statistics, geography, geology, and cultural history. Her continuing engagement with cartography at higher and more responsible levels occurred sometimes in spurts and sudden announcements. When the leading historian of cartography Professor Ernst Bernleithner died in March 1978, Freitag (2004: 5) relates the following event (translated from German): "One day, a few weeks later, Mrs. Kretschmer works—as usual—at the desk in her office at the Institute. Suddenly the door flew open. In the doorway stands Erik Arnberger,[15] quite ordinary, and stops there. Looking up, he looks surprised and says with definite certainty to her, 'Bernleithner is dead. Now you take over the whole thing!' Speaks, turns, and closes the door."

CHAPTER 7

Ingrid Kretschmer died in Linz, Austria, on 22 January 2011.

BARBARA A. BOND (B. 1943)

Barbara Bond worked for the United Kingdom Hydrographic Office, rising to one of their highest administrative positions.[16]

Born in 1943 in Manchester, Barbara Bond was the daughter and first child of Jack and Hilda Garside. Her father was the middle son of an impoverished Lancashire family; he left school at thirteen and worked as a telegraph boy in the Post Office. It was from him that Bond learned the value and importance of education: he was an autodidact and very well read. His view was always that education is the one thing that can never be taken away from you. He knew that education was the only passport out of an impoverished background. Barbara Bond informs us that he always encouraged his daughter to read and bought her a globe when she was eight, opening her eyes "to our remarkable world." It was from that globe that her interest in geography and cartography grew. She won a scholarship paid by the local authority at the age of eleven to attend an excellent school, Hulme Grammar School, Oldham, Lancashire, as her family could never have afforded the fees. She then studied at the University of Leeds, graduating in 1964 with a combined honours degree in geography and history.[17]

Her contributions can be found in the fields of mapping and charting. Starting with technical cartographic posts with the Ministry of Defence, Directorate of Military Survey, she later served in the United States as the British Liaison Officer with the then United States Defense Mapping Agency, where she "gained considerable experience in a high-profile liaison role involving day-to-day contact with both national and international organisations in the USA, Canada and Central America" (Department of Geography, 2010: 10). It is striking to see the professional friendships that developed among some of the personalities in the vignettes.[18] Not only were Eila Campbell and Helen Wallis good friends, but, though Barbara Bond was much younger, they took her under their wing and offered much support and encouragement during Bond's career. Both stayed at the home of the Bonds as house guests during their lecture tour to the United States in 1986, when the Bonds lived in North Virginia. Also, Barbara Bond met Carol Beaver in 1965 on her first trip to the United States, as they were doing the same work for their respective defence mapping agencies and became firm friends. Barbara Bond met and became great friends with many other women in the United States cartographic world, including the late Barbara Petchenik, Alberta Auringer Wood, and Patricia ("Trish") Caldwell

Lindgren, all of whom at various times held significant and influential positions in the cartographic world.

Bond returned to the United Kingdom in 1987 and was appointed in 1991 as director of Hydrographic Charting and Marine Sciences in the United Kingdom Hydrographic Office. In 1993, she became its deputy chief executive, "responsible for the overall strategic direction of the agency, with particular responsibility for the delivery of analogue and digital marine navigational charts, publications, and services to the international merchant marine and defence markets" (Department of Geography, 2010: 10). The web page for the United Kingdom Hydrographic Office—the world's most prestigious navigational charting organization, dating back to Britain's colonial past (since 1683)—describes its mission as follows: "We provide nautical charts and navigational services of the world's oceans and ports to support world shipping, including the Royal Navy. Our Admiralty products and services have been developed over 200 years, and we use the very latest techniques to continue to help protect lives at sea today" (http://www.ukho.gov.uk/Pages/Home.aspx).

Bond chaired the International Hydrographic Organisation's Antarctic Commission between 1992 and 1997.

Bond has a long-standing interest in higher education (Department of Geography, 2010: 10). She has chaired Graduate Fast Stream recruitment panels for the Cabinet Office and acted as an "external assessor and adviser for graduate recruitment in the Foreign and Commonwealth Office." She was invited to join the University of Plymouth's Board of Governors in 2000. She served as "chair of the University's Employment & Remuneration Committee and was appointed chair of the Board in January 2006." She retired from the Board in December 2008, but has continued her association with the university in her new role as pro-chancellor. She has served on the Higher Education Funding Council for England's Leadership, Governance and Management Strategic Advisory Committee and was a member of the Equality Challenge Unit's Project Group, "considering the need to increase diversity in university governance" (Department of Geography, 2010: 10). Bond was awarded the Member of the British Empire in 2009 for service to higher education; the University of Plymouth added "its own recognition of her achievements and her continuing commitment to the university" (Department of Geography, 2010: 10) by awarding her an honorary Doctorate of Science later that year.

The challenges of being a professional woman in the United Kingdom, especially in the 1960s and 1970s, were many. According to Bond, she "always found the best way to deal with sexism, which was rampant, was with a sense of humour—laugh and the world laughs with you, cry and you cry alone!" It is a testimony to her fortitude that she did not receive much encouragement from her colleagues. She says that "only one of my male colleagues ever understood why I wanted to pursue a career." Fortunately, he was her director and became a friend and mentor for many years. He supported her nomination as the first woman ever to go to the United States as a British Liaison Officer (Survey) for the Ministry of Defence. Her career took off from that point.

Bond is now retired and lives in rural Somerset with her husband of almost forty years. Their two adult children, Abi and Adam, live and work in North America, which Bond still regards as her second home. She continues her involvement with the University of Plymouth as its pro-chancellor and is also European consultant to a global GIS company that has its headquarters in India. She has been heavily committed over the past two years to helping them establish a corporate Training Academy, recruit a president for the Academy, and create the first training courses both for the company and for the wider international market. This has taken her to India twice in the past year (see also Mason, 2009).

Her personal interests included escape and evasion maps in World War II (Bond, 2009), and she is continuing her research in this topic. This research started in the early 1980s, when she was employed in what was then Military Survey. In 1983, she published an article on maps printed on silk (Bond, 1983). Francis Herbert (2009) found that in 1990, she presented a paper on escape and evasion maps at the International Map Collectors' annual symposium (Bond, 1990). She confesses that she now has "one of the most comprehensive carto-bibliographies of this largely unknown set of British military maps" (cited by Herbert, 2009).

Barbara Bond received a number of recognitions, including being invited Fellow and past-Council member of the Royal Geographical Society, and became Fellow and past-president (1991) of the British Cartographic Society (Pye, 1991: 351; Mason, 2009). In 1997, she was awarded the Prince Albert I Medal by Prince Rainier of Monaco (Plymouth University *Newsletter*, 2009). One year later, she received the British Cartographic Society Silver Medal in the service of international cartography (Mason, 2009).

Barbara Bond delivered the Helen Wallis Memorial Lecture, "Communicating with Maps: Did the Message Get Across?" at the 46th British Cartographic Society in June 2009. For Bond, the final and most important acknowledgement goes to Roger, her husband of almost forty years. Without his unstinting support and constant encouragement, she believes, she could never have succeeded.

ASIA

LIQIU MENG (B. 1963)

A renowned scientist from China, Liqiu Meng has specialized knowledge in emotional requirements of map users, map recognition, geo-data generalization, remote sensing and geospatial data mining, geo-data integration, and mobile navigation.[19] Meng is now vice-president of the Technical University of Munich (TUM).

Liqiu Meng was born in October 1963 in Changshu, near Shanghai, China, into a modest teacher's family; she belongs to the twelfth generation of teachers in the Meng family.[20] She has one son.

Beginning her studies in 1978, she obtained a B.Sc. in cartography in 1982 at the Institute of Surveying and Mapping, Military University of Information Engineering in Zhengzhou, China. She followed this with an M.Sc. degree in 1985, also in cartography, at the same university. She remained at the Institute until 1988, at which point she left China after receiving a scholarship from the Chinese government and entered the Ph.D. program at the Institute for Cartography at Hanover University, where she obtained a Ph.D. in geodetic engineering in 1993.

While moving from China to Germany was a big adjustment, Meng was determined to make the change work. She says, "It wasn't easy. Although I had taken a German course in Beijing and was able to communicate, everything was still completely different: the food, people's behaviour.... From the very beginning my attitude was: If I want to study in Germany, I have to adapt to the country's culture and not vice-versa. And I've always stuck to that" (DAAD, 2009a).

Between 1985 and 1988, she had been a teaching assistant at the institute in China where she had obtained her B.Sc. and M.Sc. While undertaking her studies at University of Hanover, she was a research assistant from 1993 to 1994.

She found the effort to apply for a visa every three months to stay in Germany quite cumbersome, and so she taught as senior lecturer at the University of Gävle in Sweden from 1994 to 1996. She played a major role in the founding of the Research Centre for Geographic Information Systems (GIS). She then served as GIS senior consultant with SWECO in Stockholm, while also giving guest lectures at the Royal Institute of Technology

CHAPTER 7

(Geomatics) from 1996 to 1998, when she obtained Swedish citizenship. In cooperation with the Swedish Armed Forces, Meng headed an important research project on the automatic generalization of geographic data (Helmholtz Association, 2011). With the completion of her *Habilitation* (post-doctoral), she was appointed professor and director of the Institute of Photogrammetry and Cartography at the Technical University of Munich (TUM), and moved back to Germany (Pressl, 2009: 3). Over a ten-year period (2001–2010), she produced ninety-three publications and fifty-three conference presentations (Meng, 2010). The eighth edition of "Cartography: Visualization of Spatio-temporal Geoinformation" (2002), which she co-authored with her Ph.D. supervisor and the latter's own supervisor, is used as the standard textbook at universities in German-speaking countries.

When a journalist from *Time* magazine asked Meng why her appointment in April 2008 as vice-president of the Technical University of Munich, Bavaria, caused quite a stir (Pressl, 2009: 1), she replied that she was an "anomaly, an unusual phenomenon." For many years, she was the first female and non-European professor at the Faculty for Civil Engineering and Geodesy, where she held the post of vice-dean. She also explained that she was the only Chinese woman in a German university at the level of management.

A strong personality who is able to bring together researchers in the Department of Cartography, her responsibilities as vice-president of TUM include internationalizing the university. Meng's role also relates to bringing in China as part of that process. In 2009, there were 824 Chinese students enrolled in TUM—the largest overseas student group in the university, even in comparison to other German universities. This process has been going on for several decades. It turns out that a number of the earlier students have returned to China and are now, as professors themselves, sending their students to TUM (Pressl, 2009: 2).

Through Meng's efforts, TUM has not only maintained but has also increased its international profile. TUM was the first German university to establish its own branch in Singapore. Her approach, however, extends beyond the usual, familiar goal of increasing the number of foreign students (TUM has a student population of 26,000) or to engage in exchange programs. Rather, she believes the presence of students from China (and elsewhere) will benefit Germany—no matter what university they study at. Moreover, German companies are so numerous in China that there ought to be a natural attraction for students to come to Germany. Once the Chinese students return home, they are likely to engage in research that will promote closer scientific ties between China and Germany (and TUM in particular) (DAAD, 2009b).

The International Association of Chinese Professionals in Geographic Information Science asked Meng, in 2007, to address a panel about Western and Eastern Cartography at the Nanjing Geoinformatics Conference (ISCP, 2007).

A caption that accompanies her profile as Senate member of the Helmholtz Association captures the spirit that moves her in cartographic work: "The wise words by Lakota Native American Sitting Bull, 'We have not inherited earth from our ancestors. We have only borrowed it from our children,' prophetically committed us to protect our earth and environment as much as possible through precautionary and provisional research and return to our children a sustainable habitat" (Helmholtz Association, 2011).

Meng is a member of the Senate of the Helmholtz Association of German Research Centres, responsible for the research sector that deals with "Earth and Climate." She is also a member of editorial boards of three international journals (ACA/SI, 2009).

In recognition of Meng's efforts to promote research on an international scale, she received the Heinz-Maier-Leibnitz Medal in 2007 and was elected to the German National Academy of Sciences in 2011 (http://www.leopol dina.org/en/academy.html). This Academy awarded her the Carus Medal in 2011 in a ceremony in the presence of German Chancellor Angela Merkel.

LATIN AMERICA

CARMEN REYES (B. 1949)

Carmen Reyes, a Mexican, is the most influential cartographer in Latin America today. The co-creator of the 2000 *Atlas Cibernético del Agua en América* [*Cybercartography Atlas Latin America*],[21] she is the founding director of Centro de Investigación en Geografía y Geomática Ing. J.L. Tamayo A.C., a research centre that is doing outstanding work on cartography.[22] She is a renowned expert in the fields of geographic information systems (GIS), digital mapping, cybercartography, and geomatics. She is the first person to have coined the term "geo-cybernetics" (in 2005), following Fraser Taylor's first elaboration of "cybercartography."

Born in 1949 in Mexico City, Reyes is very attached to her mother, Angela Guerrero. In a conversation with me, Reyes said that she cannot talk about herself without talking about her mother. When her mother was fourteen years old, Reyes's grandfather died. Her mother started work as

CHAPTER 7

an executive secretary, moving into high positions in Mexico and claiming to be a feminist. It was a close family, and Reyes felt equal to her brother. Reyes's mother urged her three daughters to go forward. Even Reyes's grandmother, who was illiterate, encouraged her daughter to go to school. Although Reyes's mother played the key role in Reyes's overall education and her view of the world, her father (who was always working away from home) was, in Reyes's views, "an outstanding man and a reference throughout my life." Both of her parents have now passed away.

Reyes completed a B.Sc. in mathematics at the National Autonomous University of Mexico (UNAM) and an M.S. in mathematics at the Metropolitan Autonomous University of Mexico (UAM), but was intending to continue her studies in Canada, the United States, or the United Kingdom. In 1972, before she made her decision, she met Tom K. Peucker (now spelled Poiker), author of *Computer Cartography*, who convinced her to go to Simon Fraser University in Burnaby, British Columbia. She had never considered doing a Ph.D. She completed her Ph.D. in GIS at Simon Fraser University; her senior supervisor was a cartographer. Her children were two and five years old when she moved to Canada.

She started her career in September 1972. Her first job was to develop a computer mapping application for a GIS to support the Planning Office of the Ministry of Education in Mexico. This was a pioneering achievement in both Mexico and the world.

Within the matrix of cartography as a man's world, Reyes realizes that the condition of women (or gender) varies from area to area. Although Mexican by birth and parentage, her own upbringing was associated with the Anglo-Saxon culture, as she attended the American School in Guatemala City and continued studying in Mexico City until high school within a bilingual (English-Spanish) culture. In her own higher education experience in Canada, she felt more discriminated against than in Mexico, where she experienced discrimination only from engineers in their fifties and sixties. Even as the sole woman among the staff of twenty-seven research centres (in 1999), she felt quite comfortable; although there have been a few gender-related issues to deal with, she has not been discriminated against. When she was invited to join the International Cartographic Association's Commission on Gender and Cartography, she did not know quite what to do, "because in my personal life," she says, "gender is not an issue." She did not face as many obstacles in Mexico as she did when she applied to come to Canada for her studies:

[In North America] I have seen some very aggressive women, because men are very aggressive. In Mexico City, there is a completely different context. So I don't think, I [will] have problems when I ask for my scholarship. When I went to the Canadian Embassy [in Mexico City], the lady in the Canadian Embassy was very tough with me, because [she asked] how come I, as a woman, wanted to come and do a Ph.D. in Canada and bring my husband? And I was going to have the primary role. And she was very, very tough. But I was able to fulfill all of the requirements and at the end I came to Canada. So that is why I tell you that I have felt worse sometimes in Canada than in Mexico [laughs]. (Interview: 10)

"To be a woman in the 20th and 21st centuries," Reyes admits, "is exciting and challenging. Many opportunities that were forbidden in the past are now open, but one does not know women who have travelled through them. One has to build these new roads and fight against prejudice, centuries of established traditions and misunderstandings. Often people are surprised that I am married, a great mom and grandma, and that I love cooking and dancing."

What allowed her to survive all the adversities found in a "world of men" was the deep love from her mother, her lessons regarding the value of work, the respect for other human beings, her own managerial skills, and, above all, the dignity that a woman has to have in a society where gender is still an issue. She believes she is "a better mother, grandma and researcher because of her [mother]."

After her return to Mexico in 1982, she worked at INEGI (Instituto Nacional de Informacion Estadistica, Geografica e Informatica) for many years, learning about map production and the use of maps. In 1998–1999, through her interaction with Dr. Fraser Taylor, director of the Geomatics and Cartographic Research Centre at Carleton University in Ottawa, Canada, and using as a point of departure his vision on cybercartography, she decided to focus her research for the first time on cartography, although cartography has been part of her career since the beginning. In 2001, she turned her attention to working on a cybercartographic atlas of Antarctica.

Reyes is the project manager of the *Continental Atlas for Latin America* under the aegis of the Pan-American Institute of Geography and History, a million-dollar project (Interview: 1). This project expresses her overall interest in collaborating with such international organizations as the World Bank and with national governments in Latin America, including El

Salvador, Costa Rica, Panama, Peru, Argentina, Chile, Brazil, and Mexico. Despite these highly normative activities, she describes herself as a "risk taker" (Interview: 12). Her interests led her to participate in at least sixty national and international projects in geomatics, involving atlases (including one for children), cybernetic documents, the management of urban green areas, water systems, air quality, electoral systems, atmospheric models, and education. These involvements have led to a long string of publications on geo-cybernetics (Reyes, Taylor, Martinez, and Lopez, 2006), cybernetic theory (Reyes, 2006a, 2006b, 2006c), environmental policies (Reyes, 2005), and electoral redistricting (Reyes, Guerrero, and Lopez, 1996). Geo-cybernetics is a field that moves the atlas from being a static element. According to Fraser Taylor, "These new types of maps are interactive. You can ask questions and get answers. They are analytical and feature a multimedia approach.... It puts all the pieces together" (Palfrey, 2000).

Reyes was the leader of a group that designed for CentroGeo a Scientific Management Model that adopts a Science 2.0 approach. The point of departure is to generate new knowledge and innovation for the purpose of filling a societal need. Using what one of her Ph.D. students has named "the Reyes method," and through interaction with society, at least three venues of research have emerged or advanced within geocybernetics: cybercartography, complex solutions in geomatics, and collective mental maps. Over one hundred projects based on this approach have had a societal impact within Mexico and some other countries of Latin America.

Reyes is one of the founding members, and former president, of the Mexican Association of Geographic Information Systems (AMESIGE). In 1999, she founded the Jorge L. Tamayo Center for Research on Geography and Geomatics (CentreGeo), supported by the Mexican National Council.

Her services include a long-standing membership of the Commission on Gender and Cartography of the International Cartographic Association since 2003. She also founded the Association of GIS in Mexico, of which she was president for six years. Reyes is one of three Latin-American representatives for the Open Geospatial Consortium (OGC) Global Advisor Committee.

From the time Reyes graduated with her Ph.D. in GIS, she always had several suppositions that guided her work and research. First, she was devoted to creating bridges between science and society; second, she was committed to making strong links between GIS and cartography; and third, her work has a populist appeal, centering on the needs and experiences of people. She sums up her approach as follows: "So what is happening simply is that knowledge has grown so much that we have sort of started dividing it—by disciplines, by sciences or whatever. But it's somewhat artificial, because my background is in mathematics, so I think in mathematics and physics—for example, in relativity theory, where does physics stop and mathematics start? You know, it is very difficult to decide. And I think something like that you can think [of] in [terms of] cartography and GIS" (Interview: 4).

While others, such as those trained only in traditional cartography or those who were trained in GIS without any knowledge of cartography, find the development of new knowledge perplexing, Reyes clearly knows where the field is going. She readily acknowledges that this is a time of confusion for many, but she sees a new cartography emerging from that confusion. Young people, according to Reyes, must be thankful to their "grey-haired" colleagues, although the young must take the disciplines ever further.

Underlying all of her work is her urgent belief that all changes in policy must come from the grassroots level. For example, she is pushing her work in geo-cybernetics to be available in schools, and points out that "the ideas and concerns of all residents of the immediate area will be vital in the decision-making process…" (Palfrey, 2000; see also Clarke, 1985: 177). Carmen describes her cartographic work on electoral redistricting as one of her

> favourite projects since I coordinated the group for redistricting Mexico at the federal level…. It was a very successful project where we introduced a Spatial Analysis approach supported by a GIS solution. It was also a turning point for Mexico from a political point of view, since for the first time in more than 70 years we had a "real" election. The redistricting was used for the 1997 and 2000 federal elections. As a scientist I feel highly fulfilled by the opportunity of having a social impact on over 46 million potential voters after the election on the whole Mexican population. (email to author, 10 June 2011)

A number of distinctions came her way. Soon after she completed her studies, the Pan-American Institute of Geography and History of the Organization of American States (OAS) awarded her with the "Samuel Gill Gamble Award for Cartography" (Centrogeo, 2010).

Through the years, Reyes has been the main provider for her family. Her family played (and continues to play) a critical role in her professional life. Her husband, Rodolfo Sanchez, whom she married in 1972, has always supported her decisions throughout their forty years of marriage (such as when she went to Canada to get her Ph.D.). He also provides his own ideas on cartography and other topics in geomatics "to challenge and help fortify my ideas," Reyes says. Her professional success is reflected in the accomplishments of her children. Her elder son, Rodolfo, an architect, is now studying for an M.Sc. at Simon Fraser University. Her younger son, Pablo, is an M.D. who specialized in internal medicine.

REGINA ARAUJO DE ALMEIDA (B. 1949)

A professor of geography at the University of São Paulo in Brazil, Almeida is a pioneer and the world's leading cartographer in tactile mapping (for the visually impaired) and developing children's maps.[23] Her other areas of expertise include indigenous mapping, perception and representation of space, geography, culture and tourism, and teaching geography.

She was born in 1949 to Dulce Araujo de Almeida and Benjamin Pereira de Almeida. Her mother died at the age of thirty-nine; Regina went to live with her grandparents. Her grandmother was a teacher, and her grandfather, a dentist. All of Regina's grandparents migrated from Portugal to Brazil. Her maternal grandmother, Eurídice, had a grandfather who was a baron at the time of Imperial Brazil (Emperor D. Pedro II). Anyone would describe her family as coming from a socially solid background, but by the time Regina was born, they had lost all the money the family once had.

Regina studied in a French school, Sacré-Coeur de Marie. Her educational trajectory[24] channelled her deep interest in history, culture, cartography, and geography. She completed a degree in history (1973–1976) at the University of São Paulo (USP), Brazil, followed by a degree in geography in 1982. In another six years (1988) she would earn a master's in (physical) geography at USP. She commenced her Ph.D. in physical geography at USP in 1989, finishing it in 1993. Her thesis, "Tactile Cartography and the Visually Impaired," expressed her long-standing interest in creating maps for non-mainstream groups. With her divorce in 1995, she began publishing under her maiden name, Almeida.

What brings Almeida's contribution to cartography to the fore is her devotion to uplifting the condition of indigenous populations, the environment, people with disabilities, pupils and teachers in elementary and other schools, people in poor urban areas, and the promotion of local culture. She normally combines these six themes in pairs of two or three. For example, in 1998–1999, her research aimed at participatory teaching, with a focus on the environment. In the same year, we first come across her burgeoning interest in tactile maps for use among Grade 1 and 2 students. In 2000, she developed a handbook for students to stimulate the learning of

Figure 7.1 Twice a year for five years, Regina de Almeida (centre, second row) would spend two weeks immersing herself in the culture of indigenous peoples while also imparting her cartographic knowledge among the Guarani (Mbyá and Nhandeva) in the Brazilian Amazonian region. *Source*: Regina Araujo de Almeida.

geography and local culture, folk and popular demonstrations, art, and artists and craftsmen in their community. Her contributions to cartography are particularly significant because she made them at a time when cartography was just breaking through traditional concepts of making and using maps, making way for subjectivity, cultural variety, and the social context (Almeida: 2001: 1).

At the same time, she expanded her vision to rely on the knowledge of the Guarani (Mbyá and Nhandeva) and several ethnic native groups from State of Acre, in the Brazilian Amazonian region. Twice a year for five years, she would spend two weeks immersing herself in their culture while also imparting her cartographic knowledge. She took great interest in their concepts of space, their unwritten language, drawings, maps, and other cultural activities (crafts, text, music, and dance, in particular). The urban areas, too, did not escape her attention. In the Vale do Ribeira and District of Bixiga (São Paulo), she developed, with several graduate and undergraduate students, "didactic material necessary to the formation of cultural and environmental monitors, providing continuity and strengthening of activities and projects that are already being developed" in these areas, highlighting natural areas, conservation areas, and rural communities with a vocation for ecotourism, environmental education, scientific research, conservation practices and sustainable management, and cultural heritage. In a term she later adopted, *Etnocartografia* became the means to use maps in indigenous schools. Her paper "Teaching Cartography for Minority Populations," originally presented in August 2006 (Almeida, 2008), contains a thoughtful and provocative set of ideas about the relevance of the topics she has been concerned about for virtually all of her career.

Almeida undertook these research activities as Professor at the School of Arts, Sciences, and Humanities from 2005 to 2007. However, the same desire of wanting to delve more deeply into the lives of the less privileged, especially youth and children, whether indigenous or not, pervades her later work.

Almeida continues to present and publish numerous papers and has six chapters in books, two of them outside of Brazil, in addition to two books. She has participated in events in Brazil and abroad to present papers. She supervised nine doctoral dissertations in ten years, and sixteen master's theses within a fifteen-year period. One of her international publications on tactile maps in geography appeared in Elsevier's *International Encyclopedia of Social and Behavioral Sciences*. She forms part of coordinating team in the Cybercartography of the Americas (a project headed by Professor Fraser Taylor at Carleton University, Ottawa, Canada, involving eleven countries. She also provided over eighty professional assessments and technical visits in twenty countries.

Almeida's international scope of service includes serving as vice-president of the International Cartographic Association (ICA), and as a key person when the ICA Working Group on Cartography and Children was established in Ottawa, Canada, during its conference in 1999. Nationally, she served on the Advisory Board of Brazil's Academy of Travel and Tourism. In 1994, the Brazilian Society of Cartography awarded her the Cartographic Order of Merit, Grade of Chevalier, which was upgraded in 2007 to Officer of the Brazilian Society of Cartography.

CONCLUSION

The historical and contemporary vignettes of the twenty-eight women cartographers (of whom twenty-three fall into the contemporary period) in Chapters 5, 6, and 7 highlighted these women's individual parentage and family backgrounds, education, interests, and accomplishments. Although we already have observed some emerging commonalities among them, it will be important to be more explicit about those commonalities (see Appendix C).

In terms of the characteristics of the twenty-three women in the contemporary period, 52 percent (N=12) were married. Moreover, there were notable discrepancies when the women began making significant contributions to the field; in general, women in North America began making those contributions about seven years earlier in their lives than is the case for the rest of women: thirty-three and forty years of age, respectively.

The following two chapters (8 and 9) are based on interviews with some thirty-eight women incumbents in the contemporary map world. The analysis of the interviews is a precursor to our exploring the social structures and potential strictures with respect to women in the contemporary map world (chapters 10 to 12).

CHAPTER 8

"Getting There without Aiming at It": Women's Experiences in Becoming Cartographers

We have now come to the point of exploring the gendered map world of contemporary cartography, offering us a chance to dialogue with some thirty-eight women cartographers about their experiences. This chapter and the following one draw heavily on the interviews I conducted with these cartographers. Where appropriate, I have also included materials I gathered for the vignettes in the preceding chapters.[1]

The formats of the previous two chapters (6 and 7) and the next two (9 and 10) raise an interesting issue. Does one learn more from the personally distinct narratives attached to each individual in the earlier chapters, or is much more gained from reading the anonymized accounts embedded in the next two chapters? Between a woman's first spark of interest in doing cartographic work and the latter-day burst of accomplishment lies a career filled with twists and turns, wider allegiances and interests, and contradictions.

The body of literature on gender and occupations is a vast one; this chapter can highlight only the most relevant works. Ellen P. Cook (1993), for example, speaks about the continuing sex differentiation in the labour market. Gerstein, Lichtman, and Barokas (1988) found that interests and aspirations in careers are still gendered. Lassalle and Spokane (1987) have found that women have a "discontinuous" career pattern, which, in effect, means that women are required to be more persistent and intensive in their involvement with their work. The source of such a career pattern is twofold. On the one hand, women see family needs as a vital component of life. Cook reminds us that it is the wife's career that adapts to meet family needs, while men follow the ethos that "making it" occupationally "requires single-minded commitment" (Cook, 1993: 231). On the other hand, women are greeted by an occupational culture, especially in those occupations traditionally held by men, that contains "stereotypic masculine characteristics," leading to an emphasis on "competitiveness, preoccupation with individual power, and acceptance of sexual aggressiveness and language" (Cook, 1993: 233–34). Nancy Johnson Smith indicates that very few studies take "the viewpoint of women themselves, about the development of their aspirations and how women arrive at the occupationally related decisions they do make" (Smith, 1997). This chapter takes up Smith's suggestion in relation to women in the map world. There are still no studies that explore in detail the career contingencies of women from their own perspective. Within the larger framework of my book, I hope in this chapter to shed light on the social processes that surround women's seeking to become cartographers.

The women's accomplishments suggest a wide range of cartographic involvement: carrying out theoretical work on graphic variables and time-lines, and on 3D maps; playing key roles in developing national atlases (electronic, historical, and demographic); developing innovative instructional programs, from elementary school to university levels; promoting innovative maps for under-represented groups, the visually impaired, and the colour blind; substantively contributing to the establishment of GIS research centres in a developing country, and forming a family of equivalent disciplines in geoinformatics. Across age, experience, and contributions, cartographers share experiences that point to sometimes long incubation periods, career detours, and wider allegiances and interests, while sorting through the ideological contradictions in cartography.

GETTING THAT FIRST SPARK
The seed of interest in maps, that "first spark," originated in early childhood experiences, and, for many, the father played a pivotal role in igniting that spark:

> My first active contact with maps was in childhood, because during the Second World War my father was working as a teacher of geography.... He was fascinated with maps and with music! These two spheres are very important in my whole life. (Marta: email to author, 2 April 2001)[2]

... it already started when I was a kid.... My father ... was involved in ... selling and buying land ... and sometimes I coloured in the maps. He asked me, "Do you like drawing? Do you like?" And then I got interested in cartography. (Els: 1)[3]

Eva Siekierska described the important influence of her father, who was "an avid painter," which most likely stimulated her interest in visual arts/cartography.

For some cartographers it was the home that generated that first spark:

[Land survey] was something we discussed very often in our family, and it was something interesting for me. The first [time] I thought about maps was when I was very little and that's when I lived in the countryside. (Helle: 1)

I used to draw maps when I was younger ... and colour them, and I was good in arts. I would make them really nice and pretty. They always attracted me. (Natalie: 1)

For someone like Kornélia, who is now a freelance cartographer, it was a map competition at her school when she was eight that got the ball rolling (Kornélia:1). For others, primary school offered the opportunity to love and make maps:

[it] was in elementary school.... And I have always been a stickler for detail. I love detail. So I would draw probably the best map in the class, with exquisite detail in it. (Lynn: 1)

... in my school life and in college life I saw this beautiful map and I ... got interested, thinking that this map had come out so beautifully and so nicely. I ... was keen to know about it and I always wanted to work on it. (Sawa: 1)

For several, there were moments in their teenage years that cultivated their interest in cartography:

When I was a teenager I was doing this orienteering, running in the woods with maps. And then it was some coincidence, friends from my parents who worked in the map shop for the National Board of Surveying, and then I got a summer job from there. And then I became interested in how they make these. (Marita: 1)

> I was competing in orienteering since I was ten, eleven years old. I knew some people that were working with the maps and I tried to find out what kind of education they got.... I had some other plans as well, but then I decided when I was eighteen that I wanted to study maps. (Inga: 1)

Evelyn Pruitt (the subject of an earlier vignette) confirmed that her first introduction to geography was anything but a "dull thing in school ... [it] was wonderful! It was outdoor mapping, you had plane tables and you paced things off, and you went interesting places, and you talked to people ... that's the best part of geography" (Walker, 2006: 432).

The early engagement with maps features an appreciation for the beauty of maps, and it is this particular attraction of maps that would define a woman cartographer's approach later in her career. The richness of aesthetic expectations, however, does not end there: many give thoughtful attention to the problem of digitization in light of the belief that maps should have an aesthetic appeal. As the following sections illustrate, the cartographer's belief that she is stepping into a man's world, and the unexpected twists and turns of her career, have contributed to her being a cartographer with a wider set of interests and allegiances, to both people and ideas.

THE FORMATIVE YEARS

While there are conscious claims to the contrary in response to the question of whether cartography is a man's world, there is no doubt that many women see themselves as "intruders." The sense of being an intruder has its incubation during the formative years. Whether or not the cartographers follow the "traditional" intent of cartography—for example, by invoking Robinson's definition found in *Elements of Cartography* (1960)—all have spent their formative years in a cartographic culture in which women were a minority. The formative period includes time spent in training, education, and the workplace.

The women's experiences of being a minority during their formative years as cartographer date from 1970 right up to 2000. The percentage can be as low as 0 (as in the case of the number of women in the geography department where one woman received her training) and is not higher than 40 (except for two cases—one in Iran, the other in a Himalayan kingdom). Although the data span many cultures and time periods, the pattern of being a minority is consistent during the formative years of women cartographers. It is striking that in some cases, the women reported being, even today, either the first or second woman student in a university department.[4]

Thus, what the cartographer carries into her work, whether as a newcomer, a mid-career professional, or a mature-career cartographer, is her place in an environment where the number of women is either very small or negligible. It is this relative absence of women that defines her current perception of herself in relation first to her co-workers and second to the discipline.

"WE ARE THE INTRUDERS"

While a number of the cartographers, such as Helle, Joyce, and Lynn, do *not* acknowledge that they were discriminated against as women,[5] there is no question that many who were interviewed acknowledged that they had stepped into a "man's world" during their formative years, either during their education or on the job. Myra's sense that she is an "intruder" in cartography is never far away. Loukie speaks of the International Cartographic Association as still such a world, where things may be changing, but only gradually. There is a heavy "but" in statements that speak of any movement toward equality: "it's open to women," says Helena, "of course, but...." Kristina surprised herself with her own observation that "It is still a man's world, more than I expected." One cartographer, who entered the field as a young woman, observes that "it was a male's world ... it was like working in a shop. The language there was like, they were swearing, they had naked women on the wall. So I had to come in and look at that ... it was a man's world. I would go to the printers, the print shop, all males, all guys would whistle, and I used to be so shy going down there" (Natalie).

For Ruth, seeing herself as an intruder during those formative years has meant having to make "an extra effort to make sure that you are understood." Cécile has "had to explode at the old boys' network" and has "had to be very cross because certainly decisions were made regarding some of my projects and I wasn't involved in them." Erika, who is still a graduate student, confides that "it's been very hard" for her; when asked what has been hard about it, she says, "I think it's more about personal relations ... and ... sometimes I had enough of being called 'girl' or something like 'chicken'" (Erika: 2). Joyce remembered an early male boss of her group of cartographers who "didn't have any experience ... and the people doing the cartography were women and the women taught him cartography."

Helen Wallis worked hard to overcome a natural diffidence with cartographer Skelton, and she radically changed her appearance. All efforts seemed to fail, however, when the time came to choose Skelton's successor at Map Library. His unfavourable assessment of her "caused a morale-sagging delay" before she was appointed as the first woman to hold that post (Campbell, 2004: 1).

For Nedelina, the unintended effects of breaking into this world during her formative years included receiving praise by men who claimed that she thinks "like a man." Ramona goes so far as to say that she "could easily become like a man." Natalie believed that she "always had to work harder or do more to be up with them [the men] and be at the same rank as them and be recognized as one of them." For Marita, it meant that she has had to be "quite aggressive" in pointing out that it was always the man in her group who was addressed by a client, although she was his superior. She is, by her own description, not an aggressive person. Inga acknowledged that she was called on the carpet for missing a particular class and that, as she was the only woman in class, her absence was more noticeable. Her professor punished her by lowering her grade, which she, all alone and quite young, challenged with an appeal to a higher board, staffed only by male academics. Erika had to struggle for her right to teach a course; she took the matter to an appeals board, with the following results: "In the end there was a really nasty fight ... I got the job, but the following weeks the staff didn't greet me" (Erika: 6).

The knowledge that they had stepped into a man's world, where, in some cases, they saw themselves as intruders and in other instances had to double their efforts to make themselves heard or acknowledged should prepare the reader for the fact that, for women cartographers, it was far more common to have male mentors than females ones (Chapter 10 delves into the topic of mentorship). One of the more telling cases is the story of Marie Tharp (mentioned in Chapter 6). Her supervisor was very skeptical of her interpretations and brushed them off as "girl talk" (Fox, 2006: 2). It took him two years to accept her ideas.

Stories about challenges that women faced are increasingly coming to our attention through a number of recent obituaries about women in the map world. Avril Maddrell, for example, informs us that Eila Campbell's male colleagues "have not been very kind to her" and recalls one comment by a male colleague who said, "there is still a lot of 'nun' about you [Eila Campbell]" (Maddrell, 2009a: 757). Apparently, when the men returned from war to resume their professional careers in geography after World War II, they did not shed their traditional image of women, and found it hard to work with them. We also learn that Eila "did not publish more in her name ... because she devoted so much of her time to work that has appeared under the names and to the credit of others" (Barber, 1995: 11). The circumstances under which she worked with Leo Bagrow did not mark a particular high point where women were taken seriously; Bagrow would almost daily badger her in his letters and mount intrigues behind her back, according to Peter Barber (1995: 11).

Working through the belief that they were stepping into a man's world, calling upon on all kinds of personal resources to overcome those inherent challenges, and acknowledging wider sets of mentoring relationships prepared the women cartographers for the twists and turns of their careers, which further enriched their personal resources and abilities. The recognition of women's contributing to cartography, even outstanding ones, would come late in life. In the case of Marie Tharp, it came in the eighth decade of her life (Lawrence, 1999: 42).

"GETTING THERE WITHOUT AIMING AT IT"
These twists and turns characterize women cartographers "getting there" in their professional work, and their belief that they may have fallen into their cartographic job "without aiming at it."

TWISTS AND TURNS
Even for those women who were able to follow through on that first spark of interest in maps and map-making, the lengthy process of becoming established cartographers was marked by detours, sudden breaks, and suddenly flung-open doors. While a good number have stayed within their general fields of study, we find many shifts within those fields, in addition to the abrupt changeovers to other fields along the way. While it is sometimes difficult to define what constitutes a field or whether a field is in fact a subfield of another one, it is clear that the thirty-eight cartographers I interviewed represent a wide variety of areas—some twenty-one fields, from architecture and art to paleoecology and pre-med. As an indication of some of these midstream changes over the collective span of training, the number in the engineering sciences has declined (from nine to seven), as has the number in the natural sciences (from seven to five). The number of people trained in the social sciences doing cartographic work, on the other hand, has increased slightly, from fifteen to sixteen, as has the number coming from "proper" cartographic training (from four to six). Four entered cartographic work via teacher training. Collectively, these cartographers have created a rich legacy from these twenty-one different fields, all flowing into doing cartographic work.

In those cases where the first spark of interest was banked and left smouldering, so to speak, the twists and turns were sudden and sharp once

CHAPTER 8

that interest had been rekindled. Similar zigzags marked both those who had no inkling, when they started their training, that they would become interested in maps and those who already had that spurt of interest. The first group of interview participants I will discuss here are those who initially had little or no interest in map-making.

A good number entered cartography quite late, after starting their work life as primary school teachers. Kristina started as an elementary school teacher; moved on to studying physical geography in university; worked successively as an editor for a national touring club, of an encyclopedia, and of a national atlas; and is now working freelance. Ettie literally moved from the school where she was a teacher, sacrificing a good position, to the field and the office, working first as a teacher, then for a mining company, and finally for a tourist publisher in the Antipodes. Ramona, who has twenty-five years' experience teaching elementary school, now teaches at a university, after being struck by the map illiteracy of her pupils and realizing that she wanted to devote the rest of her life to eradicating that illiteracy. Agnes ended up heading a sales department in a cartographic agency, first starting as cartographic editor, then moving up to the position of production manager for foreign and domestic markets. Her first job after university was as a primary school teacher. Zia, who also started her work life as a teacher, had this to say when reflecting on her current university posting, working on maps and the environment: "I didn't like the university and I never thought to be a university teacher [laughter], but I am and I like [it] very much, because, I think, every time for me that type of thing happens and I say, 'wow, and now what am I going to do?' And now I take the risk and … if I don't like it, I go out" (Zia: 5).

Others initially worked in fields such as mathematics and computer science. Marta's story starts with mathematical modelling and progresses through working in an agricultural university to linking social problems and their representation on maps. Liisa was active in computer science and surveying when she developed a love for maps and read everything she could about them in a local library, only to find herself later teaching cartography at the university level. Natalie started as a pre-medicine student but ended up taking a cartography course at a local community college. She was hired by her current employer before she had a chance to finish university. Saskia had very little interest in cartography—she was a "GIS person"—until her supervisor at a Canadian university convinced her, through his beautiful maps, to take up the interest. Now, she heads numerous GIS offices in a Latin American country where she is trying to instill in her colleagues a deeper, more inductive approach to map-making. Lynn's family encouraged her to become a lab technician; when that job came to an end, she went back to university and took a "smattering of everything," not intent on taking cartography until later in her undergraduate studies. Alenka avers she "knew nothing" when first hired by a cartographic institute; in fact, she started to cry when asked to do hill shading on her first day on the job. Now, she is a widely acclaimed figure in that area of cartographic work. Finally, we learn that Hadley had wanted to become a teacher, first going into mathematics but finishing with degrees in geography and library science, as she could not tolerate the courses required for teachers. While an undergraduate, she worked in a fine arts and architecture library and became a map librarian after finishing her degrees. Along the way, she worked in a public library and has since worked in a non-governmental organization, university libraries, a public library, and a national library.

A second group of interview participants is characterized by their interest in doing cartographic work from their first stint at work and onwards (unlike the above groups). Sawa and Helle, for example, were hired in a cartographic governmental office without any formal training (although Helle had some practical knowledge about maps, she had no formal training). Once they were hired, their employers sent them for cartographic training in conjunction with their work. Sawa's employer sent her to India, where she took a basic cartography course. Loukie found herself taking evening courses for a diploma as a geography teacher while working at a travel agency. Eventually, she did a Ph.D. in cartographic animation as a result of a major reorganization within her research and teaching institute. Ruth, who trained as a cartographic technician, ended up teaching cartography. Helena's career has spanned the whole arc of cartography, starting with photogrammetry and on to thematic maps, archeological data maps, and now children's maps. Marita's challenge was deciding between mathematics and graphic design; she found cartography "a good place to combine those roles." The turns in Inga's career can come only from the fact that she was the first woman cartographer in her country: starting in a technical school and then an agricultural school, where she took up surveying, she became a regional planner and municipal administrator. She followed this varied career with a stint as the director of a land-mapping agency; her current position is international geomatics advisor of her country. In the midst of these shifts, Inga also managed to edit a book, travelling around the country collecting information on maps. Although Amanda was first interested in surveying, she took courses in architectural drawing, followed by a degree in geography, with a specialization in mental perceptions of space. Her work now involves the production of maps. Judith, in southern Africa, worked in a geography department without any cartographic skills but with a great deal of enthusiasm: she then moved to a local government department, shifted to a surveying department, and went to a local college for training and then to an international research and training institute in Europe.

For all these cartographers, the main effect of undergoing these twists and turns in their training, education, and professional lives has been an unfailing appreciation of the wider relation of all the fields that relate to cartography—a theme to which we shall return later. The twists and turns, however, are most typical of mature-career cartographers. Pregnancy is still another factor that inevitably shaped the career paths of these women, either by temporarily slowing down their progress along whatever career paths they took or by complicating family arrangements. Myra works in one country, while her husband and child live in another. She says that "he's a good mother" (Myra: 6).

In the case of Barbara Petchenik, we noted in a previous chapter that although she had initially intended to do a Ph.D. in physical geography, in the hopes of teaching at the university level, she admitted that "a whole lot of things happened that changed my course." For Eva Siekierska, her plans to study oceanography never materialized, but they stimulated her to learn Russian well, because one potential location for graduate work was St. Petersburg (formerly Leningrad), where she was accepted to study (after being repeatedly refused a passport to study in Canada), but the whole program would last seven years. In any event, she did not go to Russia and ended up in Canada after all.

FALLING INTO THE JOB

The ups and downs that have characterized the process of finding their current work result in a unilinear professional career. Cécile, who at first "hated maps," emphasizes the importance of not speeding into postings and "not jump[ing] into things." In reflecting on her career, she admitted that "it wasn't the plan—it just happened!" Even after these twists and turns were exhausted, the dominant sense of cartographers is that they fell into the job by accident. Inga admits that she was asked to apply for the position of director of a national mapping agency, even though she had "not been working in cartography for several years." Helena, Cécile, Gabrijela, and many others describe this experience as "getting there without aiming at it." Helle says, "I never looked for a job," while Hadley sees that process as "a series of things, including a visit of a cartographer/librarian from a national library." Nedelina says that she had never expected to become a university teacher, while Ettie "never aspired for an award." Natalie speaks of doors being opened for her and "being pushed in"; it is quite common for these cartographers to be asked to apply for jobs, as Hadley and Inga were. Zia mentions that the senior professors gave her "opportunities to speak" at conferences and that "I think they look at me as I'm a woman, almost a young woman, do you understand? ... [yes] ... So, I think they feel a little friendly, you understand, that this is sympathetic ... [*maybe like a daughter?*] ... Yes, yes, maybe like that because in fact they are very, very kind to me. [But with respect to] the women: no. Not so" (Zia: 8).

We also remind the reader of Ingrid Kretschmer's account in a previous chapter of how she fell into her position when the door of her office flew open and, upon hearing of the death of the person she would replace, was simply told, "Bernleithner is dead. Now you take over the whole thing!"

Ruth's account is not atypical: "I fell hard [laughs] ... [F]riends of mine just mentioned to me that I should teach because maybe it's just my personality [that] was natural for this kind of profession, but I fell into it. Because I was unemployed at one point and someone phoned me and ... asked me to teach part-time and it progressed from there" (Ruth: 3).

Natalie's opinion that she "was very lucky" (Natalie: 2) is echoed by many, many others as they reflect back not only on their varied careers, but also on how they managed to land their first and subsequent jobs working with maps, as teachers, administrators, map-makers, computer specialists, and so on. But at what point did they arrive at the realization that they were, indeed, cartographers?

IDENTIFYING MOMENTS

Every occupational incumbent arrives at a point when her identity as incumbent becomes crystal clear to herself. These moments usually occur spontaneously and unexpectedly, either when one has been in the occupation for any length of time, or long before she has entered the occupation. Sometimes, these moments hinge on formally completing training or education, but usually are independent of such formal benchmarks. Kathy Charmaz uses the term *identifying moments* to refer to "telling moments filled with new self-images" (1991: 207). Such moments "spark sudden realizations, reveal hidden images of self, or divulge what others think." Although Charmaz derives this concept[6] from her study of people with chronic illnesses, we can very easily transfer the concept of *identifying*

CHAPTER 8

moments to other settings where someone has a sudden awareness of the new self, whether it involves a widow whose new status suddenly hits home after checking "widow" on the vital-statistics form (D. van den Hoonaard, 1997), or a new university student's realization that he or she is no longer a high school student. It should come as no surprise that a cartographer also undergoes an identifying moment when, unheralded, her status as a cartographer is unquestionable from her perspective. Interestingly enough, only a few of the women I interviewed associate their identifying moment with producing their first map. Shpela (9), for example, speaks of a national map on a scale of 1:400,000 where "the contours were very, very close one to each other." Sawa (5) produced a district map; when that map—her first—came across her table, she "really realized I'm a cartographer [laughter] … I'm really proud of my work."

Indeed, what became the identifying moment for Eva Siekierska, that is, what made her turn to cartography, was a combination of nature and art. When she was doing fieldwork in the early part of her studies, one of her professors asked her to demonstrate her drawing skills. She surprised herself at how well she did and, more significantly, realized how that talent could be incorporated into cartography. The story of Phyllis Pearsall is iconic: she hit upon the idea of creating her now famous *A–Z Maps* when, in 1935, she got lost on her way to a party in the Belgravia area in London and came up with the idea of creating an accurate map of the city.

For the majority, the most telling occurrence involved others, such as recognition from supervisors, attending a cartography conference, or publishing an article or book. Two assigned their identifying moment to what a university president and academic peers said about them: "You see, other professors at my university say that I am all-cartographer. Everything of mine is in cartography. I think only of this when I am not in my family. I think only of production and of learning, of teaching, only of cartography" (Nedelina: 5).

By far, the "doing" and the "participating in" cartographic work constituted the principal means by which they achieved that salient identifying moment. The process can be a gradual one, as some testify:

After three years I had experienced so much, I ventured to say that I'm a cartographer. (Gabrijela, email to author, 2 April 2001)

I had a professor [who] spoke [over] most of our heads. Sometimes I was afraid that it was because I was a girl that I didn't catch him. However, when

I dared to ask my student colleagues, I realized that it was the same situation for all of us. Very early in my career I was elected to the board of chartered surveyors and also in other organizations and I realized that I could contribute as well as my colleagues around the table and even sometimes [was] more dedicated. (Inga, email to author, 3 May 2001)

Natalie gradually reached that identifying moment while working on a national atlas: "I worked for the *National Atlas,* and they were so picky and everything had to be perfect ... revision upon revision upon revision. I think after a few years of working on the *Atlas,* I thought that I was a really good cartographer because the expectations were so high. You had to be good to be able to work there" (Natalie: 4).

The identifying moment for Amanda came only very recently, after she had been doing cartographic work for twelve years: "Again, it is [mentions name of a cartographer]'s influence on me and discussion with her over time ... and while dealing with people on an international [scale], you broaden your definition.... I don't think I woke up one day thinking that [I am a cartographer]. I just come more and more to the realization over time. And actually, at conferences like this is what drives it home" (Amanda: 4–5).

For yet others, the identifying moment came rather suddenly. Several used such an evocative term as "joy" to describe that experience: "Partway through the third year I suddenly realized I was *enjoying* the course. I had found the whole experience back at university very hard work—I had never studied so hard.... The second defining point was when I was awarded the prize for the fourth-year student with the best academic results. Having never aspired to gain any such award during my school days, this was very important to me" (Ettie: 3 [emphasis mine]).

Helle describes a particular technical accomplishment as her identifying moment: "It was when I understood I was set in a surveying project ... surveying and catching all the coordinates ... and I could count different coordinates for a very large field ... the beginning of a very large plan for some buildings, and that's when I saw my work in the larger ... scale" (Helle: 6–7).

Echoing the same sentiment as Amanda, Liisa believed that fellowship and association with like-minded people made her realize she was a cartographer (although she never received formal training in that area and just read as many books as she could in the field): "Maybe when I came to this ICA conference [I understood I was a cartographer]. Because when you work at some program, well, most probably I would have said that I am a GIS person, but when I came to the ICA conference, and I saw, okay, these are also cartographers, I could also be called a cartographer. So maybe, just when I saw other people who called them cartographers, and I saw that I belonged to this group" (Liisa: 4).

Cartographic work is a literate culture, and, therefore, for some the realm of publication occupies a significant place in achieving the identifying moment. One person cited work on her thesis as the instance of importance for her: "I really didn't identify myself as a cartographer until probably about the end of the first year of my graduate work, when I was actually working on a transportation thesis" (Joyce: 4).

Having an article published was a turning point in their identity as cartographers; some even remembered a particular comment from a manuscript reviewer that sparked that turning point: "We sent a paper to *Cartographica* and we had really nice comments from the reviewers. Then we felt, you know, that we were doing something special; there was one comment that was saying, 'This point of the paper is a contribution to theoretical aspects.' So we felt like we're [cartographers]" (Helena: 7).

And here is a comment by an author of a book: "I truly felt I had come to a point of realization ... when my publisher sent me a draft of the cover of the soon-to-be-published book ... and when I saw my book in a bookstore!" (Loukie, email to author, 28 March 2001).

SUMMARY

For these cartographers, the formative years of their training further deepened their potential contribution to the field. As intruders, they stepped into a man's world, but also took on the responsibility of transforming that man's world onto their own shoulders: they saw it as the woman's task, not the man's. My findings parallel those of Nancy Johnson Smith (1997), who, in her work on young women undergraduates, found that her interview participants expressed a life story that was not linear. Smith relies on Ochberg (1988), who interprets "the male plot as [a] continuous, relentless career movement—a culturally prescribed obsession with moving ever upward from adolescence onwards" (Smith, 1997: 3). The women she interviewed, however, saw the development of occupational aspirations as a response to shifting expectations set by others, involving respect, friendship, and acknowledgment. They saw their future occupation as a "fantasy," which cannot be attained by a linear process. In a set of interviews for another study, Carol B. Warren (Warren and Hackney, 2000) investigated academic women's experience of their career trajectories. She was interested in finding out whether other women's experiences had been like hers:

characterized not so much by goals and planning, but by drift and adjustment to circumstances. She found that for these late 1960s to early 1970s women professors, even the purportedly goal-organized career path of academia, especially in the earlier stages, is characterized by drift and adaptation rather than goals, planning, and choice. They described entering graduate school in terms such as "getting away" from their household duties or "having nothing else to do" in the college town where their husbands had chosen to teach. They had been brought up to adapt to circumstances, often to men's plans, rather than to make plans and goals for themselves.[7]

Although I agree with Teixeira and Gomes (2000) that studies on career changes of women (and men) are scarce, one cannot but note that career changes seem more typical for women cartographers than for the economically active population at large.

The dilemmas that characterized these women's training did not stop when they became the incumbents of their cartographic occupation, but rather continued through the twists and turns of their career. Some would arrive at that point fairly late in their natural life; even others, who had an early start, confessed to shifts in training and work. It makes sense that such experience would frame their cartographic work in a different way for women than for men, who are more likely to have taken the "straight" path.

CHAPTER 9

"We Are Good Ghosts!": Orientations and Expectations of Women Cartographers

By the time a woman enters cartography as a fully working professional, the data indicate that she stands out in many significant ways from her male colleagues (van den Hoonaard, 2000c): she brings with her an education or training in a wide variety of areas, and her career takes on the form of an occupational zigzag.[1] Her cartographic work reflects this wide disciplinary and occupational experience. This chapter explores the outlooks she brings into that structure. Her outlook shapes the way in which she participates in that social structure—it both affirms and rejects the occupational norms. More specifically, these norms centre on the purpose of maps and on the relevance of her wider allegiances and interests: personal, professional, and disciplinary.

This chapter closely examines the occupational attributes women cartographers bring into their field. These attributes pertain to resolving the tension while creating maps that are both beautiful and useful, and to the role of relationships, personal and collegial, on one hand, and disciplinary, on the other hand.[2] This chapter also speaks about the sentiments and ideas that the interviewed women have expressed about their work in cartography and on cartography-related work.

CHAPTER 9

RECONCILING THE BEAUTY AND USEFULNESS OF MAPS

The question of the beauty of maps, while at the same time privileging their usefulness, surfaces frequently in the interviews. By the time she reaches that moment, a woman cartographer has already worked through the seeming contradiction of whether beauty or utility should be emphasized in maps: that is, whether maps are an end in themselves or are tools for others. In many respects, the issue of beauty versus utility has come to the fore because of the digitization of maps. Both men and women express deep concerns about the poor design on some maps created by GIS. The late Borden D. Dent produced the standard text in cartography on the subject, *Cartography: Thematic Map Design* (1996), and Terry A. Slocum devoted a large proportion of his book *Thematic Cartography and Visualization* (1999) to design. Even from the perspective of GIS specialists (such as Ed Madej in *Cartographic Design Using ArcView GIS* [2001]), there are growing concerns about the struggle between beauty and usefulness. Judith Tyner's *Principles of Map Design* (2010) shows how beauty can still manifest itself in very useful maps.

There is clearly a struggle about the role of aesthetically pleasing maps in the face of digitalization. Nuvyn is so entranced with making maps that she confessed that "if I cannot create a map, I will die with my eyes open"—an expression in her language that indicates a single-hearted devotion to a cause (Nuvyn: 1).

What stands out from those early experiences when a woman first became interested in maps (van den Hoonaard, 2000c) is the attraction of "beautiful" maps, involving graphics that would define her involvement in cartography many years later:

> *What brings you joy about cartography?* For me, the beauty of the maps. The beauty of the maps. And the graphic designs of them. (Liisa: 2)
> [Y]ou need to find some beauty [in maps]. (Ramona: 1)

> [T]hey [in other countries] made so many beautiful, so many nice maps. So why are we are not doing so? When I go back [to my country] I really want to put the same as what the people are doing here [at an international training institute]. (Sawa: 2)

Nevertheless, men criticize the women's interest in cultivating beautiful, aesthetically pleasing maps, as one cartographer reported: "One of the teachers ... looked at one of the products that I made and said that was a very pretty map and I said, 'What do you mean by that?' And he said, 'Men make maps and women make pretty maps.' And I was quite offended by that comment at that time" (Ruth: 6).

However, the desire to combine mathematical precision and graphically beautiful details is the most important element that would guide her and others through their careers: for some, it would become a mission of sorts:

> And the artistic sense is the combination with the science part, which is important as well. (Ruth: 3)

> *What do you enjoy most about cartographic work... what gives you the most satisfaction?* ... When the beautiful map and math map come out and when the work is well done. (Alenka: 1)

It must be remembered that the first spark of interest in the beauty of maps often occurred long before digitalization or computerization become the norm in map-making. The arrival of the digital map meant that certain aesthetic qualities would have to be sacrificed, as a number of cartographers aver. There are sharp criticisms about the lack of beauty or the aesthetic quality of maps that characterizes the digital era:

> Everybody and their dog who owns a computer can make a map. And they are making garbage, most of the time. And we have lost control of doing it. No one is asking us to teach it. And it really distresses us as people who make maps. (Lynn: 9)

> I can see as a problem that people think that handling the computers and the software is cartography, which [it] is not. (Elg: 4)

However, a number also believe that the artistic, beautiful components of maps may, in fact, be less relevant today in light of digitalization, although creativity still retains its important place in map-making: "because it is the [artistic] side of mapping that women often respond to the most, I think.... And like myself, I respond to the artistic side. But I guess with the technical emphasis, women are not interested in that side" (Ruth: 7).

While some admit it is a struggle to make the beauty of maps count in the face of digitalization, there are other facets of maps that the women cartographers underscore: namely, the importance of making maps relevant for users, rather than mere displays of technical accuracy or beautiful graphics. Maps must be taken off the walls and must be seen as tools, rather than objects, even objects of beauty.

It thus seems ironic that while the interviewees largely emphasize the importance of cartographic beauty, there is an equally strong demand to "take maps off walls." Unasked, interviewees remarked that "maps are not for walls." In several respects, the imagined duality between beauty and usefulness reflects the traditional beliefs that men do "art," while women do "crafts."[3] I was first drawn to this issue when a presenter at CARTO 2000, the annual conference of the Canadian Cartographic Association, expressed frustration to her supervisor because he believed that "maps are Picassos on the wall!" when she was trying to convince him to see maps as tools (Selley, 2000).[4] I was surprised to discover that it was not uncommon for interviewees to speak in a similar vein:

> [M]aps are for people, they are not for animals and they are not for walls, they are for people. (Cécile: 3)

> [O]ur [women] students, the ones I'm involved with in teaching, don't display their maps on the wall. (Minhui: 3)[5]

Instead of displaying topographic ingenuity and precision on walls, there is now a call to make aesthetically pleasing maps also geared toward the needs of "users":

> [M]y task is to prepare good maps to be easily used by other people. (Nedelina: 6)

> [B]ut now you must be aware not only of the related technical disciplines but also societal needs, client needs, and so on, so you have to be a very open person ... very perceptive, and you have to listen well. (Myra: 2)

> I need to make maps that are directly linked to who will use them, because I feel that the importance of that contact with a person is important. (Ruth: 6)

> It's not the aesthetic quality, but I think the message, the information that is portrayed in our maps, is what a cartographer has to have in mind when he or she produces the information.... If we produce things that are as aesthetic as they can be, but useful, I am willing to live with less aesthetic but more useful. (Cécile: 7)

In reflecting on the relationship between creating beautiful maps and the need to make maps for the benefits of users, one interviewee had this to say: "You want to have more than just the data coming at you. You want a pretty picture, a pleasing, eye-pleasing picture. It helps to communicate the information that much more efficiently and quickly" (Amanda: 3). One Himalayan cartographer explains that she is driven to make beautiful maps to represent her country, and that it is part of her nationalist sentiment (Sawa: 6).

The lack of unanimity about the importance of beauty in maps is reflected in how the cartographers define cartography, although, as the reader sees, there are overarching principles about the nature and intent of making maps: as tools for communication, focusing on the needs of users. The lack of unanimity also reflects the period of great uncertainty of cartography: is it about describing landscapes (Helle) or about "data sets" (Minhui)? The uncertainty has led Joyce, a seasoned cartographer, to exclaim, "I don't know what I am anymore!" and Ettie, who is fairly new to the field, to say, "I haven't got a definition at all!" (Ettie, email to author, 1 May 2001). Others (e.g., Helena and Gabrijela) rely on the definition and purpose set out by Arthur H. Robinson in *Elements of Cartography* (1960: 11), that cartography is about "the presentation of spatial relationships."

Some women cartographers, like Barbara Petchenik, go a long way to find out how maps can be made more relevant to users—in this case, children. We noted in an earlier chapter that she interviewed over a thousand children from all over the world about their cartographic knowledge, leading her to conclude that children prefer clear and uncluttered maps without too many extraneous elements. Patricia Caldwell lectured worldwide on helping audiences understand rapidly evolving mapping-science technology. The potential loss in having beautiful, aesthetically pleasing maps has already been mourned by Eva Siekierska, partly because with the introduction of computers in cartography, fieldwork lost its charm and she could no longer exercise her drawing talents. Fieldwork is not practised anymore.

Despite the emotional importance of making her map work relevant to others, the intellectual flair of converting an abstract idea into concrete reality still remains high. It is not uncommon to attach such words as *creating* (Ruth: 1), *transforming* (Cécile: 2), *solving a spatial problem in a graphical format* (Ettie: 1), and *something concrete coming from the mind* (Marita: 11) to this process of conversion. An additional challenge is to create something beautiful (Nedelina: 2; Liisa: 2), especially while keeping up with new technology and "finding ways to show things" (Natalie: 3). An employee of an Eastern European cartographic firm talks about the freedom of the new technology, which allows her to become even more creative than when she worked on a table (cited in Agnes: 11).

This process of conversion can be summed up as a "brain–hand" connection. This connection goes a long way back in the memory of the many seasoned cartographers I interviewed for this book. The "table" was, literally, the place where the cartographer expressed her cartographic design on acetate sheets long before today's technological revolution. Doing "table work" (e.g., Sawa: 2) is an evocative term Sawa used to explain how she translated ideas into something solid, without the computerized intermediary. The love for symbology, graphical design, and minute detail, and the sense of creating something "concrete," both stem from that earlier phase of cartographic work which, for many interviewees, goes back to the 1950s, 1960s, or 1970s. Yet another facet of cartography that evokes commitment is the idea of "travelling in the room."

"Travelling in the room" is how one Eastern European cartographer describes the reward of her cartographic work and illustrates how that work is itself a fine substitute for international travel (Kornélia: 1). Helena (1–2) declares that, as a cartographer, she can identify herself "at every time, in every space, [and] know where you are." However, they, like several others, undertake international travel and become well acquainted with even the most remote corners of the globe. There is nothing selfish about this love for travel by proxy, because their international travel and knowledge are coupled with improving the fate of people elsewhere, especially in the so-called Third World. For Myra (3), the international character of her work always entails doing something "useful" for others. Bringing others into her work, moreover, applies particularly to personal and disciplinary relationships, to which I now turn.

THE IMPORTANCE OF RELATIONSHIPS

The contributions that women cartographers believe they have made, or are making, to their field are inextricably bound up with their wider allegiances and interests. Some, like Ruth, do not think about what their own contributions to the field might be. Besides, Ruth says, "it's part of human nature ... not to toot your own horn" (Ruth: 3). Ruth identifies her ten years of work on the *Historical Atlas of Canada* as a particularly interesting contribution she has made to the field (Ruth: 5).

Their perceived contributions to the field usually entail other people and are not focused on the self. For one thing, the users of maps spell out the extent to which women see their contributions as effective. While no doubt most textbooks and classes stress the importance of the map user, some women cartographers do say that that is one area where the contributions by women stand out more clearly. For example, as Lynn (8) avers, "we have the greater interest [than men] in the user of maps." One's contribution is being able to make maps that people understand (Lynn: 8). One Scandinavian cartographer articulates that the main principle is what the user wants, and this principle never goes away. The cartographer must have contact with the users (Marita: 6). To that end, it is signally important to know how to apply technology to "real-life situations" (Minhui). A South American thought that one ought to make the production side of maps more in tune with the needs of children (Agnes: 1). She has devoted her life to promoting cartographic literacy among primary schoolchildren. If teaching is the primary realm of activity of the cartographer, there, too, the "other" is kept carefully in mind and any new teaching ideas are about involving students in the cartographic project. Elg notes that her distinctive contribution to the field consists primarily in having taught cartography and cartographic editing, especially in the realm of thematic maps in atlases (Elg: 1).

Some women cartographers, like Evelyn Pruitt, in response to the resentment and prejudice that greeted her at most steps of her career, not only provided the institutional means to help upcoming women, but also supported them through her later largesse, notably the Pruitt National Fellowship for Dissertation Research through the Society for Woman Geographers.

BALANCING FAMILY AND WORK

When we peruse the demographic status of the twenty-three contemporary women cartographers featured in chapters 6 and 7, twelve were married and eleven were single at the time of their work. (One became a widow, and one was divorced.) A number of contemporary women in *Map Worlds* felt compelled to resolve the issues of balancing family and work. Choosing to never be married (or "partnered," in modern parlance) was an option.

A field as vast as the new cartography is making room for women. The opportunities, however, are filled with barriers (see, e.g., Beaujot, 2000: 66–81 on the familial obligations mentioned by many of the women interview participants). The need to balance family and work is something that women in many occupations, whether in retail sales, medicine, house cleaning, and so on, must attend to. In that sense, cartography is no

CHAPTER 9

different from these other occupations. While the following section presents the perspectives of women cartographers about balancing family and work—and these perspectives can be germane to many other fields—it is also clear that in a number of countries, such as the United Kingdom, cartography offers many opportunities for its practitioners to go freelance or set up their own cartographic firms.

The women feel keenly that the world of work and of family occupy two different mental niches—something that is also anticipated by young incoming cartographers, as we shall see later:

> I was able to make a Ph.D.... but then I had my first baby ... so it took me a few years.... The wife of one of them [professor-colleagues] told me once ... "I don't understand how you can make it in there with the other three because I know how my husband works hard during the night and he's very busy; he usually has something to do and I know that you have the kids and your husband.... (Helena: 9)

> [Here] we are in 1999 and I still think that women are still responsible for the education of their children and the running of the house, unfortunately. [*One sociologist called this the second shift*]. Yeah, well then there is the third shift. When my kids are in bed, then I get back to work in my basement on my home computer. You know, it's a challenge, it is a challenge. (Cécile: 9)

> The work is something and life outside is something else [laughs]. (Hegner: 6)

The responsibility of her doing work falls on her own shoulders, not those of the whole family, and not on those of her co-workers, who are often not aware of these third shifts of work:

> But I was telling my daughter that my mother didn't work, my aunts didn't work, my grandmother didn't work, and I was the first generation to work outside the home. So that was a big deal for me, and not only was I going into a male-dominated type of work, but I was getting out of the house, going to work, and I felt guilty about working outside the house and not being home with my children.... All my cousins stayed at home with their children. And my husband said, "Well, listen, if you want a house, you want this and that, you have to work." Well, I don't regret it, but I mean it was a real challenge, a real debate inside of me, between traditional and the new trends and between what I wanted and what I was forced to do, more or less. (Natalie: 13)

ORIENTATIONS AND EXPECTATIONS OF WOMEN CARTOGRAPHERS

The existence of two such differing worlds has had a serious impact on where women choose to do their cartographic work. In England and in some other Western countries, women cartographers are founders and heads of their own cartographic firms. In Canada, however, women seem to have chosen government service over private enterprise: "I can see all sorts of good reasons why they [women] would be in government, especially those women with children: regular hours, good maternity leave benefits, furloughs, leaves, all sorts of ways of coping with demanding lifestyles. You know, private industry, especially the geomatics industry, is not a very attractive place to be, because it is high pressure and lots of hours …" (Amanda: 9).

Husbands can make a tangible difference—for better—in the lives of these working cartographers:

I have a very understanding husband, who [however] understands only so much, but I am very lucky. I could have married somebody who was much less supportive. (Cécile: 9)

[M]y husband is a marvellous and supporting individual. He's an excellent organizer. He's a very good mother and father [laughter]. He took excellent care of my daughter while I was away.… He's really amazing, I mean, without something like that I would never have done it.… (Myra: 6)

Or for worse:

But it is a real problem in my generation where our husbands are very old-fashioned. They have the old-fashioned education at home and they really can't cook, they are unable to use washing machines, or cleaning machines, or anything.… (Liisa: 12)

And he [my husband] wouldn't help around so I had to do it all myself. And you know, I had this job that was very demanding—I was a supervisor during the day—so I mean … it was hard, I had to work really hard. I was really tired, I had to keep pushing through, pushing, pushing, pushing. (Natalie: 11)

Some put a positive spin on such struggles. The act of balancing these obligations with the demands of work makes for maturity, as one woman claims: "[W]e are more mature and more settled down, less emotional.… But you are

CHAPTER 9

more, more settled down and calm, like I am not.... [I say] who do I have to kill to get there, you just say that I want to get there" (Saskia: 12–13).

There are drastic consequences when the world of work and the world of the family cannot be kept separate. Aside from the sense of guilt, some people deliberately turn down opportunities to assume supervisory work out of a grounded sense that it will be difficult to maintain that balance between work and family:

> [I]t is rather difficult to say that the participation of women and men in management of cartography is the same. No, [they are] not the same. The ten-year barrier connected with the care of children is significant for the development of women, and the state of women is not the same as the state of men. (Marta: 3)

> [Y]eah, and of course some of them [women cartographers] don't want to be directors. From my position now I have small children. I want to have more freedom than ... when [I am] a leader. (Helena: 9)

Women also face opprobrium when a stage of their life demands more attention from them:

> [A] colleague in the school of architecture wrote her autobiographical sketch and she wrote down that for two years she didn't have any research done or any work done, just the educational obligations, because she had two children at that time, and they were young. There were many people saying that it has nothing to do with her career, but she didn't have to mention that. I still remember it.... Of course, they never asked me such a thing, but I'm pretty sure, you know, they're thinking about it.... (Helena: 12)

And when she tries twice as hard to overcome her familial obligations—in other words, when she tries to compensate—she is not trusted by her colleagues:

> I guess at that time [1973] women were starting to work. If you had children, you were supposed to be at home with your family. And you know, you ... got looked at kind of funny. And as I said, I was in the world of men. And [with me] being a perfectionist, they thought, "Hey, she is trying to take our job, she is trying to take our place, you know." That it was difficult. [But] I had to earn their respect as a cartographer. (Natalie: 9)

For someone else, it is a question of de-emotionalizing oneself:

And I said to her [the Director-General], I said, "You've had kids. How did you deal with that frustration?" And she was in an even more male-dominated organization, which was even worse. And she said, "The only thing that I have learned in the past is that you have to de-emotionalize the situation." (Cécile: 8)

Young cartographers, however, are particularly apprehensive about any sense of balance that will be ruptured by the difference between the world of work and family:

How has science helped women? In a sense, yes, we see [many] more women in congresses and conferences working, but what about their own lives? What about their private lives, their family lives, their own dreams? Are they changed? I was in one conference … and sometimes I found myself thinking, "Well, would that be my life in the next years? Would be it like that? Now I'm here. Next week, I'll be in Finland, and then at the end of the month I'll be in Portugal. What about my personal life? I'm thirty-two. Am I going to have children? And I was very sad. (Zia: 4)

She added: "I'm a little afraid, really … because for me work is not everything and I love my work, but it's not everything and that frightens me a lot" (Zia: 5).

Especially for the upcoming generation of young women, there is no realization about the intrinsic problems that women will face when choosing a career such as cartography. Denying there is a problem is the usual approach taken by younger cartographers: "I really feel and really think that we don't see the problem. We like to hide the problem. Whenever your daughter gets her degree and gets her job, [we believe that] her problem is solved: she is independent, she earns money, but nobody, nobody thinks or speaks of the problem that she [has to do] double, double work. Everybody is so happy for her professional career, but nobody thinks that when she comes home, she has to do her work" (Liisa: 12).

Nevertheless, there are moments of intense appreciation of the family's attitude that sometimes relieve the burden of work outside and inside the home:

CHAPTER 9

[Do your kids know what you do?] Oh yes, they saw me on TV the other night and they were quite excited, without saying that they are proud of their mother. My daughter who is eight is starting to understand a few things, and she helps me dress in the morning. "You know, Mama" [she says], "You have to be really beautiful for your launch of your big project." (Cécile: 11)

PERCEIVED RELEVANCE OF OTHERS AT WORK

The importance of others is a key element in the cartographer's decentred self. The opportunity for her students to "win awards and jobs" and "to attend meetings where people think that thematic maps are important" says Ruth (2–3), a teacher of cartography, is very valuable to her, as is simply a chance to "work with students" for Joyce (2). Several describe the excitement that policy makers get when being inspired by her map (Cécile: 6) or when a sudden realization dawns on the map reader ("Oh, I see!") (Lynn: 1), or when "everyone says this is a good map" (Shpela: 2). Amanda (2), who has produced a national atlas, gets a "great deal of satisfaction out of seeing people understand complex things quickly" through her atlas.

Yet, even stronger is the desire among the interviewed women to work in a team, which constitutes the highest form of reward for the woman cartographer. Group projects, whether she is an organizer of such projects or a regular participant, stand high as an intrinsic reward to making maps (Marita: 1), because "when we're developing something together as a team ... when you see that people develop and grow ... that's when you find solutions that you know are beneficial for society" (Inga: 7).

The very fact of organizing a team toward completing a project is a source of immense satisfaction:

I'm given the responsibility of organizing a team and then I have the liberty of choosing whom I like to cooperate with and then I allocate the tasks.... So I go around ask my colleagues and I say, "Can you write that part and you write that part?" "I got this part and you write that part," and then from time to time I go around and I talk to them to see how ... progress is going, and if they have difficulty, I try to help them and also I like them to be interested in the subject. (Minhui: 6)

Extending the idea of the family, cartographers treat projects in familial relationships. Sometimes a project is the equivalent of giving birth to a "child" (Marita: 9–10). Bureau chiefs see women co-workers as their "daughters," or want to be seen by these co-workers as "mothers" (Saskia: 9). Here is how one director put it: "I find that I could never achieve like what we have achieved here ... without the support of my group. And actually somebody had referred to me as a mother hen, but to me it's my extended family, and their welfare—mental, physical, financial [needs] of my staff—is to me a gauge of success.... I think we can be more creative, we can be more flexible with people and therefore have larger successes" (Cécile: 10).

Seeing oneself as part of a family is a key to success: "We have been working like this for a few years, maybe seven or eight years. We are a family. We are working together. We are all professionals working toward a goal and I don't feel that there is jealousy. Everybody is working together, giving the best of themselves, toward the same goal. And you know, not pushing each other, just working together toward one same goal. And I think that is the best thing" (Natalie: 7).

Even if a chief cartographer does not conceptualize her role as a mother, she does visualize the life of her charges in the context of *their* familial relationships: "In the case of students ... I try to support them also in the private programs if they want to be supported ... in this holistic way, I try to take my students as persons, not only students, but persons who have parents and maybe children and husband or wife. And I try to help them to plan their whole career of life ... I not only make them know something about cartography, but I try to help them to plan their life" (Liisa, 5).

The idea of a team is paramount among women, says one cartographer, who finds that working with women has been a very, very positive experience, because of the way that "you make decisions and the way your team and the way you react" (Saskia: 13). For a teacher of cartography, the ability to work in a team is the most important thing for keeping her going: "I think for me, it is the ability to work in a team. I know that myself, I need the contact with fellow workers, fellow cartographers, now fellow teachers. And I think being able to support each other in the cartographic industry is very important to me" (Ruth: 5).

As in a family context, creating the right atmosphere is an important goal at work: "Women care more about atmosphere.... [We] do observe the atmosphere as women, more than men. Men say, 'As long as I get what I wanted, I don't care how the other people feel,' but women ... like to have [a] more happy working environment" (Minhui: 7).

Another cartographer put it this way: "What mattered to me were the relationships. Now the group that I work for is very close. Most of us have been together for eight, ten years. And we work very well together and that is extremely important to me in making my environment a happy place to work" (Amanda: 6).

The process of creating something is as important as achieving the goal. In fact, the goal is the process itself: "I have always thought of myself as a creative person and in the last few years I have been extremely interested in process, and process of production.... But the intricacies of actually getting something through a lot of people, there is a lot of types of people involved in producing a map and how do we move something through all of those people as smoothly as possible?" (Amanda: 2).

For someone like the young Dutch cartographer Emmer (4), it is "sharing thoughts and so triggering new ideas" in a team that is so very important. Myra, a Greek cartographer, talks about her love of "throw[ing] ideas to other people and see[ing] what they think ... having peers around that I respect and have a different point of view and I will take their idea and look at it from a different perspective. This I find very rewarding" (Myra: 6).

The source of new developments, according to a Scandinavian, entails her being "very open [to] new ideas," something she cannot achieve without peers or co-workers (Helle: 2). The key is not competition, but cooperation: "I find lately that we have been working in teams. So everybody brings their expertise to the project. And we succeed together. It is the sharing of the success.... And [the project] came to a peak [last night] when we had that launch and we went out to dinner last night and everybody was crazy and happy. And that is our ... family" (Natalie: 7).

Still, the vestiges of male privilege remain in some work settings. An earlier chapter already referred to this quote: "There are some, I am sorry to say, old men ... who are keeping the power ... but now I have a supervisor from [the] geology department and she's a woman ... I am doing fine now" (Erika: 2).

In such a context, cartographers like Helle (1) and Ettie (4) give due emphasis to liking the colleagues around them who are also supportive. One bureau chief explains it this way: "[We are] more balanced ... we all have husbands, or fiancés, and we have children. And we are professional, very, very good, and we do our best in whatever. But we are not there to kill, but rather to enjoy work and to do things together. So competition is very different" (Saskia: 10).

For another, "It is not *my* atlas, it *our* atlas, really. And I really do believe that everybody in my group—and that includes computer people—are as passionate about this.... If I remember one thing, only one thing of all of the things I said to her [the Director] yesterday, *we* will have succeeded" (Cécile: 15).

Under circumstances where cooperation, rather than competition, is valued, it is natural to acknowledge the work of others. In such group work, the workers often use "we" (e.g., Amanda: 6; Marita: 1) to describe their collaboration on particular projects. As a consequence, a few of the interview participants mentioned that women disavow formal positions of leadership, such as this one: "I don't like be chair of the Commission. The woman is rarely very satisfied by being a leader ... we are a ghost, a good ghost" (Marta: 3).

The helpful work of Judy Olson expresses an invisible approach that has helped many a cartographer, especially when Olson was editor of *American Cartographer*. There, as we saw in an earlier chapter, through a process of her asking a potential author questions, she managed to convince him to submit the answers in handwritten form. In the end, he had an article that was "eminently suitable for publication." On an entirely different level, we already related the story of Phyllis Pearsall: when she came upon staff who were inconsolable because they were about to lose their job, she hit on the idea that the staff in her company would henceforth hold a hundred percent of the shares, and she gave up "any legal right for a dividend, a pension, or even employment" (Knowles, 2003: 19).

As we noted in an earlier chapter, we can find the same spirit of taking human concerns to the top of the list in the way Helen Wallis dealt with all those who sought her help. Whether it involved her own staff or younger scholars, she was natural and friendly. Peter Barber, the current head of Map Collections at the British Library, simply stated that she "loved people and maps—in that order," and she was able to extend that approach to national and international collaboration (Barber, 2005: 195).

Besides fostering and maintaining personal and collegial relationships, these cartographers also contextualize their work in terms of disciplines: their own and others. Specifically, they value the importance of other fields that might have a bearing on their cartographic work, both practically and theoretically.

"WALKING ON BOUNDARIES"

The cartographer's distinctive zigzag career would lead us to believe that she might see cartography as an endpoint.[6] For her, it is the beginning, not the end, however. Her unique take on the field is expressed in several ways. First, she firmly believes that the science of cartography is still an immature science. Second, the "narrative" plays an essential role when teaching or talking about cartography. Third, for cartography to advance, she also believes that the field must be open to many different specialists and disciplines. Finally, many cartographers identify themselves as risk-takers. All of these four elements shape her participation in her chosen field.

CARTOGRAPHY AS AN IMMATURE SCIENCE

The rapidly evolving state of cartography is described as "very frustrating" and "very shocking," still marked by a "lot of confusion" and "worrying" (Saskia: 5). At the same time, the last ten years have been "very exciting," involving "changing our ways of thinking" (Saskia: 5). Despite the move forward conceptually, one cartographer believes that not many people are working on "very fundamental issues." As a consequence, or perhaps because of it, the science is not mature: "But I find that the science is not mature enough at this point. We are just walking, we are being dynamic. But at this point, there is no clarity" (Saskia: 5).

However, she does expect that "more and more people are going to be working on the very fundamental issues related to both cartography and new cartography, I don't even know how to name it, but I think that it is a completely new era" (Saskia: 5).

Another cartographer relates these future prospects of cartography to her own expertise in three-dimensional cartography: "*How do other cartographers see your work? ...* It is a very interesting question because nobody believes [in 3-D maps] ... that this is the next step of cartography. But I am sure that this is the next step of cartography because the reality has three dimensions" (Nedelina: 3).

THE PLACE OF NARRATIVES

It stands to reason, according to a number of cartographers, that the only way that a dynamic field such as the new cartography can be captured is through employing the narrative technique of teaching. "Male colleagues," according to one, "may just want to put a package of knowledge in some structure." For her, building a story into cartography is a must (Myra: 10), an essential element to make it all work. The idea of combining and presenting ideas in an old form (through narratives) and making them somehow work in the context of the latest techniques, often cutting-edge ones, is simply another way of "walking the boundary." A synergy, a dynamism occurs when they combine the old and the new in this manner. A package of knowledge does not step outside its safe confines ... and it is less daring. Looking at it from the other side, it is about contributing to the *process*, rather than being preoccupied with the technology and even the solution. When not enough women get involved in cartography, the narrative and the process will not come through. Without their voices, the narrative is silenced. "Their voices need to break through," says one (Inga: 11).

Technology exercises such a strong gravitational pull that problems in cartography are defined by that technology. Moreover, the application of technology "in real life in the real projects" (Minhui: 1) is largely overlooked. Learning something new, "the new difficulties," "the new problems," and "the new challenges" (Minhui: 1) is solely defined in technical terms. They become divorced from their intended use, to solve everyday problems in the everyday world. Having entered "this terribly stifling digital world," says one (Amanda: 2), the artistry takes the backseat.

BEING OPEN TO DIFFERENT DISCIPLINES: WALKING ON BOUNDARIES

There is an overwhelming desire to walk the boundaries, to welcome different specialists into cartography. Kornelia (3) believed it was important for her to understand other specialists at a young age. There are others who are of the same opinion:

> *What really counted as you went about learning the tricks of the trade?* The education and that I am a generalist, not only knowing the field of cartography and geography, but also having a big interest in such other fields as history, biology, environment, literature, and so on. (Kristina: 2)

For Myra (2), the old cartography required only tunnel vision. But now the needs involve understanding "technological disciplines," societal needs, client needs: "you have to be a very open person ... very perceptive, and you have to listen well and so on and so forth" (Myra: 2).

Nothing short of meshing personal goals with the objectives of cartography can achieve the desired result: to increase the multi-disciplinarity of cartography: "[O]ne of the big [personal] goals [of mine] would be the increase of multi-disciplinarity. Although we are international and we are trying to work together, I think we are still lagging behind ... in inter-disciplinarity and understanding other fields, for example, sociology, anthropology, and bringing all this together, not assimilating them into geoinformatics, but forming a family of equivalent disciplines that work toward the same goal" (Myra: 3–4).

It is not only she, but it is her gender that can achieve the ideal of multidisciplinarity: "[Women have] more sensitivity for the big picture and not for the detail … and the women can be the glue for integration with other disciplines. I believe they are better at walking on boundaries" (Myra: 8).

In line with the belief that the next phase of cartography involves walking the boundaries among the disciplines, a number of interview participants see themselves as risk takers:

> I am a risk-taker. [laughs] I thought, what does it mean to do a Ph.D., you know? And he explained to me that it was going to take me longer than a master's degree, and he explained to me what it was all about: doing research, contributing in new ways to knowledge, and whatever. "Oh," and I thought, "that sounds quite interesting," and he said, "Let's do a Ph.D." (Saskia: 12–13)

> I had gone against convention by resigning a good position in the teaching profession after five years and headed back to university for four years of full-time study. I had taken myself out of my comfort zone to do this, and I stubbornly was determined to complete the course. (Ettie: 3)

Women, according to the editor of *Progress and Perspectives* (a journal devoted to promoting equity in surveying and mapping organizations), often come from "well-rounded mapping backgrounds and tend toward a more ecumenical approach toward the profession" (Straight, 1991a: 6).

New conceptions are increasingly appearing on the horizons of cartography. Margaret Pearce and Michael Hermann (e.g., 2010) have done some fascinating work bringing narratives into the picture of understanding how maps were created. They embarked on a mapping study of the seventeenth-century travel journals of Champlain in the 1600s—a record of his mapping project from the Gaspé Peninsula to Georgian Bay, Canada. By co-privileging Champlain's record with his own travel narrative, Hermann and Pearce were able to contextualize the primary design element with Champlain's words, descriptions of the geography, and experiences "to bring the reader into the landscape of the map" (2010: 41). These experiences included the conflicts and stories of Aboriginal people and Europeans. The sequential panel in their study conveyed the depth and diversity of experience in the places outlined on Champlain's maps.

Another good example of walking the boundaries comes to mind when we think of Evelyn Pruitt's talents in being able to actively engage with researchers and scholars outside her realm of work. Apparently, the invention of the triangulated irregular networks (TIN) arose out of a conversation she had with T.K. Peucker (now spelled Poiker) back in early 1972. This active engagement is a characteristic of other cartographers, such as Eva Siekierska, who was project leader of Visualization within the Knowledge Integration for Sustainable Development Program and of Visualization of Urban Archetypes. This main innovation was web-based and audio-tactile haptic feedback, relying on the work of Regina Almeida and her colleagues.

Numerous are the other examples where women cartographers naturally extend their interests to other areas: in the United States, Judy Tyner drew heavily on the social-constructivist and subjective nature of maps to develop her term "persuasive cartography." In England, Helen Wallis's approach was interdisciplinary long before it became fashionable (Campbell, 2004: 2). Barbara Shortridge (in the United States) easily extended cartography to food routines and traditions, farming, stimulus processing models, the discrimination of town size on maps, and map-reader discrimination of lettering size. Eila Campbell (in England), as was noted in an earlier chapter, created bridges among geography, cartography, and history. Carmen Reyes of Mexico expanded her areas of interest to children's atlases, cybernetics, the management of urban green areas, water systems, air quality, electoral systems, atmospheric models, and education. No less interesting is the work by Brazil's Regina de Araujo, who is now the world's leading cartographer in tactile mapping. Her other areas of expertise include indigenous mapping, perception and representation of space, geography, culture and tourism, and teaching geography.

SUMMARY

The woman cartographer's identity is not a static, unchanging process. The ongoing process of charting her career through cartography is marked by many forces that also shape her identity, partly driven by technology; by her orientation toward others, both personal and professional; and by walking on disciplinary boundaries. Exploring these dimensions of her identity as a cartographer allows us to know how she might experience the social structure of her profession.

Social structures are not far removed from the ongoing concerns of people who experience that structure. As symbolic interactionists claim, the reality of social structure is embodied in the norms and values that shape (and are shaped by) everyday interactions. These interactions are usually not problematic when participants share the definition of the situation.

This chapter's exploration of the orientation and expectations of women cartographers shows that women do experience that social structure as problematic. Most notably, the cartographers emphasize the importance of making maps more user-friendly, the struggle between work life and family life, the significance of valuing co-workers where cooperation (rather than competition) is the norm, and the need to incorporate many disciplines to move cartography beyond its current state of "infancy" (a term used by one interview participant).

It is clear that the heightened awareness of the importance of creating beautiful maps which are, at the same time, useful to others has given the cartographers the opportunity to gain a more profound understanding of their work. Moreover, because of the importance of both personal and conceptual relationships, the cartographers can place their work in a larger context. While personal relationships and obligations can exact a toll on their career, they also provide the higher road to the production of maps. These cartographers "make good ghosts," always ensuring that their team moves forward in as harmonious a manner as possible. The conceptual relationships they bring into cartography entail an opening up of cartography in ways that lead to new possibilities. At the same time, "walking on boundaries" results in a characterizing of cartography as an immature and incomplete field. All kinds of possibilities extend into the future.

The following chapters explore the social structures in the map world that canalize the experiences of women cartographers.

CHAPTER 10

Educational Opportunities and Obstacles

Regardless of the personal and career contingencies of the individual woman cartographer, she will come face to face with the educational agencies that formally launch her career in the map world. As far as the men are concerned, their acknowledgment of "gender" varies no less considerably, but is muted by the privileges they take for granted, especially in the reproduction of power in the map world.

This chapter explores the general situation of training women cartographers, but first takes a look at the participation of women in technical fields in general so as to offer a comparison with women in cartography. It then takes an in-depth look at the participation of women in a cartography teaching institute, followed by a discussion about the expectations and rewards, and the role of mentorship in the education of women cartographers.

Given the broad scope of the map world, one would expect to find many variations in the field of education (and training). Indeed, such education occurs in many different contexts. In the United States, it falls on departments of geography to teach cartography; in Canada, one finds departments of geodesy and geomatics engaged in education. But there are also departments of technology, laboratories, and specific institutes that partake in this process. Some institutes are national, some are international. All these variations point to the complex nature of pinpointing some general patterns of education in the map world. One thing can be stated with certainty, however: women are under-represented.

The situation regarding the participation of women in technical fields varies greatly with time, by national and local opportunities or obstacles, and by institutions and agencies that offer training and educational programs. No doubt the role of national professional associations also has a hand in (re)shaping educational opportunities for women (and men).

It is no secret that the participation of women in engineering and technical fields remains relatively low. According to Maryse Demoor (2000), there are still relatively few women in science and particularly in the higher echelons of academia. In Canada, where there has been active recruitment of women going on in post-secondary education, 60 percent of all students in 2010 who received post-secondary degrees, certificates, or diplomas were women (N=146,721). In 1992, it was 56.4 percent (Statistics Canada, 2010: Table 477-0014). These figures are far lower for such fields as engineering, applied sciences, technologies, and trades, where 11.9 percent of all degrees are held by women (Statistics Canada, 1998). However, there are places where the percentage of women, such as in geomatics, is considerably higher, as at the University of Calgary where, in 1999, 34 percent of the students were women (*Progress and Perspectives*, July–Aug. 1999: 4). When one narrows down the focus to engineering, only 21 percent of engineering graduates in Canada were women in 1995 (Finnie, Lavoie, and Rivard, 2001: 10).

The United States has a comparable situation, where 19.4 percent of students in engineering faculties are women (*Progress and Perspectives*, Mar.–Apr. 1999: 5), slightly up from 16.5 percent in 1996 (*Progress and Perspectives*, Jan.–Feb. 1997: 4). In 1994, there was an active discussion among educators as to what surveying nomenclatures would attract women students. The term "geodetic engineering" was the most appealing to female students (86 percent) in a survey among high school students in Maine and a few in Massachusetts, while "geographic engineering" appealed more to male students (64 percent). The term "surveying engineering" came third among female high school students (68 percent) (Schweik, 1994: 1–2).

The next section examines the participation of women in an international cartographic educational institution situated in Europe.

IN AN INTERNATIONAL CARTOGRAPHIC EDUCATIONAL INSTITUTION

In the map world, there are a number of important institutes that teach cartography (along with geodesy and related fields). It is important to look more closely at a particular European institute, because a number of the interview participants received their training there and spoke about the critical role the Institute had played in their cartographic lives.

The percentage of members of the Institute's staff who are women has remained fairly constant over the years, but there has been a strong increase in the proportion of women students.[1] There has not been much variation in the number of women staff since 1991, when women comprised 27 percent of all teaching staff. By 1998, this figure had barely changed. As is expected, the proportion of women on the part-time staff (45 percent) far outweighs the number of men (12 percent) on the part-time staff (1993 was the latest year for which these data are available). The Institute itself recognized the problematic nature of the gender issue and reported in 1997 that "no female staff was appointed to senior positions and the percentage of women occupying such positions even decreased" (Annual Report, 1997: 17). The Annual Report offers its readers a long list of teaching and administrative staff, but there is an unevenness in how men and women are designated. Men without an academic degree are not likely to be indicated with "Mr.," while women without an academic degree have "Ms." up front.

In terms of students, the Institute has experienced impressive growth in the number of women enrolled in its programs. In 1986, for example, only 12.4 percent of the 467 students were women (i.e., fifty-eight women), but the proportion nearly doubled, to 24 percent, in 1998, enabling ninety-seven women to attend the courses.

I shall look at student composition and the Institute's main vehicle of presenting itself to the outside world, its annual reports, from 1993 to 1998. A content analysis of the photographs contained in the annual reports shows that they do not do justice to the increasing levels of participation of women in the Institute's training program. Of the identifiable people in the 406 photographs in the annual reports (between 1993 and 1998), fewer than 20 percent are women. As well, men are more likely to be portrayed as faculty or as officials. Overwhelmingly, the photos of the Institute's staff members in the field are men. Women seem to fare somewhat better in photos of official settings (such as graduations), but the proportion of men rather than women is still higher. Moreover, women are more likely to be portrayed as students, rather than as faculty, and less likely in official

CHAPTER 10

functions than in in-course situations. What is noteworthy is that the only area where the proportion of women outweighs that of men is in the "entertainment" category—the annual international function the institute holds for its staff and students. Looking at it another way, more than 75 percent of all photographs only or primarily feature men, and fewer than 20 percent feature only women (the remainder of photographs feature men and women equally).

As a content analysis of an annual report is barely adequate to provide a fair understanding of how women experience educational opportunities and obstacles, it is important to consider the words of the women themselves, in terms of their expectations and rewards.

EXPECTATIONS AND REWARDS

We already know (from Chapter 2) that there are uncertainties that attend to defining who is a cartographer. There is even a problem when it comes to defining the field itself. It is therefore not surprising that in answer to the question of whether their work has lived up to their expectations, interview participants were more likely to talk about the things they did *in general*, rather than in terms of their activities within cartography. Marito (10), a freelance map-maker from Scandinavia, says she "didn't expect something more for my work." Not a few indicated their absolute surprise and even wondered that they had become (university) teachers:

> I didn't expect to be a teacher. (Nedelina: 7)

> [C]ertainly the big shift from mapping the maps to teaching about how to make mapping has been the biggest shift. (Ruth: 3)

> The political situation was such that I would never think of being a member of a university. (Helena: 6)

> We started [with] just the first step. We don't understand anything about methodological ways [of making maps] ... but we will research it and we discovered which way to make, to help students to have ... map reading [skills]. (Ramona: 4)

As another form of personal revelation of expectations, "Joyce" (5) finds that her cartographic work has moved her from her "grandiose self to things you no longer think are important." It is remarkable that cartography has led some women to exceed any expectations they might have nurtured (such as becoming a university teacher), while for someone else, doing cartographic work was a means to become more humble about one's earlier expectations!

However, there are a number of cartographers who spoke of their cartographic (rather than personal) experiences. For example, Kornélia (5) had not expected to make military maps during her career, raising fundamental questions in light of her interest in wanting only to make maps that would benefit the public. Loukie (2, 4) who is now a program administrator, says she was attracted to the aesthetic part of map-making, but not, to her disappointment, the push by technology. Ettie (4), who makes tourist maps, had "never thought I would be working on a computer 100 percent of the time!" Ending up as manager was not part of the expectation either, especially in sales ("marketing"). While expectations were minimal, the rewards of doing cartographic work are both profound and extensive.

The perceived rewards of making maps reflect, to a large extent, the philosophy that "maps are not for walls." Of the twenty-four interviewees who had an opportunity to speak about the rewards of making maps, 54 percent indicated the importance of others in map-making, either in a work team or as a user, and 42 percent spoke of the importance of converting an abstract idea into concrete reality. Only 21 percent admitted that what they like best about map-making is the idea of doing the technical part of the work, while 17 percent confessed that international travel or knowledge constitutes an intrinsic reward when it comes to map-making.

The development of expectations and rewards often happens in interactional situations rather than as formal statements issuing from educational or training establishments. Moreover, once cartographers have achieved a level of accomplishment and recognition, we wonder about the extent to which they make themselves available as mentors to junior colleagues.

MENTORSHIP

The interview participants offered valuable insights about mentorship. They spoke about "distant" sources of mentorship, such as books that had inspired them; more direct and personal mentor relationships; the praiseworthy qualities of their personal mentors; the differences between women and men mentors; and the extent to which they themselves bring all of these experiences into nurturing the next generation of women cartographers.

CHAPTER 10

"DISTANT" SOURCES OF MENTORSHIP

It is noteworthy that the interview participants mentioned books as sources of inspiration and encouragement. Two mentioned Arthur H. Robinson's book *Elements of Cartography* (1960), which served as a publication-mentor.[2] In the lives of cartographers, there were also other published works, including Neil Postman's *The End of Education* (1995), James Carson's *Finite and Infinite Games* (1987), and Mark Monmonier's *How to Lie with Maps* (1996). Others cited Jacques Bertin (1918–2010), a seminal French cartographer who edited *Semiology of Graphics*, 1967), and Jean Piaget (1896–1980), a leading Swiss developmental psychologist, as their "fathers" in linking perception to the teaching of map-making to children (Ramona) and as major sources of influence and inspiration. In one instance, Barbara Petchenik was a deep influence on a person's life, even though Petchenik was still a graduate student at the time. Marita stepped enthusiastically into cartography after reading a book by John Keates (*Understanding Maps*, 1981), liking what he said, though she has met him only twice since then.

PERSONAL MENTOR RELATIONSHIPS

The women in our study could not rely on very many men within their organizations to support and nurture the efforts of women, but when men did so, the benefits extended far and wide. However, many women in this study were only too pleased to point to mentors, mostly men, who helped them along the path to cartography. Most commonly, the mentor was their thesis supervisor or "the boss" at work. In North America, for example, women cartographers have benefited immensely from the personal mentorship of the kind described above from particular men who also deeply influenced cartography itself. According to Judith Tyner (1999: 27), there were young geographers and cartographers in the latter part of World War II and thereafter who "began teaching and changed the face of cartography." There is a fairly large group of men who were all early mentors to women, especially among the early cadre of women who obtained a Ph.D. in cartography—women cartographers who became noted in their fields.

Other than some of the men referred to earlier in chapters 6 and 7, there were so few men who were mentors that it seems relatively easy to list them by name. Ed McKay, for example, actively recruited women into geodetic organizational leadership (*Progress and Perspectives*, Jan.–Feb. 1993: 6). Stephen G. Brush took upon himself the task of outlining the obstacles women face in education (*Progress and Perspectives*, Mar.–Apr. 1992: 4). Richard Elgin, the president of a survey company in Missouri, also actively promoted women in the field (*Progress and Perspectives*, July–Aug. 1992: 6). The name of Porter W. McDonnell, an active proponent of women in surveying and mapping, was given an award in his memory (*Progress and Perspectives*, Mar.–Apr. 1994: 1). There were others, including Herbert Stoughton, a physicist and professor (*Progress and Perspectives*, Jan.–Feb. 1998: 4), and John E. Foley, another physicist (*Progress and Perspectives*, Mar.–Apr. 1998: 5).

PRAISEWORTHY QUALITIES OF MENTORS

Impressive knowledge (Ruth), an "infectious" enthusiasm (Myra, Ettie), and loving "new ideas" and being open to new people (Liisa) seemed to be the three personal ingredients that made these men promising and effective mentors. Along the same lines, Zia observed: "I think for me it was the human dimension that counts [in her mentor-teacher]. The human dimension ... is the relation with our form of conceptualizing the subjects. It's a very, very interesting form of conceptualization and that fascinates me, of course ... but then the human dimension for me is fundamental to work" (Zia: 5–6).

It is striking that the "mentee" also wanted to be "challenged" (Natalie) and demanded "100 percent of herself" (Cécile) or wanted to be "marked tough" (Lynn). The good qualities of a mentor included his or her belief that "I was good" (Inga) or the fact that he or she "believed in me" (Natalie). Minhui praised her mentor because "he always appreciated my approach" and liked the fact that she was direct. For each one, the mentor was able to translate her commendable qualities into concrete action: for instance, by "opening doors" of opportunities at work (Inga, Natalie, Gabrijela) or urging her on to write books (Ramona, Inga). For some (e.g., Amanda), the value of a mentor was in waking her up to an international perspective in cartography. One cartographer had two mentors; one did not allow her to become "lazy," while the other one was "flexible enough for her to be lazy," allowing her creativity to emerge (Cécile).

A particularly telling account comes from Nuvyn, who describes her experience at a well-known international research and teaching centre in cartography as follows: "They treated me like a very important person. They made cartography appear to be nice work. One day, I was asked to be something like a class president. When I tried to beg off because my English was not very good—poor, in fact—the instructor said that his [a non-European language] was also poor. Thus, I became president of the class and received much encouragement in that way" (Nuvyn: 2).

The bonds of mentorship continued long after the formative years. Some have kept in touch (Ruth), others became close friends (Ettie), while still others managed to exert a deep influence (in one case, about every ten years) in pushing the cartographer into new jobs and challenges (Inga). Zia, who is just completing her Ph.D., assigns an enigmatic difference to the way women treat her, as opposed to men, as follows: "… it's completely different. Here is a gender difference [that is] completely different. Well, they [the women] treat me with more distance. They speak with me … [but] only [what is] necessary. [However,] they don't [pose] difficult questions to complicate my presentations. They hurt me, [though]…" (Zia: 8).

The following extract from my field notes taken during a cartography conference typifies the motivation, interest, and interaction between a mentor and upcoming women cartographers.

> I asked him what gave him the idea of promoting women in cartography. He said that he himself came from a humble background and thus has always felt naturally inclined to help those on the outside. I also asked him how he managed to promote the idea of gender. He said several times, "Perseverance!" He said he had a hard time convincing his colleagues, including some women, except for [mentions name of a senior woman cartographer], but eventually he did, and was then asked to convince the Geomatics folks in the United States to set up a committee, too. The group did, but it also took perseverance. As we were waiting for the pedestrian light to change, I had an opportunity to observe the manner of his encouragement of women. He had mentioned that one of his efforts resulted in his creating travel awards for the better students from developing countries (in fact, the Chinese student who approached me during registration was one of them). In any case, he warmly greeted two young women students from Argentina, one of whom (or maybe both?) received such an award. Then, he pulled out of his briefcase a photo of the student with others. (The following morning, Liisa mentioned that [the mentor] had come to her and her student's presentation and how he had congratulated the student.) (Field notes, 16 August 1999: 3)

DIFFERENCES BETWEEN WOMEN AND MEN MENTORS

Of the twenty-four interview participants who said they had had mentors, only six reported that those mentors were women. In all, there were more than three times as many men mentors (twenty-eight) as women (eight)—some participants mentioned more than one mentor.

Despite the important role that mentors had played in these cartographers' lives, one is struck by the fact that relatively few women have chosen to mentor other women in the field.[3] Only two of the interview participants mentioned that they were making a special effort to mentor other women.

For some, it is a conscious decision not to imply that incoming women cartographers need any assistance. Loukie, for example, believes that although the number of women cartographers is small, this is not because of inequality; Joyce, who is an academic cartographer, does not "see any necessity for this [mentoring]." Minhui shies away from going "around my colleagues, saying, 'I'm a woman so you pay attention to me' … I don't want to do that." Saskia, who heads several research centres, says, "I have to be honest. I don't feel particularly … attached more to women than to men." For others, there were practical difficulties that stood in the way of mentoring women. Kristina, for example, is a freelance cartographer but does not have the time or the money to do "gender work"—much like Liisa, who holds a teaching position.

NURTURING THE NEXT GENERATION OF CARTOGRAPHERS

Still, despite their reluctance or lack of opportunity to mentor women, there is plenty of advice they would give to incoming women cartographers. The proffered advice falls into several categories. The first concerns *"the Self"*— an interesting choice, in light of the women cartographers' general proclivity to take others into account in their work. Kristina captures the sentiment of many on the subject of "the Self": "Stand up to your opinions. In the long run, it is important that women tell men that they react to attitudes and special actions when it happens, instead of trying to manipulate the men to the way you want decisions to take. Be aware that you get the credits for your work…. Try to enlighten gender aspects in a positive way without lying. Do not underestimate yourself and get used to standing up and arguing for your work."

They advise that women should be "more convincing" (Els), be "bolder and more confident" (Amanda), "do what's best for you" (Helle), be "prepared to question and challenge conventions" (Ettie), "trust yourself" and "don't wait for anybody to thank you"—in fact, "thank yourself" (Marita). Inga advises that women should not be "modest about accomplishment." These comments speak to the fact that these interview participants have, in the past, not put themselves on the map, but hope that the future generation will be less modest about their own accomplishments.

There are many fewer statements about *"the Other."* Helena, for example, declares that although women do not "have to be stronger than [chauvinist] man … they have to declare that [such a man] is not going to stop it [the advancement of women]." Still, the responsibility shifts from "the Other" to the women: Liisa feels at fault for not insisting that the departmental secretary do as much work for her as for a male colleague. Minhui offers this striking perspective: "I want to mainly change myself. If I change myself, I think the reaction from the men will also change … slowly, not at one time.… So I think the gender issue is more a woman's concern than a man's concern. I don't think it's practical to go to educate men.… It's more practical to change women.…"

Even though a larger number of the interview participants refrain from seeing themselves as mentors for future generations, the path of mentorship is strewn with many "how-to"s—that is, bits of practical advice. Cécile advises the women to take "one step at a time" and not to "shoot for the stars too quickly." Still others, like Natalie, who now heads a marketing division of a major atlas firm, would tell her "mentees" to walk through "any door anybody opens for you." The approach taken to mentoring is not "packaged"—that is, with routines; it is built around the "mentee," shaped around that person's narrative, so to speak. The cartographers carry this grounded approach through, even in their teaching. Myra, in speaking about the importance of having a narrative for a course, says that "male colleagues," on the contrary, "may want to just put a package of knowledge in some structure and the structure is fine and that is logical and so on." Ruth, a cartographic instructor, teaches by example, by giving anecdotal information to her students.

Finally, we hear of the *specific things* that the interview participants would encourage the novices to study. Myra, who teaches GIS, finds it important that women be oriented toward solving the problems of their own country. In the same vein, Liisa wants her young women students to

> study technical things, study mathematics, study computer-information technology and this and that. So what does it actually mean? It means that I suggest them to study subjects which are considered as male oriented: technology, mathematics, and things like that. And why I say that is because I am afraid that female students go to more soft studies. Okay … why [should] I try to push my young girl students to subjects which they are not maybe originally interested in? Okay, I know if they learn those things, they have better possibilities in the[ir] working life. (Myra)

Many others agree with this recommendation, whether it comes from Lynn, a recent Canadian graduate in cartography, or from Marita, who is a senior editor in a Scandinavian cartographic firm.

SUMMARY

Our analysis of how the map world is organized includes a brief examination of the relatively low proportion of women in the engineering sciences and in geodesy. The proportion of staff and students at an international cartographic institute reflects such low participation (with 27 percent of full-time staff and 24 percent of students). In exploring the expectations and rewards of women's training to be cartographers, one discovers that occupational expectations were low and that women saw the rewards as intrinsic, not necessarily reflected in salary or status. Finally, we considered mentorship as an integral part of education. It was notable that many women drew inspiration and knowledge not only from very established books in the field, but also benefited from personal interactions and sharing of expertise with personal mentors. We also noted that among the interviewed women, they reported relatively more men than women who had mentored them. With the rising generation of women in the map world, there is, however, some concern that the gains of the past will be either taken for granted or dismissed as irrelevant in the light of the appearance of contemporary equality. The following chapter delves more deeply into this topic.

CHAPTER 11

The Gendered Social Organization

So far, our discussion of contemporary women cartographers has brought to light not only their individual lives through vignettes, but also many of their common perspectives and experiences. We now turn our gaze to the social organizational aspects of the map world. They describe, in short, the cartographic culture that gives rise to those perspectives and experiences. This particular chapter discusses the map world as a gendered social organization.

Many complexities attend to the situation of women in cartographic organizations. Social, economic, educational, and technological factors drive these complexities. There are other layers that work themselves into these factors: namely, the diversity of organizations in the map world, regional and national peculiarities, social policy, government reorganization, as well as the role of the International Cartographic Association (ICA) in stimulating the advancement of women. "Developing" countries seem to have a higher proportion of women cartographers than "developed" countries (Chapter 12 elaborates on this finding). In some countries, social policy has resulted in more women cartographers who are active in government service, while in others, traditional outlooks, where both governmental and corporate sectors do not welcome women, have resulted in women heading their own private cartographic firms. All of these create a rather diversified landscape of women's engagement in the map world.

PROPORTIONAL NUMBER OF PARTICIPATING WOMEN CARTOGRAPHERS

It is hard to estimate both the number of people and the number of organizations devoted to cartographic activity worldwide, let alone trying to configure the proportion of participating women in that activity. Worldwide, the proportion of women in surveying and mapping was 5 percent in 2000, even though at the auxiliary and administrative personnel levels the proportion is somewhat higher (see also Brandenberger, 1993), but it is quite difficult to estimate the actual total number of women involved in cartographic work. In 1990, the ICA International Task Force on Women in Cartography was planning to distribute a survey questionnaire to two thousand women in sixty countries (International Task Force on Women in Cartography, 1990). We must presume that these women were recognized as active in the field of cartography. When one includes technical assistants (who are women, by and large), drafters, and the like, we must quadruple the figure, reaching perhaps eight thousand women as being active in cartography. If we were to include map librarians and map archivists around the world, we might add another 510 women. Thus, we end up with no more than 8,510 women who are routinely connected to map worlds. Still, given the dated information and major technoluogical developments in the field, one would be hard pressed to arrive at an accurate count of women in contemporary cartography.

Based on figures derived from a survey conducted by the ICA in 1995 (ICA, 1995), I would estimate that the developing countries contain at least 15 percent of all cartographers, surveyors, and GIS specialists. At the same time, a survey published in 1991 (Siekierska, O'Neil, and Williams, 1991) notes a fairly large difference among women with wide-ranging experiences in cartography (see Table 11.1).

There are proportionally fewer cartographers with less than five years' experience in cartography itself (41 percent) than is the case with those with six to fifteen years' experience (57 percent), and even fewer than those with sixteen years of experience or over. The reverse is true for both GIS and geodesy/geomatics, the two main affiliated occupations. Twenty-six percent of the youngest group are involved with GIS, while only 16 percent are the mid-level experienced group who work with GIS. In geodesy/geomatics, the proportion of women is still quite low, whether for the younger or mid-level

THE GENDERED SOCIAL ORGANIZATION

Table 11.1. The participation of women cartographers in cartography and related fields

Level of Experience		Cartographers	GIS	Geomatics	Other Fields	Total
0–5 years	– %	41%	26%	18%	15%	100%
	– N.	1,073	680	471	392	2,616
6–15 years	– %	57%	16%	13%	14%	100%
	– N.	1,829	513	417	449	3,208
>16 years	– %	69%	—	4%	27%	100%
	– N.	1,341	—	78	525	1,944
Total	– %	55%	15%	12%	18%	100%
	– N.	4,243	1,193	966	1,366	7,768

Source: Adapted from Siekierska, O'Neil, and Williams, 1991.

Note: Excludes the 2.9 percent of respondents who did not indicate level of experience; all percentages are rounded. "Other" includes remote sensing/photogrammetry, and other. In this 1991 survey, the researchers had included geodesy and geomatics separately from surveying. Given the current global trend to rename and redefine surveying departments as geodesy and geomatics, I have combined these three areas as geomatics.

group (15 percent and 11 percent, respectively). In effect, one could say that the introduction of certain kinds of "table" technology (i.e., GIS) has altered the profile of women participating in cartography. Table 11.1 alone suggests that there is a higher preponderance of women associated with more traditional forms of cartography than with either GIS or geomatics. In comparison to women in related fields (such as in surveying in Canada, where 4.3 percent of the 8,065 surveyors are women), there is a larger proportion of women in fields that have an immediate bearing on cartography. In fact, in 1996, some 1,165 (28.4 percent) workers among the 4,100 "mapping and related technologists and technicians" in Canada were women (Statistics Canada, 1996: Table 12.1, in Chapter 12).[1]

When we take into account our estimate of 8,510 women in fields related to cartography around the world and combine this estimate with the results of the 1991 survey (referred to in Table 11.1), we can make the following provisional guess of women in these fields: 4,400 are involved directly with cartography, 1,200 work in GIS-related occupations, 960 work in geodesy and geomatics, 510 are librarians and archivists, and 1,440 work in other fields (such as remote sensing/photogrammetry and in "other" [unspecified] areas). Perhaps just over 1,300 women cartographers are located in Europe.[2] With a range of roughly between 4 percent and 25 percent of cartography-related fields consisting of women, it is important to consider how women participate within international, regional, and national cartographic organizations themselves.[3]

Based on these preliminary observations, this chapter discusses the International Cartographic Association, regional and national cartographic organizations, and map librarians and map archivists. With a focus on international, regional, national, and topical issues, we can confidently surmise the nature and rate of women's participation in the contemporary map world.

THE INTERNATIONAL CARTOGRAPHIC ASSOCIATION

The International Cartographic Association (ICA) is the organization that captures the activities of men and women cartographers around the world and is therefore an inviting one to study. I mounted intensive fieldwork on both the 1999 (Ottawa, Canada) and the 2011 (Paris, France) conferences, with the aim of assessing the extent to which women participated in those gatherings (see Appendix A). In the process, I also learned a great deal about trends related to gender in a large number of countries. I have organized my findings under several categories: (1) membership on the ICA executive, (2) participation of women in the sessions (both as presenters and as audience members), (3) the work of the ICA commissions (including the Commission on Gender and Cartography), and (4) the nature of commercial and non-commercial exhibits at both conferences.

ICA EXECUTIVE COMMITTEE MEMBERSHIP

A closer look at the structure of the ICA reinforces one's belief that the advancement of women shows relatively little progress and even a decline in some parts of that organization. The 1999 ICA international conference in Ottawa, Canada, shows an active concern with that advancement; its 2011 international conference in Paris—fifty years after its establishment in 1961—demonstrates no such advancement in many areas of the ICA. The ICA's 12th General Assembly, held in August 1999 in Ottawa, reveals a similar proportion of women's participation in that gathering. Although the proportion of women attending the General Assembly was as high as 41 percent, the percentage of women voting as country delegates was 14 percent. On the executive dais, the proportion of women was practically the same, at 13 percent (Field notes, 21 August 1999).

What might indicate the participation of women includes their election to the ICA executive; the gendered composition of ICA commissions and working groups; women's participation in the sessions themselves, whether as presenters or in the audience; and their appearance as chief editors of atlases, which were placed on display at these international gatherings. The 2011 international conference was notably different than the 1999 event.

The gendered membership on the ICA executive committee after the 2011 conference in Paris resembles that of the executive committee before the 1999 conference in Ottawa; out of a membership of eleven, there was one woman. However, some shifts seemed to have been under way during the elections to the executive at the 12th General Assembly,[4] when there were seven men and four women. Then women occupied 36 percent of the executive membership. This proportion compared very favourably with the proportion of women delegates in 1999, which was 14 percent. The organization would be leading the way in the advancement of women in its leadership with the election of the new executive in 2003 in Durban, South Africa, resulting in the election of two women, representing 18.2 percent. By 2007, the ICA conference, held in Moscow, produced only one woman member on the executive (9.1 percent)—the same as the 2011 conference in Paris. Thus, the 2011 elections of the executive represent a significant decline in the proportional number of women.

Virtually all women cartographers I spoke with during a reception at the ICA conference in 1999 must have anticipated the rough road ahead for women's involvement in the ICA and spoke pessimistically about change in the organization and structure. One attendee mentioned that "behind every older person here, there are twenty younger ones who are unable to come to the conference and who have asked her to keep an eye out for this or that at the conference" (Field notes, 15 August 1999: 3). Others spoke of the threatened viability of the ICA, "when it costs so much to get here, especially for women and young professionals" (Field notes, 15 August 1999: 2). I heard similar comments throughout the remainder of the conference (Field notes, 16 August 1999: 2). The underlying message is that the administrators within the ICA should reduce their administrative costs perhaps by meeting less often, and thereby making it financially possible for women and young professionals to attend the conferences.

Any change in executive membership is inherently a slow process. Within the International Cartographic Association, it would take at least six years to effect any change on the executive if the membership decided to have a woman president: one typically serves several years as vice-president before coming on as president. Before being able to serve as president (two years), one generally has to serve on the executive as a general member (another two years).

PARTICIPATION IN ICA SESSIONS AND EVENTS

Learning about the proportion of women who are delegates to ICA international conferences is not a straightforward task. At the 2011 gathering, officials declined my request to see the list of participants, and I was unable to figure the gender composition of the delegates (interestingly, the ICA conference in Beijing in 2001 did publish the names of all participants). I thus resorted to other measures to provide what I hope to be a reliable estimation of such composition (see Table 11.2). The participation of women (and men) in the program was made invisible by the fact that the conferences program never listed a participant's full first name; it was always an initial.

Regarding the number of women from countries attending the ICA 2011 conference, one should note the complexity of calculating their gender composition. When I asked members of countries with a large number of attending cartographers to tell me that composition, it was difficult for them to assess the number of people or the gender composition from their own country. At the other extreme, one wonders whether it makes sense to include the statistics of countries that have only one cartographer attend the conference. Taken together, the proportional number of women is 36.8 percent.

If we take 36.8 percent as the approximate proportion of attendees who are women, we can already see (Table 11.2) disproportionally low numbers of women who participate as chairs (12.5 percent) of the sessions and who are the least likely to ask questions or make comments (23.6 percent). As presenters (38.2 percent), the women are equal to their overall attendees at the Paris conference, although as audience (31.6 percent) they seem again under-represented. It should be noted that China had a conference delegation of one hundred, of whom thirty were women.

ICA COMMISSIONS AND WORKING GROUPS

The history of women in the International Cartographic Association merits a fuller treatment than is possible within the confines of *Map Worlds*. The bulk of its practical work rests in its commissions and working groups. There are wide variations in both the number and size of the commissions.

Table 11.2. Gender composition of selected events at ICA 2011

	Total Number (women and men)	Percentage of Women
Direct observation of delegations from 42 countries (out of 50)	429	36.8%
Direct observation of 16 sessions at ICA 2011 (including General Assembly)*		
Chairs	24	12.5%
Presenters	51	38.2%
Audiences	994	31.6%
Asking questions	89	23.6

Source: Field notes, Paris, 2011.

* During the Paris conference, I made sure that I took a random sample of all sessions and technical interests to ensure that I would catch all corners of cartographic activities and interests. The organizers divided the program into eleven categories. Each category had multiple sessions. With a few exceptions, I was able to observe one or more sessions in all categories. The statistics in the above table are limited: (1) some presenters did not show up; (2) audience attendance shifted in the course of a session; (3) sometimes, no time was left for questions. The calculations separated chairs and presenters from the audience. I normally took the count 10 to 15 minutes after all had settled in.

In 1999, there were seventeen commissions, but in 2011, there were twenty-eight. To gauge the participation of women in the commissions, I tabulated the proportion of chairs and executive memberships occupied by women. Most recently (2011), of the twenty-eight commissions, three were headed by women (10.7 percent) (see Table 11.3). The current (2011) proportion of women chairs is the lowest in more than thirteen years. The proportion of women who are the remaining members of the commissions is 25 percent (that is, 119 women out of 478 members).[5] However, there were women on all commissions, and a number were corresponding members. It should be noted that the role of corresponding member is a very active one, keeping in touch with all members. Commissions can add members without

Table 11.3. Chairs of ICA commissions, by gender, 1999–2011

	No. of Commissions	No. of Chairs Who Are Women	Percentage Who Are Women
1999–2003	17	3	17.6
2003–2007	19	3	15.8
2007–2011	22	4	18.8
2011–2014	28	3	10.7

Sources: ICA web page and *ICA Newsletter*.

waiting for a meeting of the General Assembly. Membership is by invitation of the commission or working group.

The Commission on Cartography and Children has the highest proportion of women (56 percent), followed by a cluster of four commissions (Education and Training, Users and User Issues, History of Cartography, and Mapping from Satellite Imagery) whose proportional number of women ranges from 38 percent to 31 percent). There are not many surprises, if any, in these data. They reaffirm women's interest in children, education, users, and history. (The reader may recall, from a previous chapter, the opinion expressed by women interview participants about the importance of the user as part of the design of maps.)

The commissions that rank lowest in terms of women's participation include those on marine cartography, mountain cartography, and spatial data standards—all falling at 17 percent or under.

In 2007, there were nine working groups. Women headed two of them (one was co-chaired with a man; the other was co-chaired with another woman). There was sufficient information available on only two working groups: Art and Cartography, and Open Source Geospatial Technologies. The proportion of women on the former is 67 percent, and on the latter is 14 percent.

ICA COMMISSION ON GENDER AND CARTOGRAPHY

Before presenting the developments of the ICA Commission on Gender and Cartography, it is important to sketch the activities of the principal women cartographers before 1989.

THE GENDERED SOCIAL ORGANIZATION

Pre-ICA Commission Days.[6] Against the heavy preponderance of men in the ICA, one finds a band of some thirty women who had already established a cartographic reputation. The reader can quickly glean from the names below that their activities constituted essentially the larger pool of women, many of whom *Map Worlds* alluded to as pioneers in earlier chapters.

From the United Kingdom, Helen Wallis, Ph.D., was one of the first women to become actively involved with the ICA; she chaired a session on the History of Cartography at the Ottawa Conference in 1972. The success of this session led to the formation of a working group (1972) and a Standing Commission on the History of Cartography (1976), chaired for fifteen years by Wallis, who has been instrumental in a number of ICA projects, and in editing and authoring various ICA publications. The Standing Commission on the History of Cartography is the interest group with the highest participation of women. Wallis was succeeded as chair by Monique Pelletier (France); other commission members and/or contributors have included Eila Campbell (UK), Olga Kudronovska (Czechoslovakia), Leena Miekkavaara (Finland), Dorothy F. Prescott (Australia), and Ingrid Kretschmer (Austria).

From the United States, Barbara Petchenik became the first female principal delegate and head of delegation to an ICA congress, at Perth, Australia, in 1984. Judy Olson (USA) was involved with the Working Group on Map Use at least as early as 1985, when she presented a report on the impact of technological change on map users and map use. She assumed the chairmanship of the Standing Commission on Map and Spatial Data Use in 1987. Currently, the commissions on National Atlases, on Thematic Mapping from Satellite Imagery, and on Population Geography have female members; Barbara Petchenik also gave a keynote address at a joint-sponsored conference on school atlases in 1986 (Calgary, Canada).

From Canada, Eva Siekierska held the position of deputy-chair of the National Atlas Commission. There were also Louise Marcotte and Lillian Wonders, who made presentations at the Calgary conference. From Germany, Helga Ravenstein, member of the publications committee since 1983, was responsible for the German-language translation of the chapter on automated cartography of *Basic Cartography* (Vol. 2). In addition to General Assembly meetings and international conferences, the ICA also sponsored local special interest seminars in which women participated. From the former Soviet Union, Irene Zaroutskaja was an active contributor to the ICA-UNESCO seminar on Education in Cartography, held in Paris in the early

1960s. And from Switzerland, Madame Kishimoto was also active at this time.

These and other women were typically active in these areas: at a 1973 international symposium on maps and graphics for the visually handicapped,[7] four working groups were established, of which three were chaired by women; and an ICA-sponsored session at the American Congress on Surveys and Mapping convention in 1984 was opened by Barbara Petchenik, and closed by Judy Olson, both of the United States.

As a later survey would indicate (Siekierska, O'Neil, and Williams, 1991), women cartographers in the late 1980s and early 1990s were more active on the national rather than international level, with at least 19 percent having more than twenty years of cartographic experience.

The Life of the Commission. The existence of the ICA Commission on Gender and Cartography is a good example of how gender issues come to the attention of cartographers and the extent to which these issues develop a "career" in the cartographic world. At the closing of the 8th General Assembly of the ICA, in October 1987, Dr. Fraser Taylor, professor of geography at Carleton University, Ottawa, noted the low participation of women in the work of the ICA (Task Force on Women in Cartography, *Minutes*, 1989); as a result of identifying that problem, in early 1989, the Presidential Task Force on Women in Cartography was established in Ottawa under the leadership of Fraser Taylor, who was president of the ICA. Regina Almeida was the ICA vice-president for this task force and played an important role. This move was accompanied by a presidential request to all commissions and working-group chairs to include women, cartographers from developing nations, and young cartographers. The chair of the Task Force was Eva Siekierska of the Canada Centre for Mapping, Surveys, Mapping and Remote Sensing Sector of the Department of Energy, Mines, and Resources Canada.[8] The Task Force was composed of members from ten countries: Canada, Australia, Finland, Hungary, Mexico, Norway, China, Sweden, the United Kingdom, and the United States.[9]

The first phase. While activities of ICA commissions or task forces are funded by ICA's regular operating budget, the survey organized by the Task Force on Women in Cartography was funded by four governmental agencies in Canada, Norway, and Sweden, five professional associations in Canada, Sweden, and the United States, and five corporate groups.[10] The ICA Secretariat also provided some monies to fund the survey, thereby giving symbolic value to the importance of improving the status of women in the ICA.

In 1991, ICA upgraded the Task Force on Women in Cartography to the Gender and Cartography Working Group. In some respects, the career of this ICA activity (as a task force, a working group, and then a commission) reflects the various strategies women used to characterize their relationship to the organization as a whole (to be discussed later in this chapter): mainstreaming equality (1989–1995), paralleling equality (1995–1999), and ignoring inequality (1999–present).

From the moment of its birth, the aim of the commission was to discover the barriers that women faced that would prevent them from participating more fully in cartographic activities. Its first aim was to conduct a survey of two thousand women cartographers in sixty countries, examining the "barriers that have contributed to the disproportionately low participation of women in cartographic activities internationally" (International Task Force on Women in Cartography, 1990).[11] The survey led to the publication of *The Directory of Women in Cartography, Surveying, and GIS* in 1991; there were 380 entries on women from forty countries (*ICA Newsletter,* [22] Oct. 1993: 7). With a vision of mainstreaming equality, it was not uncommon for a member of the working group (renamed as such in 1991) to organize and to take a number of group members to regular workshops, such as the first national Mexican GIS conference in 1992 (*ICA Newsletter,* [22] Oct. 1993: 7). The working group also prepared a background document for the 13th United Nations Cartographic Conference for Asia and the Pacific, held in Beijing, China, in 1994 (Siekierska, 1994). Another theme of this first phase of what became the commission was developing a "partnership of women and men" (Siekierska, 1994: 2). Reflecting this vital interest in partnering with men, the working group's wording of the title in the commission would later be "gender," rather than "women."

The second phase. The phase of creating a parallel structure for women cartographers emerged after the commission had completed its survey: the barriers were deep-rooted, and themes of discouragement and lack of nurturing emerged from the survey's findings. Women cartographers, moreover, seemed quite unfamiliar with the workings of the ICA—and this seemed to be the main focus of the survey. The lack of travel funds and the lack of time (while balancing "family responsibilities and professional life") acted as deterrents as well (Siekierska, 1994: 5).[12] As the membership of the commission expanded (in fact, doubled), it became clear that what was needed was a forum, an intimate and personal forum, where the members could share their experiences as women cartographers in their home countries. In her role as chair of the commission, Siekierska played a crucial role in allowing women to share, at first hesitantly and then more frankly, their experiences, but also to hear the stories of other women in other countries.

Some ten years after its founding (during which time the task force had now been permanently renamed the ICA Commission of Gender and Cartography), the commission[13] consisted of members from eighteen countries and was able to demonstrate a wide range of activities it had undertaken during the previous four years. It included updating and publishing the *Directory of Women in Cartography, Surveying, and GIS,* and establishing and maintaining links with other organizations, including the IGU Commission on Gender and Geography, the FIG Gender and Surveying Working Group (of which the first chair was Kirsi Artimo, who was also a member of the ICA Commission on Gender and Cartography), and several departments at the University of New Brunswick, Canada (ICA Commission on Gender and Cartography, 1999). The commission also developed and maintained a website pointing to many cartographic activities on behalf of women.[14]

The third phase. The third phase was to start in 1999 with the selection of a new chair for the commission, Ewa Krzywicka-Blum (from Poland, like Siekierska) whose interest in working and mapping under-represented groups moved the commission in a rather different direction. In this phase, women's inequality became a less recognized part of the commission's work, and the mandate was expanded to include minorities (of which women were one) and even touch base with children and mapping. In August 2002, the chair of the Commission on Gender and Cartography issued a call for papers for the 21st International Cartographic Conference, to be held in Durban, South Africa, the following year. The list of recommended issues for papers did not contain the words "gender" or "women" (Krzywicka-Blum, 2002).[15] The second portion of this communication, however, asked the members of the Commission on Gender and Cartography if the name and profile of the commission ought to be changed. If no changes were suggested, would the commission concentrate on

1. developing an "inventory and trials of solutions of the problems connected with barriers leading to gender inequalities in education, accessibility to labour market activity in social and political life; or
2. studying the historical role of women in cartography; or
3. studying differences in map-making and map users between men and women groups?

THE GENDERED SOCIAL ORGANIZATION

Alternatively, [with the name change] should the Commission consider:

1. editing and publishing the ICA World Demographic Atlas (in connections with under-represented groups and groups with special cartographic needs); and
2. the participation of other ICA Commissions, such as the Maps and Graphics for Blind and Visually Impaired People," etc.? (Krzywicka-Blum, 2002)

After 2007, the commission changed its name to the Commission on Under-represented Groups and Cartography.

COMMERCIAL AND NON-COMMERCIAL EXHIBITS AT ICA CONFERENCES, 1999 AND 2011

Cartographers attending the ICA international conferences spend much of their time visiting parallel exhibits that include the Petchenik International Children's Map Competition, maps produced by participating countries, the commercial exhibit, and the exhibit of national atlases. This chapter analyzes the latter two exhibits. What stood out at the 1999 conference in Ottawa was the commercial exhibit, while one of the main exhibits at the 2011 conference in Paris was the display of national atlases. These lend themselves to conducting a gender analysis of the prevailing map world.

COMMERCIAL EXHIBIT (OTTAWA, 1999)

Corporate sponsorship of the sort described above is a fairly recent phenomenon.[16] Some 14 percent of the fifteen-page program of the ICA's 12th International Conference (held in Perth, Australia) in 1984 was devoted to advertising by cartography-related companies (R.A. Skelton Papers, 1984). However, what is striking is the lack of corporate sponsorship. In those earlier years, the exhibit was called the "technical exhibition." The description suggested that "more than 60 percent" of the space was booked by "a wide range of firms both Australian and international." Morning and afternoon teas were served at this venue "to give maximum exposure" (R.A. Skelton Papers). Even the 1991 ICA conference in Bournemouth did not yet contain corporate sponsorships as such. There was not even a description of the "technical exhibition," which would later be so common for "commercial exhibits."

Other changes also reflected the trend toward commercialization of cartography. One participant noted that some thirty years ago, the non-profit booths tended to be at the centre of the exhibition hall (Field notes, 18 Aug. 1999). Another said it was also more typical to find cartography exhibits at the centre; these have gradually been replaced by GIS-related booths and exhibits (Field notes, 18 Aug. 1999).

The mapping trade fairs were a regular part of these gatherings, represented by both the established mapping software developers and more recent dot.com and other companies. One participant at the international conference dubbed the presence of the commercial booths at the conference "the courtship of commerce," involving free gifts, scavenger hunts, and other enticements to lure conference participants to the exhibitors. As the conferences progressed, more and more participants were carrying free items, including maps, often rolled up in large cylinders. It is a social world wholly deserving of its own study. The courtship of commerce at the international conference involved two important ingredients that set it apart from any other aspect of a gathering of map-makers: it was a layout that was highly stratified and gendered. What was salient was the relationship among the cost of location of the booths, the specializations they represent, and gender, including the dress code.

As a general rule, women closer to the inner circle were more likely to be dressed in what one would describe as conventional business clothes, comparable to the suits that the men wore. However, we should understand that the gender makeup of individual booths did not necessarily reflect the makeup of the workforce of the company. For saleswomen with family responsibilities, it was quite difficult for them to leave home for a week to stand in booths, because, according to Marlene, "the logistics to leave home is a little bit more difficult" (Field notes, 16 Aug. 1999: 3).

Finally, one might refer to the fact that many employees in major software companies were "ex-military type people," according to Rebecca, a relatively new entrant into the field (Field notes, 16 Aug. 1999: 6). The hiring of former military personnel was (and still is) particularly prevalent in the larger companies, and therefore women seemed to have a bigger chance of being hired by the smaller companies, as Rebecca explains: "So I think that there is a little bit of the old boys' club that is still in this industry, and so I think that is probably one reason that you see a lot of men at this. So small companies tend to give women a bigger chance than larger companies would" (Field notes, 16 Aug. 1999: 6).

Another dimension that might explain the preponderance of men in the inner ward is that when women become experienced specialists in cartography, they are more likely to go into sales. One interview participant in a booth believes that some 80 percent of people in sales are women (Field notes, 16 Aug. 1999: 3). As she said, "So the exhibitors at the conference are actually more reflective of sales and marketing for us in the company, rather than the actual people doing the mapping" (Field notes, 16 Aug. 1999: 3).

Alongside the organizational developments and the mandate to make the setting more welcoming for women, attitudes among members were changing toward sexist advertising, such as what was found in the National Society of Professional Surveyors in the United States (NSPS) in 1988 (*Progress and Perspectives*, Jan.–Feb. 1989: 1) involving "mini-clad" women in surveying advertisements, "shirtless" women advertising surveying equipment by Apollo, Inc., "semi-nude" women in Nikon Instruments Group and Geotronics of North America calendars (*Progress and Perspectives*, Mar.–Apr. 1989: 1, 2; Jan.–Feb. 1989: 1). Geodimeter of Ontario and Ashtech Inc. were other companies that either distributed literature "offensive" to women or were subject to lawsuits (*Progress and Perspectives*, Mar.–Apr. 1993: 7, 8). Vendors of survey equipment continued to use these illustrative materials and "girlie" calendars until 1990, and some even until 1993 (*Progress and Perspectives*, Nov.–Dec. 1993: 1). Even technical journals used sexist illustrations well into the early 1990s (*Progress and Perspectives*, Mar.–Apr. 1993: 8).

EXHIBITION OF ATLASES (PARIS, 2011)

One additional gauge informs us of the gendered nature of cartographic work: namely, the editing and production of atlases. I took the opportunity at the 2011 ICA conference to avail myself of the eighty-four atlases from twenty-six countries displayed at this conference and conduct a systematic analysis of the extent to which their editorships were gendered. I relied on the conference catalogue, compiled by François Lecordix (2011), to gather the basic information; when that information was missing, I approached the participants from the relevant countries if they could help me. In a few instances, I sought out the atlas editors themselves. I have seen and checked every individual atlas. And whenever the information was not available, I wrote the atlas publishers. In the end, I had reliable information about seventy-three atlases from twenty-four countries, representing 87 percent and 92.3 percent of atlases and countries, respectively.

I included only the principal editor(s) of general atlases, not the educational atlases, unless there are no general atlases indicated for that country. Aggregately, thirty-five women and 110 men were responsible for the editing and design of the seventy-three atlases. In short, women constituted 31.8

percent of this work, although women supervised only ten atlases, or 13.7 percent. The former proportion is not out of step with the women's participation in the conference and in the sessions in particular.

REGIONAL AND NATIONAL CARTOGRAPHIC ORGANIZATIONS

Cartography adumbrates many activities, and cartography itself is part of a larger set of professional and technical work. It is therefore difficult to sort out what kinds of organizations are typically involved with cartography. For a number of organizations, there are no membership records that indicate whether a member is male or female.[17] Some international organizations, moreover, consist only of national organizations and have, as such, no individual memberships, as in the case for the International Society for Photogrammetry and Remote Sensing (ISPRS). It is not unusual for some cartographic organizations to fall under the aegis of another organization. The Cartography Speciality Group, for example, is part of the Association of American Geographers. The National Society of Professional Surveyors (NSPS) in the United States has members preparing data for maps or creating maps, but women's participation is rather uneven, despite its efforts to include them. A recent short film presented by the NSPS (*Spotlight on Surveying*) for the United States Public Broadcasting System shows three women and forty men (http://www.acsm.net/, accessed 8 Dec. 2011). The NSPS executive lists thirty-one members, only two of whom are women. Moreover, those in drafting or digitizing positions may have no interest in belonging to any of these organizations. Some areas, like hydrography, are still primarily male areas in map-making, while in the case of map librarians and archivists, one can clearly see how even this very small but vital corner of the map world is clearly marked along gender lines.

REGIONAL ORGANIZATIONS

A cartographic organization may well represent members of a variety of disciplines (such as GSI or spatial-data management), but may also function as a national chapter of an international or regional organization. For example, EUROGI, the umbrella organization founded in 1993 to develop a unified approach to the use of geographic technologies in Europe, holds within its ambit twenty-two national groups, ranging in membership from twelve people (Luxembourg or Switzerland) to seventeen hundred (Norway), for a total of 5,080 individual members (http://www.Eurogi.org/). However, none of these groups is similar to each other in terms of type of

THE GENDERED SOCIAL ORGANIZATION

membership or organizational template under which they function. Some are circumscribed to ministerial delegations, while others are wide open to a broader membership. Private corporations intersect as well with these national groups. Headquartered in Amersfoort (The Netherlands), EUROGI also has among its members EuroGeographics (situated in Brussels), which is the association of the heads of European Mapping Agencies.

There are data pertaining to other cartographic organizations that indicate that women cartographers, generally, do not constitute more than 25 percent of membership. For example, the EuroGeographics offices in Southampton, Paris, Frankfurt am Main, Helsinki, and Brussels (the association of the heads of mapping agencies in European countries) has a total of sixteen staff members, including four women. There are no women serving as president or on the management board, and one finds women in the regional offices. The one woman who is serving at the Paris head office is its secretary (http://www.eurogi.org/index_1024.html). In Italy, in 1990, there were no women serving on the National Council of Surveyors (Terenzoni, 1990), and at the provincial level only a few women.

In North America, women were heading some 21.5 percent of the sixty-five university cartographic laboratories (NACIS, 2005). Of national cartographic societies where women's membership is above 25 percent, two belong to Canada, two to the United States, and one each to Sweden and North America. What is striking is the fairly low proportion of women in United States cartographic organizations: only 11 percent in the American Cartography Association (1993) and 4 percent in the American Congress on Surveying and Mapping (1993). In any case, the rate of increase of women members averaged from 0.2 percent to 1.2 percent per year; for some organizations, these incremental rates cover as long as a thirty-three-year period (Canadian Association of Geographers), while for others it is just one year (e.g., National Society of Professional Surveyors in the United States).

NATIONAL ORGANIZATIONS
What seems like an increase in the proportion of women members may, in fact, be due to an overall decline of an organization's membership, but one where women have managed to stay in. For example, the Canadian Association of Geographers (CAG) had 802 members in 1996, but 770 in 1999 (Shoffey, 1999). In other instances, not all practitioners in a given field are members of the relevant organization. For example, at the National Geodetic Survey in the United States, there are about thirty women involved in geodetic data collection, analysis, and distribution, either as professionals or as

technicians. This figure represents about one-third of personnel involved in these activities in this agency. However, none of the women are members of the American Association for Geodetic Surveying (AAGS), while at least ten to fifteen of the sixty men are members (Doyle, 1999). Finally, *de facto* membership of women (and men) in organizations does not necessarily indicate active participation in the organizational work. Again drawing on the experience of AAGS, it seems that such participation has never been high (1–2 percent) and that at least since the mid-1970s, one seldom has seen more than two or three women participating in these meetings (Doyle, 1999). In the late 1970s, a number of women in AAGS who were eager to see women advance in its leadership managed to get Evelyn Pruitt (see Chapter 6) on the ballot as a vice-presidential candidate, but a more traditional man was elected (Janice Monk, email to author, 13 September 2011).

THE CARTOGRAPHIC SOCIETIES IN THE NETHERLANDS AND THE UNITED KINGDOM

One can also find huge variations in the participation of women in cartographic organizations along geographic lines. In European countries, such as the Netherlands, for example, a large proportion (30 percent) of the thirty-six women cartographers who belong to the Netherlands Cartographic Society (NVK) are in research and academic cartography, while some 20 percent are students.[18] Two women are heads of a department. Of the remaining women, nine (12 percent) specialize in particular map types (such as school atlases), eight (17 percent) work in methodology (such as GIS), five (13 percent) are involved in map production, and three (8 percent) work in mapping techniques (i.e., desktop and digital mapping); the remaining two (6 percent) work as a multimedia developer and as secretary. On average, 14.1 percent of NVK members were women (*NVK Membership Directory*, 1996–1997).

In the United Kingdom, the Oxford Cartographic Conference in August 2000 attracted 289 participants, 264 of whom were from the United Kingdom. The average percentage of women participating in the conference was 39.2 percent. A careful check of the reported affiliation of the participants (thirteen had not indicated any affiliation) reveals striking differences among women and men (Oxford Cartographic Conference, 2000). Nearly 73 percent of freelancers and nearly 55 percent of those working for nongovernmental organizations were women. While just over 41 percent of those working for government were women, only 36.4 percent represented universities. Significantly, only 23.5 percent of those working for private firms in the United Kingdom were women. In the world of an allied field—geography—Maddrell (2009b: 315) reports that until 1913, women may have been excluded from the Royal Geographical Society (with a few earlier exceptions). Between 1951 and 1970, in another British organization, the Geographical Association, fewer than 10 percent of the executive were women. In 1970, when Eva Taylor, the first woman professor of geography, was appointed, she was one of ten on the executive (Maddrell, 2009b: 324–325).

MAP LIBRARIANS AND MAP ARCHIVISTS

Map librarianships and map archival activities constitute a distinct branch of cartography. (Chapter 6 is devoted to a historical discussion of map libraries.) These activities are certainly connected to historical cartography, but some librarians are also involved with academic cartography and, in specialized libraries, with geodesy and geomatics. The engagement of map librarians can be extensive in the educational field as well, teaching cartographic literacy to children and students. Let us look at the gender division among map librarians and map archivists.

While it is not easy to pinpoint the precise number of map libraries around the world during the nineteenth century, it is clear that some of the most notable libraries achieved their start during this time. For example, the collections of the Société de Géographie commenced in the nineteenth century (Brooks, 2001: 28). A momentum had been building up since the establishment of the British Library in 1753 (Dubreuil, 1993: 215) to establish cartography as an intellectual discipline. Various libraries saw the intellectual relevance of collecting and preserving maps. In 1805, for example, the British Library established the India Office Records Map Collection (Dubreuil, 1993: 215), and in 1828, the Bibliothèque nationale in Paris became the legal depository of maps (Dubreuil, 1993: 73; Pognon, 1979: 195). In 1851, the collection of maps and the like became part of the mandate of the American Geographical and Statistical Society, founded that same year (Drazniowsky, 1979: 127). One of the richest depositories of maps in Russia, the former Lenin State Library, began receiving maps in 1862, although the Imperial Public Library did so in 1810 (Kozlowa, 1979: 205, 206). The Public Archives of Canada has been collecting maps since 1872, just five years after the founding of the Dominion of Canada (Winearls, 1979: 167). The Geography and Map Division of the United States Library of Congress was formally established in 1897 (Wallis, 1979a: 13). It would take at least another half-century before map libraries would become a more widely accepted fact of public life in the United States. In 1937, for example, there were fewer than thirty map libraries in the United States with one or more full-time employees (Wallis, 1979b: 107). Still, map libraries did not come into their own until after 1950s.

Today, map collections worldwide have grown considerably. In preparation for her third regular survey on map collections around the world, Lorraine Dubreuil (1993: vii) sent out questionnaires in 1989–1990 to 950 map libraries. She excluded collections with fewer than a thousand maps unless they were a national library or archive. With a final listing of 522 map collections, just over one-fifth (22 percent) had been founded in the nineteenth century alone (around 9 percent were founded before the nineteenth century, and around 69 percent were established since the nineteenth century). To be sure, many of those founded in the nineteenth century are located in Europe or in English-speaking countries; there were also a number in Latin America.

Map librarians in the nineteenth century seem to have been men, with female assistants. It was only after the nineteenth century that several notable female map librarians rose to the top: Clara Egli LeGear, a carto-bibliographer, served forty-seven years at the United States Library of Congress (1914–1961) (Wallis, 1979a: 68); Ena L. Yonge was hired in 1917 by the American Geographical Society when it was still located in New York (Joanne Perry, email to author, 17 September 2002). However, it was not until 1935 that the Library of Congress issued the first publication by a woman, a catalogue of two hundred Hispanic-American maps, atlases, globes, and so on (Modelski, 1979: 96). Her named is embedded with those of two men authors. For the first time, too, the American Geographical Society hired a woman, Mrs. Mary E. Harzell, as the first full-time assistant in its Map Department in 1944, ninety-three years after its founding (Drazniowsky, 1979: 136). Significantly, the first publication at the Library of Congress in 1949 by Clara Egli LeGear dealt with "maps, their care, repair, and preservation in libraries" (Modelski, 1979: 96). Before that time, all publications of the Library of Congress were catalogues and the like: there had been nothing about the care of maps. LeGear had already, in 1947, organized, with Walter W. Ristow, the New York group of the Geography and Map Groups of the Special Libraries Association (Wallis, 1979b: 108).

It is difficult to make national and international comparisons regarding map libraries and archives. There can be large differences in how a library or archive functions, and some agencies carry more weight than others. What is more, libraries and archives are organized somewhat differently in various continents or countries. In North America, for example, most of the

Table 11.4. Map librarians, archivists, and related positions by proportions of gender, 1993 (in percentages)

	Archivists	Librarians	Other	Totals
Women	6.3	29.1	4.1	39.5
Men	16.4	30.2	14	60.6
Totals	22.7	59.3	18.1	100.00

All figures are rounded. N=444 (excludes unidentifiable gender of nineteen people).
Source: Dubreuil (1993).

map libraries are found in university settings (and there are very few listings for map archives), but in Europe, many map archives are organized mostly under municipal or state-level libraries. In Latin America (and elsewhere), map "libraries" are normally organized under the military or departments of defence. Finally, deciding what constitutes a map library or archive can be fairly subjective. For example, Dubreuil (1993) lists only ninety-one such establishments, but Carrington and Stephenson (1978) have a listing of 743 map collections in the United States and Canada alone. The discrepancy becomes more obvious when we realize that there is a gap of fifteen years between the publications.

If we exclude the unidentifiable gender of eighty-four people from the list provided by Dubreuil (1993), we see that nearly 40 percent of people who work in map collections around the world, in sixty-nine countries and dependent territories, are women (see Table 11.4). While the categories of "archivists" and "librarians" are usually quite clear, the category of "other" is not. In the latest category, we include people who work, for example, in bureaus of specific agencies, such as statistical, public records, and surveying offices. However, there is a far higher proportion of women working as map librarians (29.1 percent) as opposed to map archivists (6.3 percent)—nearly five times as many.

CHAPTER 11

SUMMARY

This chapter has led us through the contingencies in the number and proportion of women in the current map world, with a special focus on cartographic organizations. It is clear that while one finds enormous variations in the participation of women in cartography, proportional equality remains an unfulfilled goal. At the international level, despite some gains made in 1999, there has been a noticeable drop in the proportion of women who participate in the various organizational and scientific areas of the International Cartographic Association (ICA). While attendance of women at the biannual General Assembly gatherings ranged between 36 percent and 41 percent, the proportion who were elected to the executive dropped from 36 percent to 13 percent. The proportion of voting members, however, hovered around 14 percent.

With respect to the participation in the scientific sessions (at the 2011 Paris meetings), 12.5 percent were chairs, some 38 percent of those who presented were women, while the percentage of women in the audience stood at nearly 32 percent. The more significant point, however, relates to the fact that the conference program has rendered gender invisible. The first names of all participants, whether as chairs or presenters, appear as initials. This designation of names has buried the fact of gender.

This chapter has also pointed to other gender divides. The commercial exhibit in 1999 shows deep gender stratification, with men occupying the centre stage of the exhibit. The 2011 exhibit profiled atlases, showing again how gender presents itself in that cartographic arena (just under 14 percent of seventy-three atlas chief editors from twenty-six countries are women). As far as some national cartographic organizations are concerned, we found a relatively low number of members of the Netherlands Cartographic Organization are women (14 percent). In the United Kingdom, we see that at least 39 percent of women members of that country's association of cartographers are self-employed or are heading their own company. In that context, the avenues of cartographic work for women do not lie in large corporations, in government, or even in universities. The gender division of labour also seems to hold true in the case of map librarians and map archivists, with women concentrated in the former.

Social trends, economic and social policy, changes in governmental structures, and technology account for these variations. National and regional differences based on culture and attitudes also contributed to these differences. Finally, the vast differences among cartographic and cartography-related organizations provide the bulk of the remaining explanation of variations.

THE GENDERED SOCIAL ORGANIZATION

The divergence of organizational policies themselves and social trends account for the uneven involvement of women in cartographic organizations. The pace of involvement has been uneven and fraught with struggles. One might even assert that the participation of women at the higher reaches of cartographic organizations has declined over the past twenty years—a characteristic that is shared with society in general.

These recent developments and proposals of the ICA Commission on Gender and Cartography placed the issue of women and gender at the forefront, at least until recently. They, in effect, constitute a lead-in into the next chapter of *Map Worlds*, which explores the varying forms and strategies of engagement by women cartographers, whether through mainstreaming equality, nourishing equality, or disavowing inequality.

CHAPTER 12

Female Pathways through the Present-Day Map World

Across the span of map worlds there are vast differences in the way that gender is a factor in the recruitment, maintaining, and promotion of women in cartographic organizations. In many respects, the map world reflects many germane issues when it comes to how women approach and solve their individual situations. Just as women cartographers are found in many countries, so does one find considerable variations among the challenges and strategies of doing one's job in the map world. This chapter considers the general challenges that women face in many of the fields allied with cartography, and looks at recent organizational trends that affect women and explores female pathways through (in)equality, taking into account their various expressions across cultures.[1]

CHALLENGES

There is very little that is new or surprising when one considers the participation of women in the allied fields of cartography or geodesy, such as engineering and science. Newly hired women may be seen as a "gimmick," a token, which lessens the credibility of both women and the employer (Scullen, 1990). However, my interview participants have also gone on record saying that being the "first" helped to establish them in their career; they took advantage of the wider margins within which to carry out their work.

CHAPTER 12

Government-sponsored affirmative action that requires public contractors to provide equal opportunity employment has given many women and members of other minorities in surveying their first break into the workforce. Although such programs are controversial, and although women and other minorities cringe at the thought of tokenism, such affirmative action programs have opened doors that might otherwise have been closed. The following response by a reader to an article that appeared in *Progress and Perspectives,* the magazine devoted to affirmative action in surveying, is an example:

> I knew when I was hired that I filled a quota, but I was also fully qualified, so I did not feel guilty about getting the job. Later, I was promoted to fill a quota, but again I was fully qualified. The same outfit tried to push me into a higher promotion, which I refused. I knew there were others with more seniority that deserved the position. I earned a tremendous amount of respect, and I think it was a good career move, as I gained the support of people who could help me later on. (Straight, 1991b: 3)

Published accounts about and interviews and conversations with women cartographers point to challenges that are unique in the map world. Particular forms of humour and language, and being surrounded by evidences of a male culture (such as photographs and engineering student newspapers that are slanted toward sexist interests) used to promote the "chilly" atmosphere for women, although this behaviour is now discouraged in the public culture of engineers. However, given the time frame of those I interviewed, many spoke of that chilly atmosphere. When Monique Pelletier was appointed director of the Département des Cartes et Plans at the Bibliothèque nationale de France, her predecessor told her, "I'm happy that you're replacing me, but it's too bad you're a woman" (Brooks, 2001: 27). This atmosphere is punctuated by the criticism that some women are just "too sensitive" about sexist language (Ballantyne and Hanham, 1993). Not being listened to is another facet of the work culture that undermines the full participation of women (Ballantyne and Hanham, 1993), while women's talk is normally interrupted by men. McCormack has found that women "feel isolated" and "excluded from the day-to-day camaraderie of the men who commute together, study together, [and] relax together …" (McCormack, 1991). While each point does not constitute the chilly atmosphere by itself, each contributes to "a ton of feathers" that women need to lift in order to survive (Caplan, 1993).

Time, professional backgrounds, types of organization or activity, regional differences, personal interests, and other factors determine the extent to which organizations deal with gender imbalances and tensions, and with the barriers women face in cartography. The wide divergence of efforts by the same organizations speaks to the general uncertainty and ambiguity of formal attitudes toward the participation of women. Sometimes, too, the uneven participation of women in cartographic work is brought about quite unintentionally, such as when Sweden decided to move its governmental cartographic service to a new location (Gävle), far from the capital, resulting in the loss of many women mid-level managers who were not able to move on account of family responsibilities. Observers have also noted that the introduction of technology has led to the removal of middle management in the mid-1990s, consisting primarily of women who were thus not in a position to be promoted (*Progress and Perspectives*, July–Aug. 1996: 4). One also notes "equity relapse" in the symptomatic decline in the proportion of women in organizations over the past twenty years. One finds the relapses in unions (Jacobsen, 2007: 41), in women's advocacy groups (Jaquette, 2000), in science (Demoor, 2000: 4), and in management (Wirth, 2002: 6). Some (e.g., Hinterhuber and Stumpf, 1990: 65) anticipated that women are now no longer exclusively claiming a professional life, but are also insisting on having a private life, separate from the demands of an organizational career. A number of interview participants remarked how the proportion of women in executive positions has been sliding over the past fifteen years or so. In Canada, for example, in the mid-1970s, some 25 percent of employees at Natural Resources Canada were women; in 1999, the figure remained the same (Amanda: 11). In Norway, there are now fewer women managers than there were in the early 1980s (Inga). There also seems to be a trend for the decline in the number of women applying for a Ph.D. One recruiter of graduate students who has been in her position for the past six years observed that while there are women Ph.D. students in the program, 'the numbers are declining"; she added, "I can't get them in" (Leslie: 2). Thus, cartographic organizations, educational programs, and conferences all experience these fluxes of gender participation, albeit with differences in emphasis.

CHILD CARE AT CONFERENCES AND NON-DISCRIMINATORY LANGUAGE

The earliest activities related to improving the participation of women in various societies had a flavour distinctively different from today's attempts.

CHAPTER 12

On one hand, the incorporation of child care at conferences could be seen as a major step forward, and its relative success might be due to the fact that this activity tended to confirm, in the minds of the organizers, the traditional place of women as caregivers of children. The American Congress on Surveying & Mapping (ACSM), for example, earlier focused on the idea of having children at national conferences (*Progress and Perspectives,* Mar. 1988: 7). However, arrangements at conferences might not be the ticket, after all. Kelly Olin, writing on "surveying mothers," notes that conferences present the opportunity to be away from children, "almost like a vacation" (Olin, 1988: 4), although it might be more problematic for single mothers to go to conferences. The 2011 ICA conference in Paris did not have child care. No one seemed to have missed them; cartographers with school-age children left them with their husbands or other family members.

On the other hand, the struggle to achieve a commodious language that expressed more openness to women's participation seems to have had a more difficult passage. For example, in 1988, the president-elect of the ACSM declared that proposals regarding discriminatory language "went to absurd lengths," and indicated the change from *mankind* to *humankind* as one example of such a change. Would such a change entail censorship? How would censorship operate, and, if censorship cannot be enforced, why bother with the language change? (*Progress and Perspectives,* May 1988: 2). By 1992, however, the ACSM adopted the policy of eliminating all discriminatory language (*Progress and Perspectives,* Sept.–Oct. 1992: 1).

TASK FORCES RELEVANT TO THE ADVANCEMENT OF WOMEN IN THE ORGANIZATIONAL LIFE

Almost invariably, organizations are interested in setting up task forces devoted, in particular, to increasing the proportion of women in their membership. The ACSM set up a Women's Forum in March 1983 (*Progress and Perspectives,* May–June 1999: 3), although it was chastised by members for being too forward (*Progress and Perspectives,* Mar.–Apr. 1992: 2). Within six years, though, it was reported that many women were leaving the National Society of Professional Surveyors (NSPS) due to their displeasure with its treatment of women (*Progress and Perspectives,* Mar.–Apr. 1990: 4); the task force set up as a result of the Women's Forum was floundering, only to re-emerge twelve years later, in 1999 (*Progress and Perspectives,* Nov.–Dec. 1989: 2; Sept.–Oct. 1998: 1; May–June 1999: 3). The Fédération internationale des géometres (FIG, anglicized to the International Federation of Surveyors) organized a task force specifically on women in 1988 (*Progress and*

Perspectives, Sept.–Oct. 1996: 3), although in 1983 it had already organized a task force on under-represented groups (*Progress and Perspectives,* Sept.–Oct. 1998: 1). The two task forces must have had shallow roots—that is, not deriving nourishment from their parent organization (FIG), for they seem to have disappeared (one reporter said, "it was stifled by a group of men who feared the enlargement and radical thinking of the Forum"), only to re-emerge in the summer of 1998 as the Task-Force on Under-represented Groups (*Progress and Perspectives,* Sept.–Oct. 1998: 1).

It is interesting that such efforts seem to coincide with the general decline of membership in such organizations as the ACSM, American Cartographic Association (ACA), and American Association for Geodetic Surveying (AAGS) (*Progress and Perspectives,* Mar.–Apr. 1993: 1). "Equity compulsion" was often born in a context of declining membership; the establishment of a task force also sometimes occurred in light of a declining membership. Still, the recruitment of women could not be realized only by special membership drives: the professional societies had to have a welcoming atmosphere, too. For example, in 1990, president-elect Charles Tapley of the NSPS said, "I saw women leaving NSPS; I don't want to shut anyone out" (*Progress and Perspectives,* Mar.–Apr. 1990: 4).

There were also developments that seem to have taken the wind out of any advances. Interweaving, or perhaps undermining, the efforts by women to secure a firmer base in occupations, the United States courts struck down affirmative action in 1989 (*Progress and Perspectives,* May–June 1991: 1).

Closer to home, family members of the female student influence her decision regarding whether she should or can participate in the sciences (Cannon, 1999: 3). Socialization can play a significant, positive role (McCormack, 1991), since the daughter of a mother or father who was in engineering are "more likely to receive encouragement from [her] parents to select engineering" (Cannon, 1999: 4). As an earlier chapter indicated, it is striking how influential a father is in encouraging his daughter to take a path in her studies that would eventually lead her to cartography.

Specific to cartography, other elements play a significant role in the way women tread the cartographic path. Traditional surveying usually consisted of small field parties living in close quarters that might "create [an] unease and awkwardness'" between women and men (Scullen, 1990: 87). Robert Foster claims that employers worry that women will not be able to carry field equipment, clear line, or generally keep up with the crew; that they might be a social distraction on the crew; that they might cause sexual harassment charges; that clients would be surprised to see a woman in charge; and that certain job sites are too dangerous for women (Foster, 1991: 7). The fact that few women cartographers take part in these activities makes them unduly stand out; it also may lead them to feel isolated.

There are many women incumbents in cartography who enjoy the security of their position and who do not question the social organizational structure (Ballantyne and Hanham, 1993).[2] Still, women and men who have explored in-depth the situation of women in such fields as engineering come up with different answers. The women are more pessimistic than men. For example, Robert W. Foster (1991: 7) claims that women are not stagnating in dead-end, low-paying positions in surveying, but have instead risen to positions such as section managers, senior surveyors, partners, and proprietors. Writing a report only a few years earlier, Anil Saigal noted that women engineers often "are disappointed with their first jobs and some even quit" (Saigal, 1987). The thirty-eight women who were interview participants are actively involved in the map world, but what can we learn about the strategies they adopted to survive in the map world *after* they gained admittance to it?

The strategies employed by women can work to transform existing power relations and to structure and restructure the social organizational aspects of map worlds. However, one noted woman cartographer stated, "There is a general consensus among women that their managerial styles are still not well understood and valued. I tell myself that I am in a pioneer position and that there is a prize for breaking new ground. This is an intellectual consolation" (Lohr, 1999: 3).

PATHWAYS THROUGH (IN)EQUALITY

Although the paths that women take in cartography speak to their variety of interests, talents, and opportunities, one can make the case that progress in the status of women has still been quite uneven. Maryse Demoor (2000: 4) makes the point that "somehow the way upward has been halted or slowed down these last years.... One of the most tenacious obstacles in the move toward a greater recruitment of women for the world of science and academia is the widespread belief that the situation now is as good as it can ever get; that there are no problems anymore."

The women cartographers I interviewed seem to walk on four pathways in coming to terms with power relations within cartographic organizations. Three of these perspectives acknowledge inequality in map worlds; the fourth denies there is inequality. *Mainstreaming* suggests that women work from within organized cartography to change power relationships.

The focus of *parallelism*, by contrast, does not seek to change the central organization of cartography, but rather to facilitate the participation of women who need nurturing and encouragement. The women who *observe inequality* acknowledge that there is inequality, but do not actively address it; they merely acknowledge its existence and observe inequality without intervening. The final category, *disavowing inequality*, disclaims there are power imbalances and inequalities. I have drawn the data to sustain these four perspectives on inequality from the interviews of twenty-eight women cartographers.[3] As explained in an earlier chapter, I have taken a symbolic-interactionist approach to understanding these interview data to highlight the concept of "definition of the situation." Following W.I. Thomas, when a social situation is defined as real, it is real in its consequences. The question of the definition of the situation challenges the idea of "objective structures" and is thus quite important. Everyone has been face to face with a situation that he or she has defined as "real," but seems to fly in the face of the "hard" realities of that situation. What strengthens my assertion about the importance of the definition of the situation is the presumption that all social situations are definitional ones. Merely stating that a condition is "objective" is simply that: making a declaration. If we look at even such seemingly enduring structures as class, gender, and race, there is very little that can be declared as concrete. After all, class is a relational concept; gender as a construct varies in relevance across time and social settings; and race is the ultimate social construct. It has no objective reality at all, but the consequences are nonetheless profound. Returning to the data from our interviews, we must assume that the perceptions and definitions of the interview participants are real to them. The fact that these participants are involved with map worlds (albeit in varying degrees) gives credence to their perspectives: they are, after all, experts about their own lives. It should be borne in mind, at the same time, that we cannot always assume that there are effects of the definition of the situation.

Lest the reader assume that the four pathways through (in)equality have clear-cut edges, he or she must aver that their boundaries can be imprecise. The cartographers do migrate from one category to another over the course of their lives. Moreover, a perspective can hinge on the age or occupational status of the cartographers. There is thus an overlap among the perspectives. What stands out in the interviews is that each interview participant is remarkably consistent in how she sees her own pathway through the map worlds. Age, experience, and occupational status seem to solidify the women's perspectives. Table 12.1 gives an overview of the incumbents of these

CHAPTER 12

Table 12.1. Characteristics of members of the pathways through (in)equality

	Number	Mean Age	Professional Status
Mainstreaming equality	7	49	Associate editor or manager
Paralleling equality	3	55	Manager, editor
Observing inequality	6	28	Ph.D. candidate, teacher
Disavowing inequality	12	41	Director, full professor

four pathways, indicating the number of interview participants, mean age, and professional status.

MAINSTREAMING EQUALITY

Seven (of the twenty-eight) cartographers abide by the perspective that equality should be mainstreamed. In other words, there is a desire to promote equality from within. In this regard, Ballantyne and Hanham's (1993) work on women surveyors is relevant. These cartographers are engaged in a process that involves reorganizing themselves and their colleagues from within by pointing to unfair and sexist practices, making men aware of women's rights, fighting the "old-boy's network," and re-educating the students and men about women's issues. The process also involves making the woman's voice heard and acknowledging and encouraging men to examine attitudes that are detrimental to the advancement of women. Women use both informal and formal means to mainstream equality.

These women, according to one well-placed woman geodesist, were trained in the 1980s. They are openly assertive and openly insistent when dealing with gender issues. It is not a side issue for them. The Montreal Massacre at the École Polytechnique in December 1989 (where a lone gunman killed fourteen women students) had a profound effect (in Canada, at least), creating a shift in attitudes among the male students and establishing various chairs for Women in Engineering (Field notes, 7 July 2000: 1).

The women in my study who have taken it upon themselves to mainstream equality are generally older than the other interview participants (the average age is forty-nine) and have achieved a measure of occupational success: they include a senior geographic analyst, a professor of cartography, a section head, an associate professor at an academy, a research coordinator, and a university professor.

FEMALE PATHWAYS THROUGH THE PRESENT-DAY MAP WORLD

But what do they see as viable ways to tread this pathway? First, they have taken on the responsibility of encouraging other women to enter the field, providing opportunities for them, building support networks, monitoring the progress of new recruits, encouraging women to study the technical, "hard" things, and making sure that the evaluation process is fair:

> My specific perspective is to try to get women Ph.D. candidates in because I run the Ph.D. program here ... and the recruitment and the monitoring and the whole thing plus the finances—that's one of the things I do. (Leslie: 1)[4]

> We had a program for developing females to more leading positions ... and then I was a mentor for two. All of us selected four women and I was a mentor for two of them ... and I also took part in arranging some conferences for female cartographers.... They also made some publications and ... within our authority we tried to make some guidelines for helping female[s] to ... advance.... But, of course, when I'm travelling in international [settings], I see still a great majority of men. I try to speak to all the females and to encourage, and give feedback. (Inga: 7)

> I have had to explode at the old boys' network, I have had to be very cross because certainly decisions were made regarding some of my projects, and I wasn't involved in them.... It is that they are doing things the way that they have always done them, [with] very little consultation. (Cécile: 8)

> *Do you see any special role for yourself in nurturing women in the field?* Yes. I think without doing it in an aggressive, or a feminist kind of way. I think we have to build those support networks, to help deal with the pressures, with the difficulties. I think when you see other people deal with the same issues you are dealing with, you can probably find solutions.... (Cécile: 11)

> [*In terms of evaluation process*:] What is best? What is quality? What is excellence? ... How do you judge someone? Is that grades only? Are there qualifications that you really think you must, should take into account? Really, none of them [men] think about that in a conscious manner and I noticed that being in this role, female M.Sc. students will come to me and ask me about opportunities and of course I don't [answer those questions], but I'm sort of the process manager. I make sure the process is fair and proper. (Leslie: 2)

> I suggest them to study subjects that are considered male oriented: technology, mathematics, and things like that. And why I say that is because I am

afraid that female students go to more "soft" studies. Why [do] I try to push my young girl students to subjects which they are not maybe originally interested in? Okay, I know if they learn those things, they have better possibilities in the working life. (Liisa: 11)

For Suzan Ilcan (personal communication with the author, 26 Dec. 2002), this interview participant's comment on "soft" studies is of interest. It is a point that reminded her of the wide range of scientific studies that are labelled or regarded as hard or soft science and of the many professional women's groups that have fought against such categorizations—not only because they are imbued with stereotypical notions of masculinity and femininity, but also because they support divisions of labour and professional training on the basis of gender.

The women of the mainstreaming equality perspective also make sure that their voices are heard, that they "de-emotionalize" the situation, that they have bright women around them as examples, and that they occasionally need to file a grievance to amend a situation:

> I know they cannot stand me and then they [men] say that I don't respect them at all and the problem is I am a person that says my opinion every time and I don't care about what happens. So I have this feeling that ... if you don't determine every moment, every minute what your opinion is about something, you are lost. (Helena: 8)

> And [my female boss] said that the only thing that [she] has learned in the past is that you have to de-emotionalize the situation, and you can never base it on the male–female ratio ... thing. (Cécile: 8–9)

> I don't believe that the glass ceiling is broken. I have countered sexual, professional harassment. I was on the executive of [she names a national organization]. I had to take vacation [time] to go to the professional meeting. [Another colleague] took her complaint to grievance, to no avail. There was another situation after her three-year term executive position was finished. She was asked to be on the editorial board, but her boss did not support this and she went to another division and got support. (Rachel: 1)

The challenges were formidable for these women; many of these challenges were, in the eyes of the men, invisible. Women face(d) "educational snobbiness," invisible "bathroom meetings," "mind games," and sitting through "dull conferences." Cécile, a forty-seven-year-old section head in a government agency, was particularly vocal about these challenges, but several others voiced opinions about these matters:

[T]here was a snobbiness about [educational] degrees and then the old boys' network was at work there. (Cécile: 9)

Unconscious corridor meetings, bathroom meetings and I am not involved in these "meetings," and it has happened one too many times. I have at least mentioned that in the last two appraisal periods with my boss, and I said, "You don't even know that you are doing it." And he said, "We will try to do better." And of course they never do better. (Cécile: 8)

I would say no mind games, no politics: frankness, I want frank people, I want honest people. No games, these are not games, put your cards away, and tell me what you really want to tell me. And if you don't want me here because I am a woman, tell me that you don't want me here because I am a woman. No games. I find that men play a lot of games. So, the communication is one thing. (Cécile: 13)

I believe from the women I know [that women] do work more, but perhaps it is a compulsion that we have, you know ... because we are intruders in a sense, eh? And perhaps consciously we want to make sure that nothing goes wrong. So we put more work into things, while men have a tradition of 2,000 years or 5,000 years of doing this kind of thing and they are much more relaxed [and] then they take it easier, while we are more ... conscious of things [like] a newcomer.... So you put [in] more effort while the men are ... well, they are relaxed because they have done it for so many centuries.... (Myra: 9–10)

[M]ost [parts] of that main conference were not so interesting, but it's more about positions and of course sometimes you have to take part in the dull part of it if you want to help female positions. (Inga: 14)

The fact of slow organizational change in map worlds is also something that women cartographers who want to mainstream equality must face: for example, in the International Federation of Surveyors (FIG), with four-year terms for executive membership, for the president-elect, and for the president, consecutively, it would take at least twelve years before an interested member could effect any policy changes (Inga: 14).

CHAPTER 12

Women with the mainstreaming equality perspective face, in fact, a double struggle. Most of them faced sexual discrimination when they entered the field; now, as they age, they are facing the prospect of ageism—prejudice against them based on their age. Older women experience more ageism more than older men.

PARALLELING EQUALITY
Parallelism represents a perspective promoted by another category of women cartographers. The perspective involves women who, on the average, are the oldest of all interview participants. In their late fifties, the women own a cartographic firm or hold managerial positions in such a firm. Alternatively, if the perspective involves a younger woman (there is one at the age of thirty-eight who is an editor of an atlas project), she was the mentee of someone who advocated parallelism. More than a belief, parallelism is a methodology toward equality.

According to a prominent woman geodesist, these are women in the mapping world of the 1960s (Field notes, 7 July 2000). They work on gender "on the side," trying to keep it private, keeping it from the view of men. They nurture other women separately from the men. If they are assertive, they are called aggressive. Instead of insisting that women map-makers join the banquet (e.g., at ICA meetings), they pull together their own resources, usually quite meagre, and have their own dinner at a cheaper place to allow more women map-makers to attend. Although there are, no doubt, also men who cannot afford such an expensive ticket, it is unlikely that they would get together to form a parallel dining group.

An exchange between two women cartographers at an international conference illustrates neatly the difference between mainstreaming equality and parallelism:

> Early on in the conference, XY and DL had a big discussion about whether the women of the commission should have their own dinner in the shadow of the gala dinner (because they could not pay for the $125 ticket), or whether the women should do everything possible to get everyone to the gala. DL has already secured $500 from [name of agency], but needed another $1000. XY offered to pay for a ticket, and they were hoping that a few others might be willing to sacrifice their ticket. It was also noted that the commission might offer $25 for the surrendered tickets of those who decided against going anyway, for example on account of illness (but had bought the ticket). In the end, nothing came off, except the enjoyable dinner at Les Fougères. (Field notes, 20 August 1999: 2)

It is a given that parallelism involves the establishment of support groups (including the ICA Commission on Gender and Cartography) as a means to nurture especially the younger women. This process is equivalent to the one described by Ballantyne and Hanham (1993) in their study of women surveyors in New Zealand "to encourage students to mingle with staff and with other students" and to have existing women surveyors "act as role models for prospective and actual students by giving career advice at schools" (Ballantyne and Hanham, 1993). Three women expressed the value of such a group in the following terms:

> When meeting in the commission and seeing members of the group, it is very inspiring to go to work at home. But when I am at home I feel no support for the work from the national association and the male members. (Kristina: 3)

> There were so many different aspects and so many different stories about how people became professors or professional cartographers. So it was wonderful. It encouraged all people to come together and if somebody seems to be a little bit lonely or maybe not so good in this thing, she always went to her and ... in most cases people only need to be encouraged once.... When they know one person, and then they know another, and then it goes. (Liisa: 9)

> Because even in a society such as ours, women do not have as much confidence, generally speaking, and it is nice to help them out. It is not that I don't like helping out men, young men as well—and I have—but I have a special sense of bringing up young women and encouraging them. (Amanda: 9)

In the practical world, parallelism places responsibility on the shoulders of the more senior women cartographers to undertake concrete steps to help the younger women professionals. For example, at a recent ICA conference, one older cartographer billeted the younger women who had travelled long distances (Field notes, 15 August 1999: 4).

Parallelism can also involve specific other actions that recognize women in particular, such as with financial incentives, awards, and scholarships (Ballantyne and Hanham, 1993). It also means drawing attention to yourself: "Stand up for your opinions. In the long run it is important that women tell men that they react on attitudes and special actions ... instead of trying to manipulate the men the way you want decisions to be taken. Be aware that you get the credits for your work.... Try to enlighten gender aspects in a positive way without lying. Do not underestimate yourself and get used to standing up and arguing for your work" (Kristina: 6).

In speaking of the sort of work undertaken by a senior cartographer, one younger woman professional expresses the value of parallel activities. My field notes contain the following entry:

> [Name of younger cartographer, XY] spoke of [name of senior cartographer, DR] in the highest terms. [XY] loves the way [DR] promotes discussion within the group, "with emphasis on everyone listening and taking turns speaking, as opposed to the men who like to hear their own voice." [DR] has "created a great influence among women cartographers." Initially, [XY] thought that there was no need to have a separate commission, because in the Scandinavian countries, women do have equal rights. But, as she began to think about this, she realized that they are still far away from equality. [She gave a few examples] She also mentioned that "equality is not about women becoming like men, but it is a question of recognizing women's unique talents." (Field notes, 15 Aug. 1999: 4)

Women who adhere to parallelism also believe that their work is of value to younger women cartographers who either believe that fences are no longer standing, or to remind "the young that maybe there are fences that they're not aware of" (Inga: 9).

Despite evidence that women cartographers are no less active in computer or digital mapping, parallelism, in the view of some interview participants, upholds the view that some areas in cartography are traditionally less open to women, brought on by the technical aspect of digital mapping (Kristina: 3), the "old playing rules" of men that are "impossible to break" (Kristina: 4), and the persistence—and re-emergence—of the military influence in map-making (Amanda: 7–9).

As a consequence of parallelism, a number of women prefer doing freelance work and have installed themselves as the chief executive officer of their own private map-making organization. It is a frustrating approach, however. Such work leaves them no time to address gender issues and leaves them out of the loop of working with other women (Kristina: 3). On the whole, there are advantages and disadvantages to freelance work.

In Great Britain, the situation of women cartographers is such that their strategies take on the characteristic of creating a parallel structure—sort of. Great Britain does not have a strong social policy for promoting the advancement of women in governmental science (as in Canada or the United States, for example), and private corporations do not seem inclined to attract women to their ranks. As a consequence, women cartographers become their own entrepreneurs, in effect heading their companies as CEOs. The *Freelance Directory* of the British Cartographic Society (2000) lists twenty-nine freelance cartographers in the country, while the Oxford Cartographic Conference (August, 2000) lists an additional eight freelancers. Among these thirty-seven freelancers, 43 percent are women (N=16)—the highest proportion of women in any group of cartographers.

The situation in Great Britain finds an echo in other countries. In Hungary, there is a thriving cartographic freelance community, where individual women are empowered to take on making maps that define their priorities, such as not producing maps for the military (Kornélia: 1). Even in Sweden, a significant promoter of equality, one still finds women cartographers who prefer freelancing rather than spending time in governmental or corporate service.

In some parts of Scandinavia, however, like Norway, there are fewer freelance opportunities, because most work is done in municipal settings or in big companies (Inga: 9). Unlike in Britain, there are no great differences in equity-hiring policies between state-run agencies and private companies in Sweden, but there is also hardly any exchange between the two. The State Topographic Services moved to Gävle, but the private companies have stayed in Stockholm (Field notes, October 2000: 1).

The downside of freelancing is that there are no automatic benefits, such as health insurance and public-service pensions. Mary Feindt, one of the most prominent surveyors in the United States, said, "Since I was my own employer, there were no benefits, except those I created" (Olin, 1988: 4). There are other drawbacks to doing freelance work. If work is contracted out—and women do manage to receive these contracts—the work is fairly routine and sometimes involves routine drawing work. Thus women are used more cheaply to do the routine drawing work (Inga: 11).

There are other disadvantages. A woman cartographer in the United States found working outside of the government to have had a deleterious effect. She says, "I gave up on my cartographic career over a year ago due to discrimination. I love maps, but found the politics of government contractors to be scandalous. There is a lot of ugly stuff going on out there outside the safety of government positions" (Straight, 1991b: 3).

The increased downsizing and privatization of a wide range of government services in many Western countries since around the early 1990s has

CHAPTER 12

negatively impinged on professional groups. Women (in comparison to men) and minorities have not fared well in this era of privatization. The politics of government contractors exercises a huge influence, not only in lower pay scales for women and minorities, but also in concentrating more work in the hands of men (Suzan Ilcan, personal communication, 26 Dec. 2002).

Kristina, however, voiced one advantage of going freelance:

> As a freelance I do not notice it as much as I did when I was employed. In fact, this was one of the strong reasons to start my work as a freelance. As a woman (and with no thesis) your ideas about small and big things are not listened to. All discussions about [traditional] cartography were neglected in favour of discussions of electronic maps, atlases, and data acquisition. It was impossible to break through the men's opinion, [namely] that "this is how it works and we have to follow the rules," which was the same rules as men make up together ... there were a lot of old "playing rules" that were impossible to break. The biggest problem with this is that it is almost impossible to pinpoint to the rules so that men also see them. (Kristina: 4)

Kristina's opinion affirms the idea that people on the margins are more likely to see the rules, while people at the centre take those rules for granted, but are less likely to see how those rules have an impact on others.

As stated above, the women following this pathway are older women on the other pathways. One has the impression that these women have tried to reach the top of cartographic organizations, failed, and have somehow "fallen from the top." The sense of rejection, of not having been able to break into the top ranks, can be quite a sobering experience. When I mentioned to an interview participant that it was ironic that the social science division in her academy had no women staff, she replied, "You know, they had a lady who was there ... and she could not stand it and she moved to a different position. She became a project officer" (Myra: 8).

And even speaking of graduate students: "They don't have any [women graduate students] like me, because they don't give them enough space. So people leave sooner or later. So now it has been sooner.... *Are you talking about mental space?* Yes ... I'm the second woman, the first one left a few years ago.... I think she just couldn't take it or ... like this is a kind of isolation to be here" (Kaisu: 4).

Working in parallel structures of cartographic organizations (such as the ICA and FIG) normally involves the routine work, usually done by women who head the commissions and committees. Membership on executive boards is only precariously left open for women to be elected to by the general membership.

OBSERVING INEQUALITY

The first two pathways I described above represent activism in promoting equality. Not all women are in a position to participate as actively, however. *Observing inequality* is a term I have set aside for these women. They are the youngest of all the categories and, with few exceptions, are in the early stages of their careers. Zia is thirty-two years old and is a Ph.D. student. Ruth is a cartography teacher at the age of forty-one, while Ramona is fifty-five years old and is a professor of geography (Ramona came into her work much later in life than is the case for others in her group). Erika is twenty-nine, and Kaisu is thirty; they, like Zia, are Ph.D. candidates. Finally, Ettie is thirty-seven years old and works in a private tourist map firm.

Two elements characterize the pathway of those who are observing inequality (rather than acting for change). They interpret their experience as one of learning and falling outside the structure. In terms of learning, they are surprised that men do not see that women are disadvantaged: "If you make comments about women, about gender in the profession, they are quite surprised. And often I think they are surprised that they don't really see the potential for disadvantage for women, and that surprises me more than anything" (Ruth: 9).

The women have learned other things from their experiences: namely, that the onus to be heard is on them (that is, the women), and that they must persist, and that they can or must lead by their abilities:

> You have to make an extra effort to make sure that you are understood. Because while you look at the podium at the opening ceremonies and you see that it is all male ... and, as women ... there is an added, I won't say pressure, but there is an onus on the person, if they want their voice heard, they have to be able to raise [it] above something. (Ruth: 8)

> *What do you do with that discouragement? What do you tell yourself or others?*
> I think persistence is certainly something that you make a conscious effort to [do to] overcome some of the resistance. Just by setting examples of your abilities, think of a better way to lead them by example. (Ruth: 8)

> It is a challenge to work in an area dominated by men! But I would say that because I come from a female dominated background—an all-girls' school,

CHAPTER 12

a teachers' college (predominantly female), [and I] taught for five years in an all-girls' school with a junior-school staff of all women ... and then my first job, where I was one of two women in a rural office of thirty men, took some adjustment!! It was easy to be the helpless female, but one has to show that you can deal with all situations! (Ettie: 6–7)

My advice to other women coming into cartography is that they should not be put off by the male-dominated workplace, [but] be prepared to question and challenge conventions where appropriate, to ask questions and maintain a sense of humour—becoming indignant over issues that are not important does not help. (Ettie: 7)

This latter advice seems to come from someone who advocates mainstreaming equality. Her comments underscore that the boundaries between the pathways are sometimes not clearly marked at all. Rather, some of the women cartographers do go back and forth among the categories.

Nevertheless, the feeling that one is standing outside the structure is quite common to the perspective that one is an observer—in fact, it underscores that status of being an observer:

But I think it [male dominance] is mostly related to the power, scientific, to the power of knowledge ... but for me ... I don't have the possibility to [reach that power], because I'm a woman. (Zia: 7)

Have you found any personal resistance to your being a woman cartographer? Personal resistance? I would say [yes]. As an example, at our school just recently [it] formed a committee that was fairly important in terms of strategy for the school and there were no women on this team and it just kind of made me wonder, you know.... That was very strange. (Ruth: 8)

Even for these women, there is no particular desire to form a sisterhood of sorts with other women cartographers, and if there is a desire, there is no time or temperament for it:

I have some colleagues in [mentions an Iberian country]. I speak a little about that [inequality] and that is, I think, a general problem.... Someone can handle it, but when you get into the subjects and if you want to [master the field], there's not even much time do any other things. (Zia: 5)

I have no problem working alone ... but I think that when I leave this place ... [I hope] I can appreciate things [that] the community [of cartographers] brings because when people work together there are always ... tensions. (Kaisu: 4)

DISAVOWING INEQUALITY

It should come as no surprise that the senior-level cartographers disavow any current inequalities between men and women. If there are any exceptions to that disavowal, they refer to traces of the past or to an unusual instance in the present. In many respects, their point of view echoes the findings of Ballantyne and Hanham (1993) and of Greed (1991). In their study of New Zealand women surveyors, Ballantyne and Hanham found that "the type of women attracted to, and accepted into this subculture were found to be those who subscribed to the values of the 'successful business woman or bourgeois feminist.'" This theme cropped up time after time. Greed is more to the point when she argues,

> It should not be assumed that women entering the profession will necessarily hold different views from the men; nor should any automatic assumption be made that they are likely to be feminist ... or that they will become radicalised by their experiences of being in a minority in [a] subcultural group.... Indeed non-surveying women have commented to me that some women surveyors appear to them incredibly straight and uncritical in their world view. Many of those [women] entering surveying today are safe women who are not going to question things as they do not wish to jeopardise the interests of the tribe to which they belong, or their wider class interests. (Greed, 1991: 10, 185)

The age of the majority of these cartographers ranges from upper thirties to mid-fifties. One woman is sixty-five. They occupy the top ranks in their cartographic fields. Saskia is director of a national GIS centre; Nedelina, Loukie, Minhui, and Joyce are professors of cartography; while Marta is a professor of geomatics. Helle is chief of production in a private company, Natalie is head of sales in a governmental agency, and Marita is a high-end editor in a cartographic firm.

The pattern of disavowal differs markedly from the approaches in the previous three perspectives noted above. This pattern rests primarily on the ideology of individualism, the uncritical acceptance of the structure of

CHAPTER 12

map worlds, and an interpretation of personal experience that belies any inequality. As one woman shared, these are the 1990s cartographers who have taken the full participation of women in map-making for granted. They expect to be heads and presidents of associations. Gender is neither a side issue nor something to focus on, but is distilled to personal relationships. There is mutual support among the women, but the structure as a whole is not flawed; what is flawed are the men (and sometimes women) who occupy decision-making positions in that structure.

A strong belief in individualism runs through the course of the interview transcripts of the women who disavow inequality. Marita says, "It depends on the person" (Marita: 11) and Els states, "I think it's in yourself as a person ... that you are career minded or not" (Els: 2). Natalie speaks not about the social-organizational structure where she teaches, but about treating everyone "individually" and finding out "what kind of personality" her male colleagues had in order for things to work out for her as a supervisor (Natalie: 13). Lynn suggests that women should ignore the "stereotypes that women do soft things in terms of design, fabric, art, and all that," and that cartography is "not a difficult field to get into" as a woman (Lynn: 11–12). Confirming her own beliefs in individualism, she says, "I really don't believe that the barriers are anywhere as high as they could be" (Lynn: 12). When asked, "Have you found any resistance to your being a woman cartographer in your unit or, you know, your profession?" Marita offered the following assessment: "Well, I don't know if it is about being a woman, or being me [laughs], because I can be quite strong headed into something. And if I am, I don't care about how other people feel then.... [When] I think we should do that, and if somebody then feels bad for it, it is not my problem, you know. But I just strongly go to it ... and I don't care about things about other people then (Marita: 13).

A similar perspective emerges from another interview: "I am a woman but I am not afraid of men and of their questions and works. Maybe I am a different woman. Most of my colleagues tell me that I think as a man" (Nedelina: 9).

Nedelina believes that when her colleagues tell her that she is like a man, it is a compliment.

Along with individualism comes the uncritical acceptance of the structure of map worlds, as reflected in some of the comments the women shared with the researcher: "I don't think that anything is difficult for women, because in [names a Balkan country], women and men should work together and it is not possible to stay at home and not to work. I know some men think as a woman. I don't want to put this distance between a man and woman in the working area" (Nedelina: 2, 9).

In reflecting on some of the young women in her charge in a cartographic firm, Helle offered: "I have three, four young girls who began the last two, three years and to compare with the men.... I can't see a difference between them ... because the women are satisfied with the situation and they don't want to climb up, and the girls—they want to be responsible for the whole project" (Helle: 4–5).

It is not because her "girls" are less ambitious. On the contrary: "they have the family ... they are looking at the whole situation for their life and this is enough, but the men are more [a pause] ... *ambitious, maybe?* ... yes, ambitious, yeah, but we ['girls,' women] are ambitious at something [else]" (Helle: 6).

In that light, there is nothing that men cartographers need to be informed about in relationship to any perceived inequality. Says Natalie: "Well, I would say in the past that we are just as capable as you [i.e., men] are. Today, I would say, well, we are buddies. We can do the work, the both of us. We can work together. And we can do something good together" (Natalie: 15).

Naturally, the numbers of women and men are proportionate, according to one professor in GIS: "*Is GIS a man's world?* No ... it's both a man's and a woman's [world]. I think really I have seen many female staff in ... teaching, but also in industry. In software development there's so many ... females, women who do very well in their jobs" (Minhui: 7).

The following extract from an interview underscores an uncritical approach to (in)equality: "Because in our country we never differentiate between men and women—something like that. Consciously, we do the same. So, whatever we get, so we offer to each other. We are very friendly. We help each other very much" (Sawa: 3).

If there are unequal numbers in a given field in cartography, it is because the field has always been "a more technical subject," such as in the Netherlands (Loukie: 3). In fact, states Loukie, one cannot really entertain the idea that women are being treated unequally. It's the fact that "there are so few women in cartography ... but it's not a matter of inequality" (Loukie: 5). A similar sentiment was voiced by Lynn, who said there are too few academic cartographers at her university to draw attention to any inequality (Lynn: 11–12). Marta, however, does admit that while, in her Eastern European country, the number of women and men are fairly equal in number, it is the men who occupy the high positions (Marta: 3).

CHAPTER 12

A third element of the disavowal of inequality arises from the fact that the personal experiences of the women cartographers affirm that equality is a reality for them. Here are some of the insights the women offer:

> So it seems to me that I was not aware that in the scientific arena there were so few women directors.... And someone was asking me, "How do you feel coming in?" And I said, "Fine." ... It is the same for me whether you are female or male. For me it is not an issue. (Saskia: 6)

> It never occurred to me that there was any problem in that the faculty were all male. Yeah, I can't say that I wear feminism on my sleeve. (Joyce: 12, 15)

> I was very lucky in our department. As much as was possible, gender was ignored. I mean, I have not had to fight one case of sexual harassment, nothing. It was actually gender blind, which I prefer. And within our organization, I really have not caught instances of it. We have had women on the executive, and again, as blind as you can be to it. (Lynn: 10)

Several suggested that they were discriminated against: not by men but by women: "I was more able to go to men to talk about it than to women. So I was kind of discriminated against by [my] own gender, too, I would say" (Els: 6).

For still others, discrimination is something of the isolated past, something that comes up only in a particular circumstance: for example, when women present themselves in a group with men, it is the men who are usually asked questions or acknowledged first (Marita: 11).

The implications of disavowing inequality involve a lack of desire to give advice to other, possibly young, women:

> Advice that would help them? Well, I think that's not really advice for only women. [When] you want to [give advice] ... I think that's very important, but that's not gender biased. It's just for cartographers. I don't know [advice] for females. (Els: 7)

> Well, I would say [to women], "I am equal." I would say I would open for somebody who has a talent. I will open the door, whether it is a women or a man. (Natalie: 6)

FEMALE PATHWAYS THROUGH THE PRESENT-DAY MAP WORLD

> *And what about nurturing young women cartographers?* Well, I must say again that depends so much on the personality. So, yes, if the person is good or worth it. But, no, if I think the person is not worth it. (Marita: 14)

The existence of a "sisterhood" (however defined) of cartographers would not serve the interests of women or of the field. In this connection, the ICA Commission on Gender and Cartography or of FIG does not seem to serve a particular purpose at all:

> It did not make sense to me [to join the ICA Commission on Gender], and it still doesn't make sense to me. I never know what to do there, because in my personal life, gender is not an issue. Then it is very difficult for me to sit down in a group and think, "Now I have to think gender...." I understand and I know the problems, and I am interested in the problem, and that is why I am there. But I never know what to say or which way to go since I don't see gender as an issue, nor age. We were laughing one day because I can hire someone who is twenty-three, or someone that is seventy-three. So it is okay, as long as they are professionally well fitted. (Saskia: 7–8)

As a consequence, the notion of gender becomes a less central concern. In fact, it disappears altogether, because some believe that women should be concerned with under-represented groups on maps—they disavow any gender differences in the social organization of the map world—and are even further removed from the concerns of "gender," rejecting the concept of gender altogether.

The idea that patriarchy is an "all-encompassing phenomenon" (McCormack, 1991) embracing the whole social structure does not seem to enter into their reflection. Their concept is far removed from the idea, as Thelma McCormack claims, that "Women engineers are about as likely to overthrow patriarchy as male engineers were to overthrow capitalism" (McCormack, 1991).

One may well wonder, gauging the implications of some of the above findings, about the field of cartography, where young women cartographers believe that fences no longer exist and where a large number of older women cartographers disavow the existence of inequality. Such a perspective underscores the *an sich* sense of women in cartography when one adds in the fact that so few women, if any, are interested in the historical role that women have played in cartography itself. The pronounced lack of sisterhood among cartographers simply underscores this fact.

CHAPTER 12

What seems salient, however, across all categories or perspectives, is the problem of older men as a category to be reckoned with by men. Of the seven interview participants, only one, who happens to be a young woman, expressed a positive idea about the presence of older men (she is a Ph.D. candidate):

> I've been very, very well treated by the oldest men. They are wonderful. The most [help] sometimes [I got] were the ones where I studied [with] the names you would recognize, for example, like [mentions name of a prominent cartographer] and all of them treat me very, very well. They give me opportunities to speak. (Zia: 8)

The others, however, expressed largely a negative opinion about the older men cartographers:

> One of the irritating attitudes is that you are always met as a woman who is not supposed to make any career in the same field as [older] men are. (Kristina: 4)

> I guess the main problem is that the staff on the third floor is almost the same people who were here twenty years ago. (Kaisu: 3)

> It's very difficult to convince senior academic staff ... that they are willing to search for women. It's very difficult to convince them that it's not just some sort of political issue [that] happens to be just on the agenda and it will disappear next year. (Leslie: 1)

What seems to help the younger generation of men to more fully acknowledge the participation of women is that they have wives working outside the home: "I think it's safe to say that for all the men it's much more difficult to accept [but] younger men who have also younger women and look at their women every day and their women are working, and they are more supportive [while] all the wives of [older] men usually don't have careers. So ..." (Myra: 9).

Before we turn our attention to how gender relations work themselves out in the practical, everyday world of cartographic conferences (see Chapter 13), I shall briefly touch upon the relativity of the concept of gender in developing and developed countries.

Table 12.2. Percentage of active women in the surveying and mapping world at all educational levels: the world by regions, 1980 and 1990

	1980	1990	Increase in %
Africa	n/a	4.4%	—
North America	14.4%	16.5	+2.1
Latin America	17.1	15.3	–1.7
Asia	5.5	7.5	+2.0
Australasia	10.0	12.4	+2.4
Europe	15.9	21.5	+5.6
Totals	13.1	14.8	+1.7

Source: Brandenberger (1997: 154).

RELATIVITY OF GENDER IN DEVELOPING AND DEVELOPED COUNTRIES

With only 15 percent of the world's women cartographers residing in developing countries,[5] and the general lack of information about the particular conditions of employment, it is difficult to give precise meaning to how the concept of gender currently plays itself out in various regions around the world. It does seem clear, though, that proportionally speaking, there were more women in cartography in developing countries than in developed countries, as compared to the overall participation of women in cartography (see Table 12.2). It is important to point out that given the dated data, the figures may not reflect the contemporary percentages in countries, but these are the latest data available.

As a United Nations expert reported, women in the surveying and mapping worlds were very much under-represented in North America (see, e.g., Brandenberger, 1997). But did the concept of gender vary across the regions? As we saw in an earlier chapter, one interview participant from Latin America (which proportionally has the highest number of women in surveying and mapping) found that the concept of gender is

CHAPTER 12

Table 12.3. Gender composition of cartographers at ICA 2011 (Selected countries, by region)

	≥ 50%	49–31%	≤ 30%	0%
Eastern Europe	3	3	2	1
South America	3	1	0	0
Southern Europe	2	1	0	0
Africa	2	1	0	4
Western Europe	1	3	1	0
Far East	1	1	2	1
Australia	0	1	1	0
Middle East	0	1	0	2
North America	0	0	1	0
Indian Subcontinent	0	0	1	0
Central America and Caribbean	0	0	0	1

Source: Field notes, Paris, 2011.

culturally, very different here [in Canada]. The perception of gender, from what I have learned living in Canada, is very different from country to country. And even from region to region.... When I went to the Canadian Embassy, the lady in the Canadian Embassy was very tough with me, because [she asked] how come I, as a woman, wanted to come and do a Ph.D. in Canada and bring my husband? And I was going to have the primary role. And she was very, very tough. But I was able to fulfill all of the requirements and at the end I came to Canada. So that is why I tell you that I have felt worse sometimes in Canada, than in [mentions name of Latin American country] [laughs]. (Saskia: 6, 10)

However, the findings seem counter-intuitive to some readers when one considers the contemporary gender composition by specific regions and

Table 12.4. Occupational niches of women cartographers, surveyors, and GIS staff in developed and developing countries, 1995 (in percentages)

	Developed Countries	Developing Countries
Government service	66.1%	28.1%
Private employment	13.6	49.4
University	20.3	22.5
Totals	100.0	100.00
(Numbers)	(59)	(334)

Source: ICA (1995).

countries (see Table 12.3). Eastern European and South American countries have the highest proportion of women attending the ICA conference, followed by southern Europe and Africa. Western Europe and the Far East had proportionally fewer women, while other regions (Australia, the Middle East, North America, the Indian Subcontinent, and Central America and the Caribbean) had no representation of women.

These findings reflect those by Maria Charles (2011: 22), who found that in science, technology, engineering, and mathematics, developing countries were far more likely to demonstrate gender equality than developed nations were. As she says, "Ironically, the freedom of choice that's so celebrated in affluent Western democracies seems to help construct and give agency to stereotypically gendered 'selves'" (Charles, 2011: 28).

In the pursuit of building their "strengths" and realizing their "true selves," (young) women in developed countries inadvertently follow gendered paths in careers and occupations (Charles, 2011: 28).

Someone who was involved in graduate student enrollment at an academy spoke of the many obstacles that women cartographers in developing countries can face when trying to secure an advanced degree in cartography or geodesy (Leslie: 3). At the same time, in developed countries, including Scandinavia, there are still many problems associated with being a woman cartographer, as a university lecturer told me (Liisa: 12).

The preponderance of women cartographers in the developing world not employed by state organizations does not contribute to their overall welfare, as is the case for women cartographers in private employment or who work freelance in the developed world (Table 12.4).

SUMMARY

The data gleaned from interviews suggest a number of pathways through inequality. Mainstreaming implies a pathway organized to change power relationships in the map world. Parallelism, by contrast, is not about changing how cartography distributes power through gender relations. Rather, it facilitates the participation of women who need the nurturing and encouragement. The pathway that constitutes principally observing inequality (rather than engaging in changing it) acknowledges that there is inequality, but the wayfarers do not actively address the inequality; they merely acknowledge its existence and observe inequality without intervening. The final pathway disavows inequality and disclaims power imbalances and inequalities. Each of the four pathways is predicated on the individual experiences, the particular social location in the map world, the applicable structural exigencies of one's education, and the organizational life in which cartographers find themselves.

The one area that may not follow these pathways is the "developing" world, where the indigenous settling in of cartography is a far more recent phenomenon than in the "developed" world. No doubt the participation of women from the former in international conferences and training in international cartographic institutes will result in the creation of new kinds of pathways in the map world.

The following chapter concludes the study. It brings the analysis back to the involvement of women in cartography in terms of the role of technology as both arbiter and handmaiden. Just as importantly, the last chapter brings the significance of our study to the fore, tying it into the larger discourse about women, feminism, and cartography.

CHAPTER 13

Gender Shifts

Our story of women in historical and contemporary cartography does not easily lend itself to creating a grand narrative, whether feminist or otherwise. Aside from the inconsistent and incomplete historical record of women's participation in cartography, cartography spans too many cultures and too many varied periods to warrant an easy explanation. Nevertheless, the reader is entitled to ask two fundamental questions: (1) What can one learn from tracing the engagement of women with cartography (primarily since the sixteenth century)? and (2) What is the relevance of our findings when we relate those findings to the feminist discourse about women and cartography?

Many scholars acknowledge that the historical record has been slow in recognizing the involvement of women in map-making. Moreover, those responsible for developing that historical record have carved out particular niches and decided which of those niches belong to map-making. If one would explore map-making in previous eras, would one include printers? Publishers? The business widows of the seventeenth century who ran mapping firms? Colourists? Engravers? The niches get more complex as one moves forward through time: globe makers, geography teachers, academics, map librarians, map archivists, and so on. How would one go about arriving at such a boundary? Those steeped in historical cartography bring in assumptions about the field that make integral sense to them, but boundaries can be arbitrary. Even when we speak of the "margins" of these niches, we unwittingly underscore, according to Adams and Tancred (2000: 122), the male discourse as to what is regarded as important, or not. Hence, the idea of map worlds is a good one. It invites us to consider wider aspects of cartography than are normally included.

THE ENGAGEMENT OF WOMEN IN THE HISTORY OF CARTOGRAPHY

Map Worlds shows an unmistakable engagement of women in cartography. It took the efforts of Alice Hudson (1989, 1999a, 1999b, 2000) and of Hudson and Mary Ritzlin (2000a, 2000b), when they published their lists of pre-twentieth-century women in cartography, to demonstrate that women's engagements in cartography were extensive and influential. Soon thereafter, other scholars (e.g., Huffman, 1997) became apprised of the works by Hudson and Ritzlin and began contemplating the implications of a feminist cartography and geography. The root argument always began with a reflection on the historical contributions of women to cartography. Hence, they justifiably turned to the lists generated by Hudson and Ritzlin, the New York City map librarian and owner of a Chicago-area map antiquarian shop, respectively. Those lists set the whole study of women in cartography, including *Map Worlds*, into motion.

The search for early women cartographers was on. We became apprised of, for example, Hoogvliet's bold claims (1996) that women produced the thirteenth-century Ebstorf map; Barbara Whitby (2005) provided details of Shanawdithit (ca. 1800–1829), a Beothuk, producing some maps in the nineteenth century. Mina Hubbard (1870–1956) mapped Labrador (Chapter 5). Numerous other studies piqued interest in the subject matter. While *Map Worlds* followed in their footsteps, highlighting specific women in early cartography, it has placed their participation in a social and historical context. Drawing on a wide swath of Dutch, English, French, German, and Norwegian sources, the book points to the deep relevance of women in the map trades: for example, when it came to creating, maintaining, and developing the famed Houses of Cartography in the Low Lands in the golden age of cartography (Chapter 3), when masses of schoolgirls at the turn of the eighteenth and nineteenth centuries created embroidered maps, or when one finds a wealth of women engravers, such as Eliza Coles, who was America's first map engraver. Moreover, in both Europe and the United States, women ran numerous map businesses (Chapter 4). Of further note is Kirstine

Colban (a.k.a. Stine Aas), who was among the first to have produced perspective maps. *Map Worlds,* it is hoped, does more than reclaim the past of women's engagement in cartography: it constructs a link between women and cartography, without which cartography might have looked quite different. It is a historiography that, in effect, entails both women and men, and a narrative that is shaped by the wider workings of society.

The *first wave* corresponds to the epoch in Western cartography when map ateliers were the currency, during the sixteenth and seventeenth centuries. Women were engaged in specialized tasks in various corners of map worlds, but generally close to home. They were colourists and engravers, among other occupations. Their participation, as far as we can see from the record, confers upon them no distinctive or unique status.

The *second wave* entails a more public role, but one that arose out of the first wave. The business widows continued the work of their husbands, often bringing the business to greater heights of prosperity. Well-placed intermarriages among the cartography Houses guaranteed the proper assembly of resources to ensure the continuance of the Houses, some of which lasted for more than a hundred and sixty years, and to permit the accumulation of wealth. In this light, one would also have to consider the role played by guilds, especially the printers' guilds, in their recognition of the fair right of women to run businesses as they saw fit.

The *third wave* is associated with the start of scientific cartography in the seventeenth and eighteenth centuries. The historical record, in particular, becomes very sketchy at this point. Scientific cartography, as an element of the Enlightenment, moved outside the confines of ateliers and map shops. The focus on measuring land (as an expression of state control) may have initiated the drive to encourage men to undertake "scientific" tasks and women to consign themselves to such "natural" tasks as child rearing, effectively excluding women from some of the main tasks in cartography. There is a paucity of records in that regard.

The *fourth wave* characterizes the mid part of the nineteenth century. Solitary women explorers, adventurers, and travellers (and sometimes others, too, alongside their husbands) took advantage of the public's growing interest in colonizing non-Christian lands and peoples, asserting empire-building dreams. This mode of participation meant that women stepped out of place and established new norms. Closer to home, too, geography teachers were women, building on cartographic knowledge and initiating novel instructional exercises, whether encouraging girl pupils to build globes or developing mnemonic devices for learning about distant countries.

CHAPTER 13

THE ENGAGEMENT OF WOMEN IN CONTEMPORARY CARTOGRAPHY

Many participants occupy today's map world. Gone are the colourists and engravers, but newer occupations gained entry into the map world, such as photogrammetrists, remote sensing specialists, and GIS specialists, to name a few. Making a record of the participation of women in all of these fields would be a gargantuan task, but it is significant to note that each of these fields involves its own unique historical and technological aspects and practices. It might be foolhardy, if not impossible, to draw comparisons among them.

The character of the *fifth wave* of women cartographers is fundamentally different from any previous wave. Caught in the love and discovery of science, there was a single-minded but unassuming pose toward one's participation in the cartographic science that lasted from the end of the nineteenth century through the major part of the twentieth century. Dismissive of self, women became invisible—not on account of a lack in historical record keeping, but because they saw science as the greater good and maps as serving the greater good of society, which often articulated a social-activist temper. Florence Kelley's maps (1859–1932) documented the distressing experiences of Chicago's immigrants. The self had no business in that engagement with science. Men achieved recognition; some women were seen as helpmates to those men.

The *sixth wave* developed during the 1970s and 1980s when the participation of women in cartography became a self-conscious act of engagement. The female pathways through cartographic institutions were diverse, and involved several strategies to cope with systemic inequality. To understand those strategies more deeply, *Map Worlds* used the idea of vignettes to honour the significant contributions of some two dozen women, starting in the latter part of the nineteenth century.

While one can portray the vignettes as elaborate and viable entries for an imaginary encyclopedia on women and cartography, they are, in effect, detailed accounts of the struggles and challenges that the women faced as they moved toward the point of making remarkable contributions to cartography and to science. The vignettes underscore the intellectual biography and genealogy of women cartographers—another instance of the interwoven nature of the work of women and men cartographers, yielding a cornucopia of rich and plentiful accomplishments. Those accomplishments spill over into thematic maps (Florence Kelley), maps that are mindful of indigenous toponyms (clearly an anti-colonial stance), earth-shaking findings by Marie

Tharp (1920–2006), portrayed through oceanographic maps that revolutionized ideas about continental drift, eye-catching city maps, and digital mapping. The innovative maps by Ellen Churchill Semple (1863–1932) laid the groundwork of the human geography movement.

The third segment of *Map Worlds* rendered the personal insights and experiences of contemporary women as they pursue their education and careers, and how those personal experiences mesh with the institutional and organizational demands of cartography. While the individual vignettes are not anonymized, the personal experiences (obtained though interviewing thirty-eight women) are, because I deemed it significant to underscore the commonality of experience of women cartographers—across the board, whether as academic cartographers, geodesists, or map practitioners. Unlike many men, most women have experienced major shifts in their varied education and careers, often criss-crossing many disciplines before falling into cartography. As a result, many disciplines have shaped the kinds of contributions that women are making to cartography.

THE ENGAGEMENT WITH FEMINIST DISCOURSE

Numerous indeed are the fields that now have drawn the interest of scholars to the participation of women in as many fields as one can imagine: music (Cohen, 1981), science and technology (Etzkowitz, Kemelgor, Uzzi, 2000), medicine (Dodd and Gorham, 1994), sociology (Lengermann and Niebrugge-Brantley, 1998), and so on. It was only a matter of time before people began exploring the role of women in cartography. *Map Worlds* offers a rehabilitative history. However, the book goes well beyond history, because the main focus of the book is contemporary women cartographers.

In recent years, a number of scholars have advocated a feminist cartography (and geography). It stands to reason that *Map Worlds* must address their concerns and participate in their discourse. That discourse stands on five ideas on whether women cartographers (1) have changed the masculinist, positivist vision of cartography; (2) have played the role of democratizing mapping; (3) have created mapping alternatives; (4) have initiated the rise of critical cartography, out of the hands of traditional control; or (5) have transformed knowledge itself while addressing power and hierarchies.

CHANGING THE MASCULINIST, POSITIVIST VISION OF CARTOGRAPHY

Like all human endeavours, cartography is part of the larger social context and is shaped as such by historical processes of discovery, exploration, colonization, empire building, and so on. The immediate context of cartography, however, belongs to technology. Sometimes as arbiter, sometimes as handmaiden, it dictates the conditions that make cartography either a hospitable place for women or a place that eschews women. The conditions were usually not clear cut.

Challenging the positivist vision of cartography has been around for at least ninety years, if not more, when Belgian surrealists in 1929 drew a world map that "prized imaginative distortions over scientific exactitude" (Bulson, 2010: 19). Others, too, have pointed to the need to deconstruct the map. In this context, there has been an explosion of interest in cartography as art. From the perspective of artists, according to Eric Bulson (2010: 19), "a majority ... are more concerned with exposing how maps fail to capture a subjective, human experience in space.... Behind it all is the desire to make maps speak about histories and stories that are unwritten or ignored."

Until the arrival of scientific cartography in the seventeenth century, women found room in cartography. There is some evidence to suggest that women were involved in creating manuscript maps—the technology and the means were on hand. The arrival of printed maps created the need for copper engraving and colouring, tasks that could be carried on at home or in ateliers close to home.

The era of scientific cartography that aimed at precise measurements of one's sovereign kingdom brought major shifts to the field and resulted in major mapping exercises beyond the atelier. No longer were the map ateliers and the written reports of explorers the primary leaven of cartography, but the theodolite and the surveyor's stick became the source of cartography in its transition to fieldwork. The sentiment of doing fieldwork as an integral part of cartography had great staying power, probably well into the 1980s—some three centuries. During these three centuries, the contributions of women coursed along the uneven tributaries of cartography, with each tributary opening up or closing down according to each technological advance. Map stitching ceased when book binding became automated; astronomy remained a course open to women because its technology allowed it to be set up on one's rooftop, closer to home, fostering the participation of women; globe technology did not need factories, but could be done in nineteenth-century schools and ensured the engagement of women in that facet of cartography. These few examples, taken from the pages of cartographic history, illustrate the varying means of women's participation through the many tributaries in the map worlds.

GENDER SHIFTS

From within cartography itself—such as Arthur H. Robinson and Barbara Bartz Petchenik (1976), Robinson (1989), John Pickles (1992, 1995), Dennis Wood (1992), Geoff King (1996), and Mark Monmonier (e.g., 1977, 1996)—there has been a growing realization that map-makers could no longer satisfy themselves with the idea that maps are objective, wholly scientistic renderings of space. At least two decades earlier, Armin Lobeck (1956) advocated the need to recognize the subjective elements in maps. A number of women cartographers, such as Judith Tyner (1982, 1987) and others, were part of the groundswell of this fundamental change and significantly contributed to that discourse. Within the confines of *Map Worlds* we already met "Kristina" in Chapter 2, who sees her task as a cartographer as involving a large variety of fields, including history, biology, the environment, and literature. In Chapter 6 we learned of Barbara Petchenik (1939–1992). One woman cartographer (Evelyn Pruitt, 1918–2000) seemed to have reinforced the quantitative, masculinist perspective, but she also managed to inspire others to create remote sensing, which she used as an important component of understanding coastal systems under duress. Eva Siekierska (b. 1945) was among the first to concretize the development of the digital atlas and seemed to extend the masculinist approach to maps. The idea of digital layering of maps allowed cartographers to overlay maps with relevant social and economic data, making them more useful for a larger audience of map readers and users.

While highly advanced in geospatial data mining and integration, Liqiu Meng (b. 1963) of the Technical University of Munich has always had a special interest in the emotional requirements of map users, but her dedication to the technical and scientific aspects of cartography mandates a special integration of these interests into something that covers a wider area of accomplishments than would normally be the case.

LEADING TO THE DEMOCRATIZATION OF MAPPING

The engagement of participants in cartography/map worlds varies in intensity from period to period. During the seventeenth and eighteenth centuries, for example, the state played a more overbearing role in the cultural and physical production of maps than perhaps in any other period. Similarly, guilds, ateliers, printers, and publishers displayed an important role at other times. The rising influence of the military in cartography (and many governmental cartographic services are still housed in military agencies) is another area that begs for more research; the invention of GPS, for example, was a military effort (Greulich, 2006: 25). While World War I signified the rise of military cartography and the possible exclusion of women, World War II created inadvertent gender shifts in cartography. The hiring of "Millie the Mapper" as an answer to the shortage of men cartographers in some departments of war constituted one of those gender shifts. The gender code in military cartography was thus temporarily broken and may have paved the way for future generations of women. The concentration of map-making companies into fewer enterprises can diminish the presence of women in those companies. Recently, an investment firm bought an important Hungarian map-making company and then divided and sold its various assets, cutting back its sizable labour force of several thousand to just a handful of employees, and no longer included a number of prominent women map-makers (Field notes, 6 July 2011, Paris). We also note that the military and larger corporations have begun playing a greater role in the contemporary map worlds, opening the field to retired military personnel. So much so that gender issues that were prominent in the 1980s are suffering a setback of sorts. Whether in the field of satellite mapping or the availability of data, the military often still holds the key to cartographic advances. In the spread of technical expertise, including computer programs, corporations have found a way to create software monopolies. The military is not only the arbiter of what data sets should be made available to cartographers (and the public), but it also provides the necessary human resources to corporations as military personnel retired and were hired by these same corporations.

The rise of American academic women cartographers since the 1960s, unlike "Millie the Mapper," gave a fresh impulse to thinking about and creating many maps that fell outside the purview of the military and the government. One has only to attend the exhibits at the International Cartographic Association to realize how deeply embedded the military and the government are in contemporary cartography (van den Hoonaard, 2000a). Patricia Caldwell has substantially contributed to democratizing contemporary cartography by focusing on how to match the interests of educational systems with the potential of graphic digital systems. Crowd sourcing, the Open Street Map, and various social networks now contribute extensively to the democratization of cartography. As a result of a work assignment by Natural Resources Canada, Eva Siekierska and others constitute a contribution to democratizing maps; the development of maps of Canada's North reflects local or indigenous designations. The Canadian Helen M. Kerfoot (b. 1938) established a worldwide reputation through her involvement in the United Nations Group of Experts on Geographical Names and indigenized countless names to counteract the effects of colonization of geographical names. The idea of place names reflecting indigenous designations has been actively developed for many years by Lynn Peplinski, The Kitikmeot Heritage Society, Gia Laidler, Claudio Aporta, Glenn Brauen, Fraser Taylor, and the Gwich'in Social and Cultural Institute, among others.

CREATING MAPPING ALTERNATIVES

Voicing the perspective of feminist geographers, Marianna Pavlovskaya and Kevin St. Martin (2007: 591) believe that a Cartesian view of space "fails to represent space in terms of relations, networks, connections, emotions, and other nonstandard patterns or movements that characterize women's life-worlds." Rejecting the Cartesian view (that Pavlovskaya and St. Martin believe is tantamount to a masculinist view of space) entails alternative ways of knowing and alternative mapping practices (2007: 592). However, the transition from the Cartesian view began in the nineteenth century. The early thematic maps of the late nineteenth and early twentieth centuries designed by Florence Kelley rendered the plight of immigrants in an acute manner.

There are contemporary examples of cartography portraying what matters in daily life, and doing so evocatively. The *National Atlas of Sweden*, for example, expresses many feminist modalities, such as mapping the daily activities of a household (Helmfrid, 1996: 153). Feminist cartography has the goal of engaging the map reader (Pavlovskaya and St. Martin, 2007: 587); the "passive reader" is a thing of the past. Long before feminist cartography became an epistemology in plain view of cartographers, there were already powerful maps in the market that induced map readers to be actively engaged. One thinks of the visceral, colourful, and engaging maps created by Eduard Imhof (1895–1986) in Switzerland and, in France, the Institut national de géographie.

These "feminist visualities" involve new kinds of research, new research questions, and diverse research methods. This "spatial turn" has found resonance in the social sciences, the humanities, and art where there are now elaborate cartographic metaphors. However, these sorts of visualities were already a common practice centuries ago. The postmodern world is just now catching up to those early ideas. A number of women in France participated in the art of making maps in a distinctive way—a manner that was to be emulated for centuries to come. One of the most interesting contributions was made by novelist Madeleine de Scudéry (1607–1701), who achieved considerable fame through her extensive literary work, authoring some twenty volumes between 1649–1653 and 1656–1660, all of which were the subject of considerable literary attention (Möbius, 1982: 152). De Scudéry had incorporated the so-called *Carte de Tendre*—an imaginary map depicting the constraints and possibilities of platonic relationships between women and men (1653–1654)—into the first volume of her romance *Clélie*. Although not the first of such maps (there was one published in Venice as early as 1499), the author's map became a huge success. Franz Reitinger attributes its success not only to the high standing of the author, but also to its "controversial subject matter" and the intellectual appeal of this new genre of "literary-allegorical cartography" (Reitinger, 1999: 109). It was a later artist, Bertall (1820–1883), who illustrated another allegorical map of de Scudéry in one of Balzac's novels on marriage ("map of the Honeymoon").

In more contemporary terms, one finds Judy Olson (b. 1944), whose major contribution falls in the field of use of colour in mapping and the psychology of maps and designing maps for people with defective colour vision. Judith Tyner (b. 1938) devoted her career to studying textile maps and promulgated her idea of "persuasive cartography." An innovative take on alternative mapping is Barbara Shortridge's efforts producing thematic maps dealing with food, restaurants, and consumption patterns. Phyllis Pearsall (1906–1996) countered the defective official map of London with her own uniquely detailed map (the *A–Z Map*) that was not only accurate but also of greater use to map readers; she created, published, and sold her maps at considerable personal sacrifice.

The rise of cartography in cyberspace has led Carmen Reyes (b. 1949) of Mexico to consider both new approaches to and topics of cartography that matter to many citizens. These approaches led her to work with at least sixty national and international projects, involving atlases for children, cybernetics, the management of urban green areas, water systems, air quality, electoral systems, atmospheric models, and education. In many respects, Reyes's interests align with those of Regina Araujo de Almeida (b. 1949), a professor of geography at the University of São Paulo in Brazil. Almeida is the world's leading cartographer in tactile mapping and children's maps. Her other areas of expertise include indigenous mapping, perception and representation of space, geography, culture and tourism, and teaching geography.

INITIATING THE RISE OF CRITICAL CARTOGRAPHY

The feminist critique contends that "knowledge is situated and implicated in the production of social power" (Pavlovskaya and St. Martin, 2007: 590). The period of colonization and global domination is an archetypical expression of such knowledge where maps served the mappers, or, more accurately, the militaries and conquering states of the world. The world thus portrayed was understood as an accurate portrayal of the world. While the crassest examples of such cartography have withered away, they have been replaced by far more subtle manifestations of domination (just think of the portrayal of the Soviet Union in American school atlases during the 1950s and 1960s, where American mappers displayed the Soviet areas in black).

Still, our story is filled with numerous examples of women cartographers embracing indigenous mapping practices, terms, and toponyms. Someone like Barbara Buttenfield (b. 1952) delved deeply into how cartographic knowledge is constructed, focusing in particular on representations of uncertainty.

TRANSFORMING KNOWLEDGE ITSELF WHILE ADDRESSING POWER AND HIERARCHIES

As indicated in the opening of this chapter, it is impossible to create a grand narrative when it comes to cartography. Some creative cartographic outlets are resisted, while other new paths open up without much ado. In the fabric of the map world, tears suddenly appear, ripping apart conventional cartographic wisdom, while other rips reveal an approach that was resorted to centuries ago, which through the passage of time had been forgotten.

The current pursuit of cartography by corporations, however, does not augur well for an era of gender equality. The proliferation of routine work in large-scale companies has a hand in producing inequitable gender relations, as the following comments by a woman cartographer testify:

> *Is there anything else in terms of gender and map-making that you want me to consider?* Of course, when it comes to females and the technical level, there are fewer occupations or fewer positions today than it used to be ... due to computerizing.... When drawing pipelines ... in most technical departments in the municipalities ... many females [were] doing the drawings. So, a lot of them [women] are now rationalized away ... you needed fewer females. Some of them started private drawing services.... But you also have that period when you had great digitizing. Then the females were doing the

> digitizing, but that also works just for a period. So, females are more used, [are] misused.... They do the routine work.... (Inga: 11)

The process of digitizing maps on a local, national, and global scale—also with the help of those corporations—suggests entirely new vistas for research. Geographic information systems (GIS) moved surveying and mapping away from fieldwork itself as the basis of maps. GIS, according to someone who was deeply involved in mapping, "was like the meteor that hit the dinosaurs of surveying and mapping organizations" (Field notes, 7 July 2000: 1).

Jennifer Brayton (2003), in her summary of research on the perceptions of the role of technology in the promotion of women, points to the seemingly unobtainable goal of using technology to advance women. (Computer) technology was greeted as a "neutral tool, created and shaped outside of societal influences" (Brayton, 2003: 8). As early as 1983, however, a number of scholars, Brayton points out, acknowledged the pervasive societal context and influence in shaping technology, both as a physical object and as human activity. Some went as far as saying that technology has allowed society to enter a female era—a belief that was particularly prevalent in the 1980s (Woodfield, 2000: 187). Modern economies were shifting to more relational, cooperative ways of thinking, as opposed to competitive thinking, and computer technology would redress the gender imbalances.

According to Brayton, cyberfeminists like Faith Wilding, Sadie Plant, and Donna Haraway began to conceptualize technology as blurring the boundaries of such dualities as human/machine, mind/body, male/female, and so on, so that women and men alike "can take pleasure in technology and share a technological identity" (Brayton, 2003: 23). A similar sentiment comes from Ursula Franklin, who speaks about a "redemptive technology"—"a technology that is holistic rather than assembly-line, participatory rather than compliant, a technology that minimizes harm and counts environmental costs, a technology that is accountable to people" (McCormack, 1991).

All of these "waves of optimism," according to Woodfield (2000: 188), have, however, failed in the "golden opportunity" that technology could have offered to women. The fact of gender alone, according to Ruth Woodfield, accounts for the failure by organizations to realize equitable gender relations. Women "were, in abstract, explicitly recognised by their colleagues as possessing good levels of interpersonal expertise [so vaunted by organizations], but were rarely, if ever, the recipients of the concretised form of this recognition: rewards, respect, and remuneration" (Woodfield, 2000: 188).

No doubt the work of Mary Adela Blagg (1858–1944) comes to mind. Virtually without much recognition and remuneration, and unaided, she developed a scheme that standardized all names that had been assigned to lunar features. Significantly, her devotion to such a task laid the groundwork for later work by the International Astronomical Union on lunar mapping. One generation later, Kira Shingareva (b. 1938) of the former Soviet Union would be part of a research team that would be the first to map the reverse side of the moon. At the height of the Cold War, Shingareva's group presented the new map of the reverse side, often challenging the Western taken-for-granted attitudes about naming practices, and signalling new approaches.

The stature of Helen Wallis (1924–1995) is more difficult to assess in light of the issues raised by feminism. As the first woman head of the Map Library of the British Museum, she aggressively acquired one of the most enviable collections of maps in the world, but did so in a spirit of helpfulness and service to all who came to seek her advice. Her own research on South Pacific maps upset conventional knowledge about patterns of discovery and exploration. One should also mention Eila Campbell (1915–1994), an eminent geographer at London University, who substantially bridged the gap between geography, cartography, and history, partly through her diligence in editing *Imago Mundi*, the premier journal of historical cartography.

UNTYING THE GENDER KNOT

Allan G. Johnston's superb 1997 book, *The Gender Knot*, speaks to the relevance of both women and men working together to create a gender-balanced society. To have just one gender pulling at the knot simply tightens the knot, rather than loosening it. Sociologists are accustomed to looking at the big picture. They deal with the study of social processes and social organization. They are less likely to study "why" people do things than ask the "how" questions, which would tell them a lot more about social processes. The study of social processes leads one to look at society, or any group for that matter, as a form of (dis)agreement on how people cooperate along the lines of mutually agreed upon meanings that lead to interaction. In reality, social reality is negotiated, entailing conflicts and disagreements. The study of social processes also allows us to explore what happens when people break its implicit norms.

The study of social organization looks at society as not merely an aggregate of individuals, but as something over and beyond the individuals who compose it. To illustrate such a point, one needs to simply think back to one's school classes: even though the number and kinds of people were the same, each class had its own culture that extended beyond the individuals that composed it. When it comes down to gender relations, as in my study, it is not so much what women and men do, but rather the larger context (in our case, the map world), a context that still presumes a male outlook as normative (see, for example, Johnston, 1997). It makes no sense to blame men (or women) for our present social organization. In some respects, social organization is like a household that both the powerful and the not-so-powerful cooperate to maintain. It is this way of accomplishing things together that garners the interest of the sociologist, without blame, censure, or praise.

One also finds that women and men participate in radically different ways in map worlds, as far as gender is concerned. We can identify several categories of female pathways through contemporary map worlds. There are those who acknowledge that power and gender imbalances exist. Some turn to the heart of cartographic organizations to mainstream equality; others create nurturing, parallel structures. Although with diminished power compared to the central organization of cartography, parallel structures facilitate the participation of those who need the nurturing and encouragement. Still others are only in a position to observe inequality. The final category, consisting of women who disclaim the existence of any gender imbalance, fosters the idea that women should be concerned with under-represented groups on maps, and even rejects the concept of gender altogether.

As far as the men are concerned, their acknowledgment of "gender" varies no less considerably, but is muted by the privileges they take for granted, especially in the reproduction of power in the map world.

Just as *Map Worlds* has taken its cues from other contexts in the study of women in various fields and occupations, so does cartography, as a whole, now provide an unerodable place for women. Still, the map world offers an ambivalent structure for women. It is not likely that that ambivalence will disappear. For one thing, there is no unalloyed interest (on the part of many women cartographers) in the history of women in cartography, as my interviews with the thirty-eight women cartographers seem to show. In answer to my question *"Do you have any interest in the historical knowledge of women in cartography?"* several offered the following in reply:

> I have never really had an interest in that [history]. (Ruth: 10)

> I know nothing about women in cartography in the historical sense. (Ettie: 7)

I would be interested in knowing, but I must say that I haven't made an effort to inform myself about that. (Natalie: 15)

I was totally ignorant of any of that stuff, absolutely ignorant. (Lynn: 13)

Our study of the female pathways through cartography's social structure also shows that there is no agreement on the degree to which women participate in that structure. It is not technology that has created ambivalence (in fact, technology has had an enabling influence), but the traditional outlooks and perspectives that still pervade society, which women and men face as they come to terms with gender relations.

Beginning with the tail end of the twentieth century, we started seeing a waning of institutional positions held by women (see, e.g., Demoor, 2000: 3–4). There are now proportionally fewer women in positions of leadership. Ruth Woodfield, author of *Women, Work, and Computing* (2000: 190), has already noted that even progressive organizations, however well intentioned, "require a firm foundation in progressive social and psychological relations if they are to avoid mutating into their opposite." The loss of women from positions of authority is due to "a cultural lag [that] exists between some key 'progressive' developments in organisations and current levels of socio-political consciousness, and ... this often renders such developments conservative rather than liberal in their effects" (Woodfield, 2000: 191). The current practice of governmental privatization of services is like an irresistible undertow that breaks apart the foundation of the movement toward equality.

Like the story of women in historical and contemporary cartography, the analysis of cartography and feminism follows a non-linear course. In some areas, such as critical cartography, the creation of alternative maps, and the transformation of knowledge, the processes must be quite satisfactory in the eyes of feminist cartographers. Perhaps less mature are issues surrounding the dismantling of the masculinist vision of cartography, perhaps made more difficult by the presence and integration of technology in cartography. Finally, it remains to be seen if or when cartography can be wrested from the control of the military, corporations, and government. Military, corporate, and governmental institutions have vested themselves seriously in maintaining the status quo. Will it take a crisis to shift the balance in cartography, or will it progress more naturally and fluidly? Was Marie Tharp our contemporary Rosa Parks in cartography, or has cartography become too routinized in its organization and structure to create a new balance?

APPENDIX A

Methodology

Given the large scope of this work, it stands to reason that I used a variety of research methods. The principal ones involved analyzing unpublished and published documents, correspondence, and interviews, and using participant observation. For clarity, I present these methods in the same order and relevance as the successive sections of *Map Worlds*. The reality of doing research involved my constant migration among the methods. Participant observation might lead to interviews which then might lead to the study of documents, and so on.

DOCUMENTS
Documents in cartography were extensive. They involved a variety of languages, including English, French, Dutch, Norwegian, Icelandic, German, Spanish, and Portuguese. My acquaintance with the latter two were more rudimentary than I had hoped.

PUBLISHED DOCUMENTS
The standard works in cartography are *de rigueur* in this line of work, and all of the published sources I consulted can be found in the References section of this book. The reader will also note references to some of cartography's key journals, including *Imago Mundi, Cartographica,* and *Progress and*

APPENDIX A

Perspectives; national newsletters; and particular reference works, with special reference to the multi-volume encyclopedic *The History of Cartography*. Trade journals in surveying, geodesy, and the like were also useful. To create the vignettes, I relied extensively on personal contact and on the friends and colleagues of those women cartographers who had already died.

UNPUBLISHED DOCUMENTS

I have made extensive use of listservs of a number of national cartographic societies, which allowed me to correspond with specialists in historical cartography and cartography, map librarians, map dealers, and map archivists, as well as with members of national and international societies, conference organizers, and volunteers. Usually, such correspondence took the shape of emails, but also of regular mail. I corresponded with at least fifty people, although the precise number is unknown. These contacts provided in-depth information to my queries about cartography in the fifteenth to eighteenth centuries. There are a large number of websites devoted to recording the contributions and families of cartographers in the golden age of cartography.

VIGNETTES OF WOMEN PIONEERS

The process of sifting through the names of many women cartographers could not have taken place without the indispensable help of Dr. Fraser Taylor of Carleton University; Ms. Alberta Auringer Wood, formerly of Memorial University of Newfoundland; Ms. Wendy Straight of *Progress and Perspectives*; Dr. Henry Steward of Clark University; Dr. Julia Siemer of the University of Regina; Ms. Mary Ritzlin of Evanston, Illinois; Dr. Mark Monmonier; Dr. Ewen A. Whitaker, formerly of the University of Arizona; and Dr. David Coleman of the University of New Brunswick. Judith Tyner was also helpful in identifying pioneers. Collectively, these pioneers embraced cartography, cyber-cartography, surveying (now called geodesy), map librarianship, historical cartography, antique map dealership, historical cartography, and seneology.

I kept a running list of potential women pioneers in contemporary cartography as I proceeded through my research. In the end, my list grew from twelve names to twenty-three. I promised that I would not only send the still-living pioneers what I have written about them, but that I would try to secure their approval for the vignette before publication in the book. In the case of one, I had to rely on publicly available documents, because I could not reach the person directly and secure more information that might be of interest to the reader.

METHODOLOGY

I aimed at creating a list that was not Anglo-Saxon in orientation, but rather encompasses women pioneers in North America, Europe, Central and South America, and Asia. No doubt the perceptive reader will spot particular gaps. In creating the vignettes, I took advantage of what the Internet could offer. I amplified my research by sorting through the publications and works of the women featured and seized the opportunity to correspond with all pioneers. If the pioneer was deceased, I contacted the person who would most likely have known about her. The final vignette was thus a collaborative effort between the living subject (or an acquaintance) and me, and an attempt to correlate the many documents available to me.

PARTICIPANT OBSERVATION

Participant observation involves the intensive study of a particular culture. Normally, the social anthropologist (or sociologist) immerses her- or himself in the ways of doing and thinking of that culture, often for an extended period of time. Such an engagement could last anywhere from several months to several years. It became clear to me that I should not only become a registered member of such professional groups as the Canadian Cartographic Association, but also attend the seminars and conferences hosted by these groups. I had entered cartography briefly in the 1960s as an assistant cartographic editor on account of my love for maps. When I chose to conduct participant observation of the map world, I had not expected to become so actively involved in mapping conferences as I did. I simply found it too difficult to restrain myself at conferences from participating fully in the proceedings by asking questions, making comments, and even presenting a paper or two.

A note about the methodology that forms the basis of chapters 8 and 9. As Appendix A tells, *Map Worlds* is comprised of vignettes of five near-contemporary and twenty-three contemporary women, as well as data derived from my interviews with thirty-eight other contemporary women. The sixty-one women reside(d) in twenty-four countries, work(ed) in a variety of positions, and represent(ed) huge career variations.[1] They are scattered all over the world, but they are not representative of countries. Absent are women from Switzerland (which has a long and substantial cartographic tradition). The relative lack of representation of women across the broad spectrum of the map world plays against a recognition of the kind of training, great cultural and political differences, and differences of resources that affect the map world in each country. Other facets of the study might highlight these apparent shortcomings: namely, the women come from very disparate countries, each involving, presumably, different challenges and

APPENDIX A

opportunities. What seems to complicate matters is that cartography does not contain enough uniformity of training. Cartography involves so many different specialties (e.g., geodesy, remote sensing). All of these variations are problematic in any study of an international nature.[2]

As a field researcher, my usual first resort involves doing participant observation research. Doing a survey using a list of cartographers provided by one of the International Cartographic Association (ICA)'s reports (Siekierska, O'Neil, and Williams, 1991) would not permit me to get data of any depth. The prospect of getting too small a response rate after all that effort and expense did not sound too appealing, either. Bearing in mind that cartography has undergone radical changes since my days as an assistant cartographic editor in the mid-1960s, I devoted a considerable amount of time acquainting myself with the new cartography. After I received a research grant from the Social Sciences and Humanities Research Council of Canada in 1997, I learned that the ICA was holding its international conference in Ottawa, Canada, in 1999. I jumped at the chance of attending this conference. In addition to "hanging out" (a favoured method of field researchers), I presented a paper on the idea of "map worlds" (van den Hoonaard, 1999). Dr. Eva Siekierska (who introduced to me the idea of gender and cartography) was most helpful in acquainting me with participants at the ICA conference. I attended as many sessions as I could, whether plenary or simultaneous ones, and took field notes of audience participation in these sessions, as well as observed how presenters engaged with the audience with a focus on gender. From these various vantage points, one could see who was central to the organizational dynamics of the conference. Throughout the conference, I was able to obtain a good number of interviews; having seen how the interview participants interacted with others at the conference, in addition to having listened to some of their papers, I was able to pursue questions and issues that seemed pertinent at that time. I also kept a pile of statements about my research for anyone to pick up if she was interested in talking with me by email. Some did.

Oftentimes, my weary moments (and a yearning for coffee and refreshments) brought me to the exhibit hall, where another facet of the ICA conference became a source of research, some of which I later presented and published (van den Hoonaard, 2000a, 2001a, 2001b, 2002a). When the ICA daily newsletter wrote a piece about my attending the conference, a few more individuals offered to talk to me. As more doors and opportunities opened up, I was able to attend other conferences, nationally and internationally (and even a cartography summer school in Norway). More observational data and interviews ensued.

Asking questions proved to be helpful in establishing my deep interest in cartography. I was careful always to start with a preamble that I was a sociologist, not a cartographer. One surprising outcome occurred when a cartographer asked me if I would run for a position on the executive of a national organization of map-makers. I declined such key involvement, but later accepted the nomination to be chair of the Historical Cartography Section of the Canadian Cartographic Association (CCA). Part of that involvement meant that the CCA required me to write three or four articles a year for *Cartouche,* the association's newsletter. At the international level, I was asked to serve as acting secretary of the ICA's Commission on Gender and Cartography in 1999, during its meetings in Ottawa.

Participant observation, however, did not end with presenting and critiquing papers at conferences and the like. I participated in other, primarily social activities of the cartographers. At international gatherings, I attended a banquet organized primarily for women who could not afford to attend the more expensive, official banquet of the conference. I also was an active participant in evenings' special outings, such as concerts. I even experienced the dismay of fellow cartographers, one of whom had her wallet pickpocketed and another whose laptop computer was stolen from under her table during lunch. At the national level, I played pool with a number of cartographers as an evening relaxation during the conference. In this manner, I sought to integrate myself in a natural way into the conference life of the cartographers whom I was studying. At my suggestion, the cartographers began to see me as the equivalent of an anthropologist's studying his "tribe," which, in turn, brought forth a variety of unsolicited comments about what was unique about cartographers. Twice I participated in the orienteering program held in conjunction with a cartography conference (in Waterloo and Regina).

My other intent was to embark on a formal study of the ICA conference as a key element in the world of cartographers. As my paper fell on the third day of the first ICA conference I attended, I had plenty of time to write up field notes of my observations and conversations with ICA participants at all levels. Their response to my presence in the midst of one of their most important rituals was gratifying: a feature article about my presence in the "People Corner" of the ICA *Newsletter for Ottawa 1999* (21 August 1999) underscored the general interest of participants in being the subject of study.[3]

The full list of eleven major cartography venues I participated in follows here: *in 1999,* the International Cartographic Association Congress (Ottawa, August); *in 2000,* the joint conferences of the Association of Canadian Map Libraries and Archives and the Canadian Cartographic Association

(Edmonton), and the Teaching Maps for Children Conference (Budapest, September); *in 2001,* the joint conferences of the Association of Canadian Map Libraries and Archives and the Canadian Cartographic Association (Montreal), the Digital Earth Conference (Fredericton, June),[4] and the GeoForum, Summer School for Cartography (Tromsø, Norway, August); *in 2002,* the Canadian Cartographic Association conference (Waterloo); *in 2008,* the annual conference of the Canadian Cartographic Association and the Association of Canadian Map Libraries and Archives (Vancouver, May); *in 2010,* Canadian Cartographic Association annual meetings (Regina, June); *in 2011,* the joint ICA symposium (Orléans, France, June), and the International Cartographic Conference (Paris, July). If one were to include my attendance at meetings of the ICA Commission on Gender and Cartography (Ottawa, 1999 and 2011) and a British Library lecture at the Warburg Institute (London, March 2000), one could say I attended fourteen such gatherings.

In addition to limited field research observations and discussions with exhibitors, I also collected and analyzed conference programs, conference newsletters, and official conference photographs. As every conference participant knows, there can be many discrepancies between what is formally touted as conference sessions and what actually takes place. One should be careful in interpreting the results: there are last-minute additions in a session, some presenters do not show up, and some papers are read by proxy. Moreover, it is not uncommon for people to do more than one presentation. Of the thirty-six women either chairing, presenting, or postering, eight women doubled or even tripled or quadrupled in any of these functions. Poster sessions in particular had multiple presenters.[5]

However, I also made good use of "semi-published" documents, such as printed conference programs, the circulation of which is restricted to conference delegates or participants, and a large number of websites (also found in the References section). In the case of past conferences or ones I was not able to attend, printed programs were sometimes available in map archives.

INTERVIEWS

Almost two years into the research, I had secured interviews and had informal discussions with sixty-seven cartographers (thirty-eight women and twenty-nine men). Once I was able to approach cartographers on the international level, it became clear—quite literally, for those who transcribed the interviews—that both the researcher and the interviewee faced linguistic challenges. The interview schedule, originally written in English, was both subtly and roughly modified by non-native English speakers whose mother tongues included Bhutanese, Bulgarian, Dutch, Finnish, German, Greek, French, Hebrew, Hungarian, Icelandic, Mandarin, Norwegian, Persian, Polish, Portuguese, Russian, Shona, Slovenian, Spanish, Swedish, and Turkish—twenty-one languages. Appendix B provides the interview schedules. From beginning to end, the schedules changed as topics become more or less relevant to explore, which is typical of qualitative research interviews.

When I began to assemble and analyze the data related to the interview participants, I was struck by similar patterns among participants in their respective map worlds. This finding did not arise from a deliberate, pre-research hypothesis, but emerged from the inductive approach in my research. I read each transcript far too many times to count, but such reading provided the main impulse of my analysis of the data. The interviews in chapters 8 and 9 represent these inductive approaches of exploration, data gathering, writing, and analysis. For many qualitative researchers, analysis is vested in "writing up the data" and involves the simultaneous process of writing and analysis. In several instances, I was able to ask follow-up questions and request clarifications by email. In one sitting, I interviewed two cartographers at the same time in the presence of their chief of section, who acted as their interpreter. The chief had arranged the meeting on very short notice.

The conference venues presented practical dilemmas of conducting interviews. Any reader who is familiar with "conferencing" knows of the limitations of time imposed on participants, especially the presenters. A planned life-history study of cartographers was, in hindsight, an unrealistic goal. Virtually the only time I was able to conduct interviews was when a cartographer was not particularly interested in a given session. Some were persuaded to go for the interview; others excused themselves for more compelling reasons and still others wanted to socialize with colleagues—friends from around the world. The evenings were usually not available for interviews. Once we were able to meet during a session, it took some time to find a corner in which to conduct the interview, but not without distractions once a friend or colleague caught the eye of the interview participant. I could hardly ask the participants to be interviewed privately in the hotel room; the interviews had to be held in public spaces, usually in a hotel lobby. As I was not able to undertake the anticipated two-hour interview, time forced me to cover only those aspects of the interview schedule that seemed most appropriate for that interview participant. To achieve this goal of enforced "efficiency," I had to devote considerable time to finding out in advance the particular strengths of the interviewee and the relevance of those strengths for my research. As conferences constituted the participant observation part of my research, it also meant that any time I spent preparing myself for each interview was taken from time I needed to devote to taking observational field notes. The process was not satisfactory.

In listening to the transcribed tapes, I am struck by the background din. As I always avoided situations that even by hint of a suggestion could be ambiguously misunderstood, I always conducted the interviews in public places such as hotel lobbies or on the street. In some instances, this situation proved a challenge to maintaining the anonymity of interview participants. Moreover, there seemed to be no way of answering this particular constraint insofar as the noise was concerned. The transcribers thus also faced a special challenge, in addition to the interviewees speaking English as a second language.

Here are illustrations of the other types of problems I ran into. It sometimes took me several conferences to catch up with a particular individual I wished to interview. In one instance, a noted cartographer happened to be free during lunch and gave a spontaneous interview. I had left my tape recorder in another room (as I had not expected to find anyone ready to be interviewed), but managed to collect several paper plates, which I used for note taking. The event was not lost on the other cartographers. One other cartographer I was interviewing had to halt midway when one of her colleagues rushed over to notify her of an earthquake that had struck her town. We all worried anxiously about the fate of her family.

The thirty-eight women cartographers I interviewed came from eleven major geographical areas of the globe (or twenty-four countries) (see Table A.1). Many interviews lasted for forty-five minutes, one was as short as thirty minutes, and a few ran longer than an hour.[6]

I tried to choose a broad spectrum of interview participants, based on region, stages of career, and involvement in cartography, from "tabletop" cartography to GIS to digital applications. The average age of interview participants was 44.5 years and the average stay in their respective cartographic careers varied greatly among these three groups: there were seventeen cartographers of mature experience with an average of 54.5 years of age and 27.4 career years;[7] the thirteen mid-career cartographers have spent between 8 and 20 years doing cartographic work (for an average of 11.8 years; their average age is 41.4 years);[8] the eight remaining cartographers were just starting out on their careers (they are, on average, 29.8 years old and have been in their careers for an average of 3.9 years).[9] The age of the

METHODOLOGY

Table A.1. Source regions of (near) contemporary women vignettes and interview participants, 1999–2012

Geographic Area	Number of Countries	Number of Vignettes	Number of Interviewees	Total Number
Africa	1	0	1	1
Subtotal	1	0	1	1
Americas				
North America	2	14	9	21
Central America	1	1	1	2
South America	2	2	2	4
Subtotal	5	17	21	29
Asia				
Indian Subcontinent	1	0	1	1
Far East	2	1	1	2
Middle East	2	0	2	2
Subtotal	5	1	4	5
Australia and New Zealand	1	0	1	1
Subtotal	1	0	1	1
Europe				
Southern Europe	3	0	6	6
Eastern Europe & Russia	3	1	4	5
Northwest Europe	5	3	10	13
Austria	1	1	0	1
Subtotal	12	5	20	25
Grand Totals	24	23	38	61

cartographers does not necessarily correspond to the length of stay in their career: some have come "late" into the field. The average ages of entry for the mature, mid-, and early career interviewees are 27, 30, and 26 years of age. I did not predefine who was a cartographer; it was the interview participants, with few exceptions, who defined themselves as such. In one instance, I left 130 copies of my interview schedule on a table near the information booth of ICA 1999 (Ottawa) and received several written replies to this questionnaire through emails.

It was in this context that I found a very diverse group of occupational incumbents in the field of cartography, all of whom actively participated in this ICA gathering. Surveyors, academics, policy makers, atlas makers, and people involved with remote sensing were united by their love of maps and their work related to maps. Up until that point, I had not realized how very diverse the field is today. To complicate matters, "my" interview participants worked in governmental agencies or the private sector or were private entrepreneurs.

The venues of the interviews were the 1999 ICA congress in Ottawa; the 2000 ICA congress in Edmonton; the 2001 Montreal joint conferences of the Association of Canadian Map Libraries and Archives and the Canadian Cartographic Association; and the Teaching Maps for Children conference held in Budapest, Hungary, in September 2000; in addition to interviews at several research institutes and universities in the Netherlands, Scandinavia, and Slovenia. Over the course of two years, two graduate students—Ms. Lenora Sleep and Ms. Linda Caissie, then graduate students in the Department of Sociology at the University of New Brunswick—transcribed 330 single-spaced pages of analyzable text.

DOING INTERNATIONAL FIELD RESEARCH

The primary challenge of researching the social world of cartographers arose from the fact that they are among the first of the truly globalized communities, revealing political, linguistic, and practical obstacles. Rather than studying a settled community—even a gathering-hunting community could have lessened the challenge—the "tribe" of cartographers is a global one. Cartography arose for the most part since the mid-sixteenth century as an occupation with national boundaries serving national political interests. It was in the 1870s that the international dimensions of cartography became formally recognized and institutionalized. By the time I conceived, in the mid-1990s, the idea of studying map worlds, that is, the community of cartographers, the tribal perspective had already been firmly internationalized and globalized. Even the small, local world of desktop cartography demanded a global perspective in light of the need to take advantage of technological developments.

The trans-political character of cartography resulted in a web of activity that spans borders. In the course of the research, I travelled to Ottawa, London (England), Edmonton, Budapest, Enschede (the Netherlands), Oslo, Tromsø, Göteborg, Linköping (Sweden), Stockholm, Oulu (Finland), Ljubljana (Slovenia), Paris, Orléans, Waterloo (Canada), Vancouver, Regina, and Montreal, where I would find cartographers from around the world. The modest research funds offered by the Social Sciences and Humanities Research Council of Canada meant that supplementary travel funds had to be sought from other sources, usually giving lectures about my other research interests while embarking on my own cartographic research trip. Not local, but international settings became *de rigueur*.

THE SOCIAL ORGANIZATION OF THE MAP WORLD

The last section of *Map Worlds* concerns itself with the social organization of the map world. In effect, one needs to find out how the vignettes of the pioneers and the interviews shape, and are shaped by, the social organization of the map world. In exploring these higher-level abstractions, it became critical to weave the data from the vignettes and the interviews into the body of literature and documents that pertain to cartographic organizations and associations. Many of these documents are not publicly available. And because these documents normally omit considerations of gender, I had to approach many ICA national organizations myself to secure those kinds of data. The ICA conference in Paris (July 2011), as well as the official reports of these organizations on the ICA website, proved to be an important source for that information. The focus was on women's participation in ICA commissions and its national organizations.

I used direct observation of the membership of the organizations attending the Paris ICA conference, paying careful attention to the organization's composition in the home country and examining whether its attendance at the ICA conference fully reflected that composition.

APPENDIX B

Topics Covered in an In-Depth Interview

The exact range of the topics will vary from one interviewee to the next. In a qualitative approach to in-depth interviewing, the researcher follows the lead of the interviewee in the topics or issues that interest the latter. The researcher explores the taken-for-granted assumptions and probes the areas to which the interviewee seems to attach particular significance. For example, the first question is very important, since it must elicit those perspectives and issues that the interviewee deems important. The researcher must then follow through on these comments.

A. The First Question
How did you become interested in cartographic work?

B. General Opinions about the Field
What do you consider as the rewards of cartography? Or, what excites you about the field? Its obstacles? Its challenges? Its dilemmas?
What do you see as your contribution to cartography? How do others see your contribution?
Have you experienced any shifts in professional interests? How did that come about?
What skills or talents do you need the most to be a cartographer?

C. Acquiring the Cartographic Perspective

What made the critical difference for you in choosing cartography as a field? Did anyone inspire you to take up this field? If so, how did that happen? Do you have any mentor(s)?

D. Getting Involved in Cartography

What led you to consider cartography as a possible career?
Can you briefly outline for me the steps you have taken in your career?

E. Achieving Identity as a Cartographer

At what point did you consider yourself a cartographer?
How do people outside the field react when you say you are a cartographer? What do you say?
What sustained you in your decision to become a cartographer?

F. Acquiring Expertise as a Cartographer

What were the defining moments in your training? Your career?
What really counted as you went about learning the tricks of the trade?
Are you passing on your cartographic expertise to the next generation? In what manner?

G. Experiencing Professional Relationships

How has your career turned out differently than your expectations?
What counts, or what establishes one's reputation as a cartographer?
What forms of cooperation in the field are the most rewarding for you?
How have the professional relationships with your colleagues, men and women, changed as you progressed through the field?
How do you think these relationships are different than in other fields?
How would you describe your involvement with the Commission of Gender and Cartography?

H. Gender in Map-making

Do you believe there is a gender aspect to map-making?
Are women inclined to do better in any area of cartography than men?
Do you think that women go into different areas in cartography than men? What are these areas?
Are men better at certain things than women in cartography?
Do you see a difference between how you, as a female cartographer, do cartography and how men do as cartographers?

I. Personal Aspects of Gender in Map-making

It must be a real challenge to work in an area dominated by men. I would like to know what it's like.

Have you found any resistance to your being a woman cartographer? If so, how have you handled such resistance?

Have you experienced any discouragement?

How do you network and cooperate with, and nurture other women in the field?

If you were giving advice to other women coming up in cartography, what advice would you give them?

What advice would you give to your male colleagues about encouraging women to undertake cartography?

J. Other Questions

As someone who is not trained in cartography, I would like you to tell me of any historical knowledge about women in cartography that interests you.

Can you send me your c.v.?

APPENDIX C

Overview of Twenty-Eight Women Pioneers in Cartography

APPENDIX C

	Home Base	Year of birth	Highest degree	Concerns	Areas	Awards
Chapter 5 (Near-contemporary)						
Blagg	UK	1858	One course	International	Lunar maps	Moon craters named after her
Kelley	USA	1859	L.L.	National	Thematic	Unknown
Semple	USA	1863	M.A.	National	Human geography	Yes
Hubbard	Labrador	1870	None	Territorial	Basic mapping	None
Pearsall	UK	1906	None	Cities	City maps	None
Chapters 6 and 7 (Contemporary)						
Campbell	UK	1915	M.A.	International	Cartographic history	Yes
Pruitt	USA	1918	M.A.	National	Remote sensing; Coastal areas	Yes
Tharp	USA	1920	M.Sc.	International	Ocean mapping	Yes (late in life)
Wallis	UK	1924	D.Phil.	International	Cartographic history	Yes
Hsu	USA	1932	Ph.D.	International	Symbolization; Chinese cartography; Projection	Unknown
Tyner	USA	1938	Ph.D.	National	Academic persuasive cartography	Unknown
Shingareva	Russia	1938	Ph.D.	International	Planetary	Yes
Kerfoot	Canada	1938	B.Sc.	International	Toponymy	Yes

Kretschmer	Austria	1939	Ph.D.	International	Historical cartography	Yes
Petchenik	USA	1939	Ph.D.	International	Spatial comprehension; Children's maps	Map competition named after her
Gilmartin	USA	1941	Ph.D.	National	Cognitive maps	Unknown
Auringer Wood	Canada	1942	A.M.	National and International	Map librarianship	Yes
Bond	UK	1943	B.A.	International	Admiralty charts	Yes
Shortridge	USA	1943	Ph.D.	National	Thematic maps	Yes
Olson	USA	1944	Ph.D.	International	Academic GIS	Yes
Caldwell	USA	1945	Ph.D.	International	Educational systems and potential of graphic digital systems	Unknown
Siekierska	Canada	1945	Ph.D.	National	Electronic, tactile, multilingual	Yes
Clawson	USA	1948	B.Sc.	National	Digital mapping	Yes
Reyes	Mexico	1949	Ph.D.	International	Cyber-cartography	Yes
Almeida	Brazil	1949	Ph.D.	International	Tactile mapping	Yes
Buttenfield	USA	1952	Ph.D.	National	Generalization	Yes
Mersey	Canada	1953	Ph.D.	National	Geomatics technologies and land management	Yes
Meng	China	1963	Ph.D.	International	Geo-data generalization	Yes

Notes

NOTES TO CHAPTER 1

1. The study of women in cartography is a relatively late phenomenon in the general study of women, work, and technology. Many of the important works appeared in the late 1970s and the 1980s. Barbara Drygulski Wright's work entitled *Women, Work, and Technology* (1987) is particularly relevant. The various contributors point out that not only will technology displace women workers, but women workers will experience technological change very differently than men. While earlier research trumpets technology as an agent for change (for the condition of women), later scholarship underscores the fact that technology brings about only superficial change (see McGaw, 1982).
2. Alice Hudson and Mary Ritzlin have produced the most systematic and comprehensive list of pre-twentieth-century women cartographers (Hudson and Ritzlin, 2000a, 2000b).
3. As many of my interview participants were trained in "surveying engineering" departments, I have retained this term in this book when I describe their experiences, rather than using the more current term, "geodesy" or "geomatics."
4. *Geographic Perspectives on Women Newsletter; Canadian Women and Geography Newsletter; Gender, Place, and Culture: A Journal of Feminist Geography; Gender and Geography Newsletter; Progress and Perspectives;* ACSM *Bulletin* (Special Issue, 1987); and *Meridian* (Special Issue, 1988).

NOTES TO CHAPTER 1

5 When using "map worlds" in the plural sense, I intend to address the general features of all map worlds, regardless of the time period. I generally use the singular "map world" when I am speaking of one within a specific time period.

6 Even without such a reference, there is nevertheless a small account of a prime minister's younger sister's embroidering maps (Thrower, 1972: 23).

7 In related fields, there are discussions about map embroiderers (Ring, 1993; Tyner, 1994, 1996, 2001). Victorian women travellers receive considerable attention (e.g., McEwan, 2000; Stefoff, 1992), as do women astronomers (Ritzlin, 1999), women scientific instrument makers (Morrison-Low, 1991), and women geographers and women geography teachers (Ritzlin, 1993).

NOTES TO CHAPTER 2

1 A small portion of this chapter was presented at the Annual Meetings of the Canadian Cartographic Association in June 2000 (van den Hoonaard, 2000b).

2 I have come across the following twenty-four occupational incumbents in cartography: academic cartographer, astronomer, atlas cartographer, cartographic technician, community college instructor, computer specialist, freelance map-maker, geodesist, geography teacher, GIS specialist, head of sales, high school environmental science teacher, map collector, map historian, map librarian, map-maker, photogrammetrist, primary school teacher, program administrator, regional planner, surveyor, teacher of cartographic literacy, tourist-map publisher, university professor. These occupations are not mutually exclusive.

3 I use pseudonyms whenever I am referring to the interview participants in my study. In this chapter, I have placed the pseudonym in quotation marks; the number in parentheses after the name indicates the page of the transcribed interview.

4 The term "tabletop" cartographers refers to the technique of applying map symbols, letters, and so on directly onto a flat surface, located on a table of sorts.

5 These artifacts and processes include extracting minerals, making the copper plate, planishing the plate with a hammer, polishing the plate with a grinding stone and water, pumice stone, smooth hone, charcoal, steel burnisher, thirteen different burnishing tools, special paper, virgin wax and feather, pencil, transparent paper, Venetian varnish, carbon paper, needle and acid, ten different engraver tools or burins, six different punches and lettering tools, roulette to create a stippled effect, sharpener for the tools, black felt, olive oil, grinding

the pigment for the ink in a medium of nut oil, iron pot and flame, ink, rolling press, paper, a smooth board, inking ball made of linen, soft rag, charcoal brazier (replaced by a steam box in 1818), and flannel (Verner, 1975: 52–53).

6 These products involve the use of tin plates, wax, silver salts, light-sensitive bitumen, benzene, copperplate, chromium salts, developers, etching acids, distortion-free lenses, wet-collodion plate negatives, stones, silver nitrate, glass negative, wax, soluble bitumen, diamond, acid-resisting bitumen, lithographic ink, acid, turpentine, paper, four colours (black, yellow, blue, red), and special frames with pins (Koeman, 1975).

7 They include preprinted letters, visibility meter table, typeface chart, tracing paper, translucent material, mechanical lettering device (including Leroy, Varigraph, Wrico), photo-lettering device, straightedge, curve, Ames lettering instrument, lettering angles, pencil lettering pens (several kinds, including Leroy), books on lettering, template, scriber, T-square, stencil, imprinter, ink, gummed stock, thin tissue, Cellophane, acetate (glossy, matte finish), flexible film, a thin, sharp knife, and official place name speller (Robinson, 1960: 245–56).

NOTES TO CHAPTER 3

1 The earth was represented by a circle ("O"), within which the mapper divided the three continents (Europe, Asia, and Africa) much like the letter "T," which marks off three spaces; Jerusalem was typically placed at the centre of such a map. According to Tooley (1978: 12), there are no fewer than six hundred of these (*mappaemundi*) maps dating from the eighth to the fifteenth centuries.

2 I rely very much on Hoogvliet's written account (1996) of the Ebstorf map, although other sources are also indicated in my rendition.

3 Tooley (1978: 14) tells us that the Ebstorf map was created in 1284.

4 Konrad Miller "redrew" other maps, such as Lambert of St. Omer's *Liber Floridus* (1112–1121 A.D.) (Chekin, 1999: 14).

5 With the expanding map world came women whose images first appeared on colourful map edges (see Traub, 1997: 18). Suarez (1997: 12) reminds us of Ortelius's title page of his famous 1570 atlas that portrays the Americas as the "head and breasts of a woman with a flame underneath, symbolizing that nothing was known of its body and, of its people, only that they lit fires." There are, nevertheless, some tantalizing bits of imagery on a number of seventeenth-century map title pages and cartouches. One image on the front page of *Atlas Contractus,* published by the heirs of Johannes Janssonius in 1666 (Shirley, 1996: 14), shows a woman guiding Gerardus Mercator's hand holding a compass. Given the great stature of Mercator, it is particularly sig-

nificant that a woman seems to guide him. She holds his hand firmly, an intimate moment in a scene filled otherwise only by men. Her clothing resembles the kind that artists imagine women in Greek mythology wear, suggesting that she is Clio, one of the Muses, the source of Mercator's inspiration.

The other image originates in a cartouche in "Prima Pars Brabantiae cuius caput Lovanium," a map by Willem Janszoon Blaeu (1571–1638) originally published in *Theatrum Orbis Terrarum* in 1635 (Duncker and Weiss, 1983: 90–91); it is found on the front cover of *Map Worlds*. The woman, wearing contemporary clothing, has actually not only taken a compass in hand, but also seems to be showing the man over her shoulder a point on a map.

6 This section relies on the following sources: Lloyd A. Brown (1949), Duncker and Weiss (1983), Hudson and Ritzlin (2000b), Koeman (1967), Chancy (1996), Cole (1996), van der Krogt (1996), Ritzlin (1986), and van den Hoonaard (2009a). I also used these Internet sources: www.vintage-map.com/en/Cartographers:_:14.html; http://new.myfonts.com/person/Christophe_Plantin/; http://www.oxforddnb.com/articles/20/20853-article.html?back=; http://stewartcollection.googlepages.com/Inthefamilywamapmakingdynasty08090.pdf; and www.worldviewmaps.com.

7 I am indebted to Mary Ritzlin, who shared her knowledge of colourists from this era (email to author, 12 February 2011).

8 The paragraphs that deal with the wealth and longevity of cartographic families rests on an article I wrote for *Cartouche,* the newsletter of the Canadian Cartographic Association (van den Hoonaard, 2009a).

9 This section alludes to a number of critical sources. Linguistic transcriptions (such as from sixteenth-century Dutch to English) created inconsistencies in spelling and genealogies. Aside from these critical sources, I have also derived benefit from the following portal: http://www.biografischportaal.nl. It derives much of that information from the *Nationaal Biografisch Woordenboek*, with links to http://www.historici.nl/retroboeken/nbwv.

10 J. Keuning (1948) provides one of the few full-scale overviews of Jodocus Hondius Jr. and his family.

11 Van der Krogt mentions 1612 as the date of Hondius's death (van der Krogt, 1996: 63).

12 I have taken Henricus Hondius's birth and death dates from de Haan (1883: 129).

13 Historical cartographer Lloyd Brown (1949: 166) only makes the point that Hondius's son (Henricus) and son-in-law (Johannes Janssonius) took over the business.

14 Pieter van den Keere produced the famed "Leo Belgicus" map.

15 Each of the nine printing presses at Blaeu's publishing house (the best in Europe) were named after a Muse; the damage was 382,000 gulden, representing a very spectacular sum. Gerald Hulst van Keulen bought up remaining inventory.
16 Ritzlin quotes J.H. Hessels (1969: xxiii, 14, 330) as a source for this information.
17 He also signed his work as Guilielmus Janssonius, and then Willems Janszoon, but after 1621 he regularly used G. Blaeu.
18 The most common references to widowhood are *Vidua* (Latin), *Veuve* (French), or *Weduwe* (Dutch/Flemish).
19 Hudson and Ritzlin (2000b) invoke the following names: the widow of H. Peetersen (1551); Paschina De Jode (widow of Gerard de Jode) (1593–1613); Catherine Morentdort/Moretus (widow of Theodore Morentdort) (1598); Coletta van den Keere (1612–1620); the widow of Jan Baptist Vrients (1612); the widow of Joannis Cnobbari (1641); Francina van Offenberg Janszoon (the widow of Jodocus Janszoon (1642–1665); and Abraham Goos's widow.

NOTES TO CHAPTER 4

1 I am deeply indebted to the scholarship of Judith Tyner, who produced a number of relevant and timely articles on the subject of embroidered maps (1994, 1996, 2001).
2 Differences between maps made at Quaker schools and those at non-Quaker schools were noticeable. While Quaker maps "tend to have few embellishments and a simple cartouche of vines and leaves," the non-Quaker maps often had "an elaborate floral border that overshadow[ed] the map itself" (Tyner, 2001: 38).
3 The equivalent of embroidered maps returned 110 years later, when handkerchief maps were included by the Royal Canadian Air Force as part of escape kits during World War II (Crawley, 1985: 55; Dancocks, 1983: 159–160; Prouse, 1982). Maps continue to be a favourite topic for cross-stitchers. I thank my daughter Lisa-Jo van den Scott, an inveterate and exceptional cross-stitcher, for finding various websites related to cross-stitching maps, especially www.cross-stitch.com.au/sitemap.html, an Australian website that specializes in maps, and for working flora and fauna into her cartographic depictions.
4 The adoption of lithographic printing was not universal. For example, the long-standing and successful Dutch firm of Covens and Mortier (1685–1866) never resorted to lithography (van Egmond, 2002: 67).
5 Mary Ritzlin's article on Mary Biddle (1990a) offers a list of sources regarding Biddle's life.

NOTES TO CHAPTER 4

6 Stauffer mentions Mrs. Akin, who was a Massachusetts engraver from 1806 to 1808, known through a picture of her showing children "with a specimen of her abilities in the graphic arts" (Stauffer, 1907: 6). Miss H.V. Bracket, one of the earliest women engravers upon copper in the United States, drew, etched, and engraved a Bible in 1816 (Stauffer, 1907: 28–29). For a brief moment, Alice Hall, born in 1847 in England, evidenced considerable talent in drawing and etching portraits of Washington, but illness in the family caused her to abandon a profession in which she had made "a promising beginning" (Stauffer, 1907: 113). Helen E. Lawson engraved illustrations in the works of Professor Haldeman and Dr. Binney, in addition to several plates of birds for a publication around 1880 (Stauffer, 1907: 158). Both Maria A. and Emily Maverick, of New York, engraved several "admirable stipple illustrations" for an edition of Shakespeare published about 1830. Emily Sartain, born in Philadelphia in 1841, learned to engrave under the auspices of her father. She eventually engraved a few mezzotint portraits and became the principal of the Philadelphia School of Design for Women. She also exhibited at the Paris Salon in 1875 and 1883 (Stauffer, 1907: 233–34).

7 The imposing paper factory of l'Anglée, about 100 kilometres south of Paris near the city of Montargis, provided, like many paperworks, housing for its workers that flanked the factory.

8 Website: http://www.worldviewmaps.com (of WorldView Antique Maps, Katonah, New York).

9 We are just now, early in the twenty-first century, seeing the tail end of this process, as evidenced, for example, by the Inuit in northern Canada, who are formally reasserting the native toponyms (Rundstrom, 1991: 9).

10 To that end, a Danish naval officer, Nils Frederik Ravn (1826–1910), introduced isolines, which he pioneered in 1857 (Thrower, 1996: 150).

11 This section draws heavily on Ritzlin (1990c), who cites a number of important and relevant sources bearing on Emma Willard's life.

12 Ritzlin (1990c) lists several sources of information on Emma Willard.

13 An anonymous reviewer made the suggestion that these atlases were sold to her students.

14 Ristow mentions that Silas Farmer purchased these interests in 1863 (Ristow, 1985: 278).

15 A reviewer of an early draft of *Map Worlds* mentioned that there is a still earlier globe, dating back to 1760.

16 Elly Dekker and Peter van der Krogt give a number of examples of globes made by girls and women in *Globes from the Western World* (1993).

17 The section on Norway is based on van den Hoonaard (2009b). I am grateful for the assistance and interest I received from Roald Aanrud, Kari Strande, Marit Moen of Lofotmuseet (Kabelvåg), and Sveinulf Hegstad of Tromsø Museum, Norway. Main sources: Aanrud (1988a, 1988b, 1988c), Cappelen Antikvariat (2009), Anon. (1997), Kari Strande (email to author, 3 May 2001), http://www.tfb.no/db/personalhistorie/3_7_20070227_220633.pdf. Permissions: Perspective Maps: Lofotmuseet, Lofoten Islands, Norway.

Photo of Kirstine Colban (Stine Aas): the source of the photo is so far unknown; I would appreciate it if anyone could locate the original. Roald Aanrud (email to author, 18 November 2009) does not recall the location. The National Museum in Oslo does not have a copy.

18 I relied on Edith G. Firth (2000) for a comprehensive view of Elizabeth Simcoe's life. Although Canada did not become a country until 1867, and the western provinces joined Confederation in the following years, I use "Canada" to describe Mrs. Simcoe's setting, since that is how the country is now known.

19 I consulted several sources on Elizabeth Simcoe. Mary McMichael Ritzlin wrote the lead-in article (1990b), pointing the reader to some original sources, such as Fowler (1977) and Innis (1965).

20 I have used Stephen R. Bown (2000), Whitby (2005), and Winter (1975) to cover the story of Shanawdithit.

NOTES TO CHAPTER 5

1 It was N. Creutzberg who coined the term "thematic map" for these types of maps (Wallis and Edney (2003: 1112).

2 The vignette of Florence Kelley is based on a book by Sklar (1995), an article by Nina Brown (2009a), and entries by Bowman (1995), Blumberg (1974), and Breckenridge (1944).

3 This biography has relied on a number of sources: Gloria Cooksey (2002), Susan R. Brooker-Gross (2000), Douglas R. McManis (1996), and Brown (2009b).

4 Susan Brooker-Gross (2000:1) states that Semple was allowed to hear Ratzel's lectures "from outside the open classroom door."

5 Janice Monk (1998) offers a detailed history of women academics at Clark University. Mildred Berman devoted an article (1974) to Ellen Semple's struggles with academia.

6 Once again, I use the term "Canada," even though the country was not officially called by that name until 1867, and a number of provinces did not join Confederation until even later.

NOTES TO CHAPTER 5

7. I have derived this number from the list of women map-makers compiled by Alice Hudson and Mary McMichael Ritzlin (2000b).
8. I relied on Anne Hart's biography of Mina Hubbard in Buchanan and Greene (2005) as the basis of this overview. Others have written about Hubbard, including Pierre Berton ("The Revenge of Mina Hubbard," 1978) and James West Davidson and John Rugge (*Great Heart*, 1997). I have not consulted these latter two accounts.
9. I derived this figure from ICA (1995), a directory of women involved in cartography, surveying (geodesy), and GIS.
10. I have used a number of sources to construct the vignette of Mary Blagg, including Hutchings (2004), the *Wiley Book on Astronomy* (Answers.com), and Wikipedia.
11. One reference refers to him as John Charles Blagg (Wikipedia on Answers.com).
12. I have based her biography on numerous other ones, including one by Leo Knowles (2003), Angie Macdonald (2008), Design Museum (2006), Sarah Hartley (2002), Geographers' A–Z Map Co. Ltd (2006), Claire Heald (2006), and Fred Bond (1996). *From Bedsitter to Household Name* (Pearsall, 1990) constitutes her detailed, and quite humorous, autobiography.
13. Claire Heald (2006: 1) claims Pearsall was on her way to a party, holding the most recent Ordnance Survey map that was sixteen years old.
14. Heald (1996: 2), like Design Museum (2006), mentions that the plane crash occurred in 1945.
15. Surveys from the air began a half-century before airplanes. Gaspard Felix Tournachon ("Nadar") was the first person to take photographs from the air, using a balloon. In 1858, a French military engineer, Laussedat, used these photographs for mapping purposes (Hodgkiss, 1981: 55).
16. I have relied rather heavily on Judith Tyner's research (e.g., 1999) for the discussion on "Mapping Maids." She is continuing this research.

NOTES TO CHAPTER 6

1. Of the eleven women in North America, seven are living; of the ten women elsewhere in the world, five are living.
2. I used a variety of sources to compile Marie Tharp's biography, but notably Margalit Fox (2006), the *Encyclopedia of World Biography* (Anon., 2004), Cathy Barton's detailed biography of Tharp (2002), and the Earth Institute News Archive of Columbia University (2006). Marie Tharp's own nine-page biography, which she presented on the occasion of the Mary Sears Woman Pioneer in Oceanography Award at the Woods Hole Oceanographic Institute (Tharp, 1999), offers a telling glimpse of her life and works.

3 I derived the vignette on Mary Clawson from detailed information she sent me on 25 March 2012.
4 Robert McMaster and Susanna McMaster (2002: 317–18) offer a bird's-eye view of women in academic cartography.
5 A reminder that names in quotation marks are pseudonyms of interview participants.
6 These other venues include the universities of Chicago, South Carolina, Northern Illinois, Southwest Texas State, Michigan State, Oregon State, Penn State, SUNY at Buffalo, Ohio State, Syracuse, Minnesota, UC–Los Angeles, UC–Santa Barbara, Clark, Georgia, San Diego State, and George Mason (McMaster and McMaster, 2002: 305).
7 An article by Judith Tyner (2006) gives an insightful overview of the professional connections between Arthur Robinson, George Jenks, John Sherman, and Norman Thrower.
8 The obituary of John Sherman can be found at http://www.accessmylibrary.com/article-1G1-53947012/memoriam-john-clinton-sherman.html and http://faculty.washington.edu/krumme/faculty/shermanbio.html.
9 His obituary is found at http://www.csun.edu/%7Ehfgeg003/csg/winter97.html.
10 His obituary can be found in *Cartography and Geographic Information Systems*, 24 (2), Fall 1997: 117, by Jim Carter.
11 An obituary by J. Jesse Walker (2006) provided the substance of this vignette of Evelyn Pruitt, in addition to Janice Monk, email to author, 13 September 2011.
12 Her six-page article (Pruitt, 1979) gives a fine overview of the activities of the Office of Naval Research and Geography.
13 This vignette of Mei-Ling Hsu relied mainly on a tribute by Zhou (2010) and by the Association of American Geographers (2009). It also relied on missives from Judy Olson to me (22 June 2011) and on a transcript from Janice Monk, who interviewed Mei-ling Hsu on 4 May 1990.
14 Some sources state that she completed her thesis in 1966, but Hsu herself (Monk, 1990: 8) says it was 1972.
15 I consulted two sources to create this vignette of Judith Tyner: *The Foghorn* (Fall 2005: 4) and her homepage on the California State University–Long Beach website (http://www.csulb.edu/colleges/cla/departments/geography/faculty/tyner). She also sent me her updated vignette on 22 July 2011, and then, on 27 October 2011, provided additional information.
16 This vignette is based on these sources: Auringer Wood (1993), FamilyTreeMaker Online (2011), and J. Robinson (1994). The Newberry Library has one

box of her papers, dated 1965–1992 (Call Number: Midwest MS Petchenik; Location 3a 41 2).

17 This all-too-brief bio of Patricia Gilmartin is based on the Department of Geography Faculty web page (http://www.cas.sc.edu/geog/people/faculty.htm) and her blog (http://patgilmartinaboutme.blogspot.com). See also Anon., 1991: 3, 6.

18 As a source for the vignette on Barbara G. Shortridge, I consulted her seven-page curriculum vitae, in addition to information she provided in an email to me (29 February 2012).

19 I based my vignette of Judy Olson on a wide variety of sources, namely Auringer Wood (2005), Anon. (2008), a website of the Department of Geography at Michigan State University, Olson (1997), and email exchanges between 2002 and 2011.

20 *Cartographic Perspectives* (2005) is a special issue of the journal of the North American Cartographic Information Society that contains highly informative details of how Woodward mentored his students.

21 Anon., 2008: 11 provided a number of details about her professional activities.

22 To create this overview of Patricia Caldwell's (Lindgren) life, I relied on Straight (1990: 5), Greulich (2006), Wendy Straight (personal communication), and emails from Caldwell, dated 21 and 22 February 2012.

23 One of the principal sources for Dr. Barbara Buttenfield is her online c.v.: http://www.colorado.edu/geography/babs/webpage/vita_jan11.pdf. I also corresponded with Dr. Buttenfield, especially in June 2011, with regard to her vignette.

24 What follows is a list of fifteen women whose work touches on cartography. Eva Siekierska and Helen Kerfoot provided me with synopses of their respective contributions or activities. I am very grateful to both of them for their help.

25 http://www.acmla.org/honorary_nomination_Welch.pdf.

26 http://www.science.ca/scientists/scientistprofile.php?pID=346&pg=0.

27 The vignette on Helen Kerfoot is based on a variety of public documents: DESA (2007), Royal Canadian Geographical Society (2011), ANPS (2004), Tierney (2008), and Rysted (2007: 6). I also relied on correspondence from Kerfoot, especially her email of 20 December 2011.

28 I have based this vignette on Auringer Wood on my extensive contact and correspondence with her between 1999 and 2012.

29 On 16 March 2011, Dr. Eva Siekierska sent me an eighteen-page biography based on a template she had prepared for the vignettes of all women pioneers. I have also had considerable contact with Siekierska since 1993, including correspondence during that time.

30 In January 2012, the Government of Canada started the process of archiving a number of websites devoted to many of the maps mentioned in this vignette.
31 I base the vignette of Janet Mersey on her c.v. and on emails she sent me in February 2012. I also added information provided by the Canadian Institute of Geomatics (http://www.cig-acsg.ca/english/general/exec_council_bio.php).
32 Today, women in cartography do not seem to be given awards later in life. We note Dawn Wright of Oregon State University, who was recently named chief scientist for ESRI, Mei-Po Kwan (also of Ohio State University), who was recipient of the AAG Distinguished Scholarly Honors Award in 2011, Trudy Suchan of the United States Bureau of Census, and Cynthia Brewer at Penn State University, who were the 2007 winners of the AAG Globe Book Award for their *Census Atlas of the United States* (see Roth, 2010), and Sarah Elwood of the University of Washington, who received prestigious grants for work related to participatory GIS/mapping. I wish to thank a reviewer for this suggested list of awardees.

NOTES TO CHAPTER 7

1 A note of thanks to the reviewers of an earlier draft of *Map Worlds*.
2 Avril Maddrell's thorough work, *Complex Locations* (2009b: 222–26), incorporates a detailed account of Eila Campbell's life.
3 Wallis (1994: 361) provided most of the information regarding Campbell's education and career.
4 I met Professor Koeman in his cartography lab at the university in 1960, when I was eighteen years old. My love for maps led me to visit this distinguished scholar who had contributed so much to cartography. He seemed bemused that someone had, on a whim, bicycled from The Hague, some 56 kilometres. The atmosphere and the physical layout of his lab (and the university) created in me such a desire that I resolved to do academic work, despite my having been kicked out of Grade 10 and never having returned to high school.
5 I am grateful to the articles by W.R. Mead (1995b), Harry Margary (1995), Monique Pelletier (1995), Helen Wallis (1994), and Peter Barber (1995), which I relied on to develop this vignette.
6 I am grateful to Barbara Bond, who supplied me with additional information about Helen Wallis (email to author, 10 September 2011). I have added "T." in the Campbell reference to avoid confusion with Eila Campbell, whose vignette follows immediately.
7 I relied on Ravenhill (1996), Tony Campbell (1995, 2004), Fairclough (1974), Thrower (1999), Barber (2005), and Mead (1995a) to develop this section on Wallis's education and her early career at the British Museum (now the British Library).

NOTES TO CHAPTER 7

8   Wallis's other awards and recognitions include the Fulbright Scholarship (1979), Honorary Fellowship (Library Association, 1985), Hon. D.Litt. (Davidson College, North Carolina), Honorary Fellow (Portsmouth Polytechnic), Honorary Fellow (Library Association), Caird Medal (National Maritime Museum), Silver Medal (British Cartographic Society, 1988), Distinguished Achievement Award (Society of Woman Geographers), Honorary Fellowship (International Cartographic Association, 1991), the prestigious Victoria medal (the Royal Geographical Society, 1995), and the IMCoS Helen Wallis Award (The International Map Collectors' Society, 1995).

9   Unless mentioned otherwise, this biographical sketch is based on my interview with Kira Shingareva, in Budapest on 6 September 2000. Dr. Ewen A. Whitaker also provided me with relevant background information in a letter to me dated 28 March 2011.

10 From 1959 on, according to Hargitai (2004: 158), "Soviet scientists had the exclusive right to name newly observed features of the far side of the Moon, which resulted in a predominance of Soviet names."

11 It was a long-standing tradition that no heavenly body be named after someone who was still alive. However, a first name was used, without the last name, if the person was still alive. Hargitai (2004) provides an interesting and useful discussion about the naming of heavenly objects, both in the past and in the present. He says that "of 5070 planetary names (excluding Lunar names), 12% are of Greek, 7% of Latin/Roman, 5% of British/English, 4% of Russian, 4% of American, 3.5% of French, and 2.5% of Norse origin. The remaining 62% is taken from 280 past and present nations, cultures, and countries whose numbers are constantly growing" (Hargitai, 2004: 159).

12 Ewen A. Whitaker's *Mapping and Naming the Moon* (1999) provides a full account of the exploration of the Moon. According to David Strauss (2002), he was ideally situated to produce this history as a participant in the Apollo missions and a member of the Task Group of Lunar Nomenclature of the International Astronomical Union. Strauss tells us that Whitaker was "directly involved in conflicts between representatives of different countries over naming newly discovered lunar features."

13 I have used the following two sources to denote the general facets of Ingrid Kretschmer's biography: Aschenberner (2004) and Freitag (2004). Peter Aschenberner was her research assistant starting in 1966. Her official obituary (in German) can be found at http://www.kartengeschichte.ch.

14 Wolfgang Pillewizer was professor at the Institute of Cartography and Reproduction Technology of University of Vienna of Technology from 1971 to 1981. He was especially interested in glacier research and participated in

NOTES TO CHAPTER 7

high-altitude and polar expeditions. http://de.wikipedia.org/wiki/Wolfgang_Pillewizer.

15 Erik Arnberger (1917–1987) was a highly regarded cartographer; the range of his work reached both national and international spheres of influence.

16 Barbara Bond was kind in providing me with her most recently updated bio, 11 May 2011, as well as some supplementary material (email to author, 28 August and 10 September 2011), including a write-up by the Department of Geography of the University of Leeds (www.geog.leeds.ac.uk/fileadmin/downloads/school/alumni/ALUMNI2010.pdf).

17 The University of Leeds School of Geography (2010) *Alumni Newsletter* and Karen Mason (2009) provided the substance of information about Barbara Bond's education and career.

18 These remarks about professional friendships came from Barbara Bond herself (email to author, 10 September 2011).

19 To develop her biography, I used a variety of sources. *Gesellschaft für Deutsche Professoren Chinesischer Herkunft E.V.* (the Association of German Professors of Chinese Background) was very useful. I also maintained correspondence with L. Meng during 2011.

20 The name "Meng" originates from Mencius (http://en.wikipedia.org/wiki/Mencius). Liqiu Meng reports that "when I tell my Chinese [colleagues and friends] that I am a teacher and my name is Meng, they would say, 'no wonder'. The name Meng is predestined to the teaching career. My birthplace Changshu also happens to be known as a birthplace for scholars, possibly due to its favourable living environment. Some fragments about this place can be found under http://en.wikipedia.org/wiki/Changshu and http://www.jiangsu.net/city/city.php?name=changshu. For these two reasons, I often tell my parents that I would do my best to keep the aura of Meng and Changshu."

21 www.atlaslatinoamerica.org.

22 Unless mentioned otherwise, I derived many details about Reyes's life when I had a conversation with her at the ICA conference in Ottawa, Ontario, on 20 August 1999. Emails to me, 20 July and 6 October 2011, also provided me with many details about her work and family.

23 I created the vignette for Regina de Almeida based on interviews with her, as well as extensive email exchanges in July and October 2011.

24 Regina de Almeida's website provided many details about her educational and career interests. See http://buscatextual.cnpq.br/buscatextual/visualizacv.do?metodo=apresentar&id=K4787992D9.

NOTES TO CHAPTER 8

1. A portion of this chapter was presented at the Joint Conference of the Association of Canadian Map Libraries and Archives and the Canadian Cartographic Association, Montreal, 30 May–2 June 2001. A significant portion was published in *Cartographica* (van den Hoonaard, 2000c).
2. Because this and the successive chapter deal with interview data, I have not used quotation marks to set off the pseudonyms of the interview participants.
3. A number following the interview participant's pseudonym indicates the page number of the transcribed interview.
4. Living in a developed country does not necessarily mean that the proportion of women is higher than in a developing country, as illustrated by the fact that 86 percent of the staff in a Himalayan office where an interview participant did her apprenticeship were women. Interestingly, the two cartographers who claim that cartography is not a "man's world" came from developing countries.
5. Several interpretations of my statement are in order. One must assume that, at the very least, the cartographers believe they were not discriminated against. Does this belief correspond to objective conditions, or do they not want to say otherwise?
6. *Working with Sensitizing Concepts* (van den Hoonaard, 1997) discusses the use of "sensitizing concepts" (such as *identifying moment*) in social research.
7. I am indebted to Deborah K. van den Hoonaard for pointing me to this particular study of women academics.

NOTES TO CHAPTER 9

1. A portion of this chapter was presented at CARTO 2002, the Annual Meetings of the Canadian Cartographic Association, Wilfrid Laurier University, Waterloo, Ontario, 27 May 2002.
2. For an interesting edited work, see John Paul Jones III, Heidi J. Nast, and Susan M. Roberts (1997). It includes a chapter by Nikolas H. Huffman, "Charting the Other Maps: Cartography and Visual Methods in Feminist Research."
3. Deborah K. van den Hoonaard reminded me of this traditional distinction of nomenclature.
4. One cartographic instructor (Minhui) commented that, in her experience, her women students are less inclined to display their maps on walls than are men students.
5. One wonders whether the above approach might not reflect aspects of humility or modesty that women bring into their contributions to their work and occupation. Nuvyn, for example, states that "I am not a good teacher, but I'm friendly" (Nuvyn: 1).

6 Future research might want to explore whether this zigzag pattern is also characteristic of the careers of men cartographers. My impression is that it is not.

NOTES TO CHAPTER 10

1 I used the annual reports of the Institute as the source of all my data in this section.
2 *Cartographic Perspectives* (2005) is a special issue of the journal of the North American Cartographic Information Society that contains highly informative details of how Robinson mentored his students.
3 It is worth noting that when I presented a portion of this chapter dealing with mentorship at a cartography conference, the audience was visibly upset about the unintended implications of discussing the role of mentors in the advancement of women cartography.

NOTES TO CHAPTER 11

1 In the United States, in the late 1980s, some 4 percent of workers in the surveying industry were women (*Progress and Perspectives,* May–June 1996: 1), a relatively low figure as compared to Canada, although women have been in surveying in the United States at least since 1918 (*Progress and Perspectives,* Jan.–Feb. 1995: 4).
2 I have derived this figure from the twenty-two national memberships of EUROGI, which total a membership of 5,086 cartographers. Using the average proportion of 25 percent to denote women cartographers, I estimate that there are 1,312 women cartographers in Europe alone.
3 I have excluded surveyors from this tally, primarily because they do not identify themselves as "cartographers," although they certainly do work that leads to maps. The percentage of women who are surveyors is still smaller than is the case for cartographers. In the United Kingdom, for example, 10.4 percent of the 107,942 surveyors in the Royal Institute of Chartered Surveyors (RICS) are women (N=11,173) (Fussey, 2002). In Japan, in 1992, 6 percent of surveying and mapping personnel out of 300,000 were women (*Progress and Perspectives,* Sept.–Oct. 1992: 1).
4 My field notes of 21 August 1999 are the principal source of information in this section on the 12th General Assembly.
5 The nature of the list of members varied considerably. Some listed only members of the executive, while others were more likely to list anyone who took interest in the activities of the commissions. In any case, I did exclude "correspondence members" and members who did not have an address or postal code (figuring that the latter group would be less involved).

6 Some of the information in this section originates from Siekierska, O'Neil, and Williams (1991) and from Siekierska (1997).
7 Judith Dixon (Library of Congress), Karen Luxton (Baruch College), and Billie Louise Bentzen (Boston College)—all thus from the United States.
8 It was Eva Siekierska's interest in advancing women in cartography that provoked my interest in doing research on gender in historical and contemporary cartography. Siekierska and I were members of a committee in a religious community in Ottawa when she shared her excitement with the establishment of the ICA Commission on Gender and Cartography.
9 Its membership was composed of the following people, in addition to Eva Siekierska, mentioned above: Edel Lundemo (a photogrammetrist working with the Norwegian Mapping Authority in Oslo; she also organized the first meeting of the ICA Task Force in 1990); Agnes Ajtay (d. 1994; in addition to being a lecturer at Eötvös Loránd University in Budapest, she worked at Kartographia, a mapping corporation of the Hungarian government); Ludwine d'Andigne d'Assis (a program specialist at the Division of Earth Sciences, UNESCO Headquarters, Paris); Ulla Durval (a marketing director of Maps International, a company located in Stockholm, Sweden, and the former marketing director of Esselte International, a major map producing company in Sweden); Natalia Guschina (editor, Map Kartografiya, in Moscow); Elri Liebenberg (professor of geography at the University of South Africa in Pretoria); Yang Jun (deputy director, Cartography Department, Research Institute of Surveying and Mapping, Beijing; now works at MacDonald Setwiller in Vancouver); Kirsi Makkonen (professor of Cartography, Department of Surveying, University of Technology, Otakaari, Finland); Helen Mounsey (Ph.D., originally worked for Birkbeck College of London University, but was employed at the private firm, Coopers, Lybrand and Deloitte, United Kingdom); Carmen Reyes (Ph.D., a very successful GIS consultant with Geographic Informations Systems in Mexico City); Sandra Shaw (deputy executive director, Bureau of Intelligence and Research, US Department of State, Washington, D.C.); and Kerry Smythe (Cartographic Division, Department of Land Administration, Perth, Australia).
10 Surveys, Mapping and Remote Sensing Sector; Department of Energy, Mines and Resources Canada; The Norwegian Association for Cartography; Geodesy, Hydrography and Photogrammetry; Norwegian Mapping Authority; Canadian Institute for Surveys and Mapping; The McElhanny Group; Rand McNally; Intergraph Corporation; Swedish Cartographic Association; Swedish Landsurvey; ICA International Secretariat; R.R. Donnelly and Sons Co.; American Association of Geographers; Champlain Institute; Canadian Cartographic Association; and Earth & Ocean Research.

11 Eventually, the Task Force sent out 1,300 questionnaires, of which 600 were returned and 412 were used in the final tabulations. The Task Force received questionnaires from women in thirty-four countries, with the largest number coming from the United States, Canada, Norway, and Sweden (Siekierska, 1994: 4).

12 In addition to the report itself (Siekierska, O'Neil, and Williams, 1991), there are at least three other reports that have dealt with this survey: Beaver (1993), Schweik (1993), and Siekierska (1995).

13 In 1999, at the height of the commission's activities, the following were its members: Seema Ahmed Ibraheem (Bahrain), Regina Almeida de Araujo (Brazil), Kirsi Artimo (Finland), Gizella Bassa (Hungary), Clara Caro (Colombia), Mutsuko Hoya (Japan), Margereta Elg (Sweden), Ewa Krzuwicka-Blum (Poland), Li Li (China), Helena Margot (South Africa), Mohamed Hassimiou Fofana (Guinée), Anna Mushi (Tanzania), Carmen Reyes (Mexico), Necla Ulugtekin (Turkey), Maria Valsamaki (Greece), Jarmila Vanova (Czech Republic), and Tamara Vereschaka (Russia).

14 The website is http://www.nrcan.gc.ca/siekiers.

15 The three suggested items include "Characteristics of demographic data collections," "Cartographic presentations of minorities or different inequalities within a population," and "Perceptual variety of map users."

16 Portions of this section on commercial exhibits have been previously published (van den Hoonaard, 2001a, 2001b, and 2002a).

17 The organizations that do not keep such a list include the Canadian Institute of Geomatics (CIG), the Association of Canadian Map Libraries and Archives (ACMLA), the Canadian Aeronautics and Space Institute (CASI), American Association for Geodetic Surveying (AAGS), Association of Canada Lands Surveyors (ACLS), and American Geographical Society (AGS).

18 All percentages in this paragraph are rounded off.

NOTES TO CHAPTER 12

1 I am indebted to one of the chapter headings in Ruth Woodfield's book *Women, Work, and Computing* (2001), which suggested "Pathways."

2 When I presented my findings (van den Hoonaard, 2002b) to a group of Canadian cartographers, outlining what the women interview participants had said about the obstacles some of them faced, the audience objected to these findings. It was as difficult for them to critically examine the map world as it was for me to convince them of my findings.

3 I conducted a total of thirty-eight interviews, but ten interviews did not provide any insights about these pathways through inequality.

4 The interviews in this chapter took place from 1999 onward.

5 I derived this figure from ICA (1995), a directory of women involved in cartography, surveying (geodesy), and GIS.

NOTES TO APPENDIX A

1 As I base my main analysis of this and the subsequent chapters on interviews with contemporary cartographers, I intend to use the present tense.
2 I wish to express my thanks to a reader of a far earlier version of *Map Worlds* for pointing to these specific methodological concerns.
3 I also recall attending the barbecue held by the ICA at the National Aviation Museum, where I saw a delegate with a camera. After I asked him if he would take a few photographs of the gathering, he fixed his gaze on me and said, "I know you. You are the sociologist!" I had never met the man before.
4 "ICA 1999" is the shortened version I will use for the 1999 International Cartographic Association Conference; also, I will use "Digital Earth" instead of "Digital Earth 2001 Conference."
5 Laura Roy, with the highest participation rate, had four poster sessions, but was also a presenter in a regular session.
6 I have used pseudonyms in referring to the thirty-eight cartographers, some of their own choosing. Persons who are in the early stage of their careers are: Elsa, Erika, Gabrijela, Judith, Kaisu, Parvine, Sawa, and Zia. Mid-career people are Amanda, Ettie, Inga, Leslie, Liisa, Myra, Loukie, Lynn, Marita, Minhui, Nedelina, Nuvyn, and Ruth. Interviewees in the mature stage of their career are Agnes, Alenka, Cécile, Hadley, Helena, Helle, Joyce, Kornélia, Kristina, Marlene, Marta, Natalie, Rachel, Ramona, Saskia, Shpela, and Tatiana.
7 The youngest of the mature cartographers is forty-five years of age; the oldest, seventy-one.
8 The youngest age of mid-career cartographers is thirty-three; the oldest is fifty-two.
9 The youngest person I interviewed who is early in her career is twenty-six years of age; the oldest is thirty-four.

References

AAG (Association of American Geographers). 2010. "Honors of the Association of American Geographers," http://www.aag.org/cs/honors.
Aanrud, Roald. 1988a. "Den første av dem?: Kirstine Colban fra Vågan in is Lofoten." *A la Kart.* 11 (Aug.): 6–8.
———. 1988b. "Kartografen ble kancellirådinne" *A la Kart.* 12 (Sept.): 13–14.
———. 1988c. "Kristine Colbans alias Stine Aas: 'Asylet' og arven." *A la Kart.* 13 (Oct.): 16–17.
ACA/SI. 2009. "Making Ends Meet: Innovative Ways of Funding International Students." *ACA/SI European Policy Seminar,* Stockholm, 15 October.
Adams, Annmarie, and Peta Tancred. 2000. *"Designing Women": Gender and the Architectural Profession.* Toronto: University of Toronto Press.
Almeida, R.A. 2001. "Cartography and Indigenous Populations: A Case Study with Brazilian Indians from the Amazon Region." In *Twentieth International Cartographic Association Conference—ICA 2001, Beijing. Proceedings. Beijing: ICA-ACI, 2001.* Vol. 1.
———. 2008. "Ensino de Cartografia para Populações Minoritárias." *Boletim Paulista de Geografia.* 1: 111–29.
Annual Report(s). [International Cartographic Educational Institute]. The Netherlands.
Anon. 1991. "Dr Pat Gilmartin." *Progress and Perspectives.* July–Aug.: 3, 6

REFERENCES

———. 1992. "In Memoriam." *Progress and Perspectives.* July–Aug.: 3.

———. 1993. "Female Army Mapmakers: 50 Year Reunion." *Progress and Perspectives: Affirmative Action in Surveying and Mapping.* (July-Aug.): 1–2.

———. 1997. "Scandinavian Literature, Norwegian Literature, The Age of Wergeland." *Encyclopaedia Britannica.*

———. 2004. "Marie Tharp." *Encyclopedia of World Biography*, http://www.highbeam.com/doc/1G2-3404706343.html.

———. 2008. "Prof. Judy M. Olson awarded the 2008 Earle J. Fennell Award." *CaGIS.* June: 11.

———. 2012. "The 3rd Reader's Report." Submitted to Douglas Hilderbrand, ed., University of Toronto Press, Toronto, 6 February.

ANPS. 2004. "International Training Course in Toponymy." *Placenames Australia: Newsletter of the Australian National Placenames Survey.* (December): 2–3, www.anps.org.au/documents/Dec-2004.pdf.

Applied Geography Specialty Group. 1984. "The James R. Anderson Medal of Honor in Applied Geography," http://agsg.binghamton.edu/nominate-am.html.

Ardener, Shirley, ed. 1981. *Women and Space: Ground Rules and Social Maps.* London: Croom Helm.

Aschenberner, Peter. 2004. "Ingrid Kretschmer." Presidential Address, German Cartographic Society, Stuttgart. 13 October, http://www.dgfk.net/index.php?do=mer&do2=kre.

Auringer Wood, Alberta, with C.H. Wood, eds. 1979. *Remote Sensing of Earth Resources*, by M. Leonard Bryan. Detroit: Gale Research (Geography and Travel Information Guide Series. Vol. 1).

Auringer Wood, Alberta, with C.H. Wood, eds. 1980. *Travel in Oceania, Australia, and New Zealand,* by Robert E. Burton. Detroit: Gale Research (Geography and Travel Information Guide Series, Vol. 2.)

———. 1981. *Travel in the United States,* by Joyce A. Post and Jeremiah B. Post. Detroit: Gale Research (Geography and Travel Information Guide Series, Vol. 3).

———. 1992. "Barbara Bartz Petchenik: Her Works, Citations to Her Works, Works about Her." *Cartographica.* 29 (2): 60–61.

———. 1993. "Notes by Alberta Auringer Wood." Mimeo. 4 pp.

———. 2005. "CCA Awards of Distinction." *Cartouche.* 58: 10.

Bagrow, Leo. 1964. *History of Cartography.* Trans. D.L. Paisley, rev. and enlarged by R.A. Skelton. Cambridge, MA: Harvard University Press.

———. 1966. *History of Cartography.* Trans. D.L. Paisley, with corrections by R.A. Skelton. Cambridge, MA: Harvard University Press.

REFERENCES

Ballantyne, Brian, and Sue Hanham. 1993. *Gender Imbalance in Surveying: Questions, Answers and Strategies.* Submitted to *New Zealand Surveyor.* Dunedin, New Zealand: Department of Surveying: University of Otago.

Barber, Peter. 1995. "The Eila Campbell Papers." *Imago Mundi.* 47: 10–11.

———. 2004. "Helen Wallis Fellowship: Invitation to Applicants." 10 February. http://mailman.geo.uu.nl/pipermail/maphist/2004-February/003067.html.

———. 2005. "In Memory of Helen Wallis (1924–1995)." *Cartographic Journal.* 42 (3): 195.

Barker, Hannah. 1997. "Women, Work and the Industrial Revolution: Female Involvement in the English Printing Trades, circa 1700–1840." In Hannah Barker and Elaine Chalus, eds. *Gender in Eighteenth Century England: Roles, Representations, and Responsibilities.* New York: Longman.

Barton, Cathy. 2002. "Marie Tharp: Oceanographic Cartographer, and Her Contributions to the Revolution in the Earth Sciences." In D.R. Oldroyd, ed., *The Earth Inside and Out: Some Major Contributions to Geology in the Twentieth Century.* London: Geological Society, Special Publications, 192.

Beaujot, Roderick. 2000. *Earning and Caring.* Peterborough, ON: Broadview Press.

Beaver, Carol. 1993. "The Role of Women in Surveying and Mapping." Paper submitted by ICA to the Fifth United Nations Regional Cartographic Conference for the Americas, New York, 11–15 January.

Becker, Howard S. 1982. *Art Worlds.* Los Angeles: University of California Press.

———. 1998. *Tricks of the Trade: How to Think about Your Research While You're Doing It.* Chicago: University of Chicago Press.

Berman, Mildred. 1974. "Sex Discrimination and Geography: The Case of Ellen Churchill Semple." *Professional Geographer.* 26 (1): 8–11.

———. 1980. "Milicent Todd Bingham: Human Geographer and Literary Scholar." *Professional Geographer.* 32 (2): 199–204.

———. 1984. "On Being a Woman in American Geography: A Personal Experience." *Antipode.* 16 (3): 61–66.

Bertin, Jacques. 1967. *Sémiologie Graphique. Les diagrammes, les réseaux, les cartes.* With Marc Barbut et al. Paris: Gauthier-Villars. (Trans. 1983. *Semiology of Graphics* by William J. Berg).

Berton, Pierre. 1978. "The Revenge of Mina Hubbard." In *The Wild Frontier: More Tales from the Remarkable Past.* Toronto: McClelland and Stewart.

Birkbeck, University of London. 2007. "Biography of Eila M.J. Campbell." Statement, http://bbk.ac.uk/events/campbell/eila_campbell.

Blair, Raymond N. 1983. *The Lithographers Manual.* 7th ed. Pittsburgh, PA: Graphic Arts Technical Foundation.

REFERENCES

Blumberg, Dorothy R. 1974. "Kelley, Florence." In John A. Garraty, ed. *Dictionary of American Biography*. New York: Harper and Row.

Blunt, Alison, and Gillian Rose. 1994. "Introduction: Women's Colonial and Post-colonial Geographies." In Alison Blunt and Gillian Rose, eds. *Writing Women and Space: Colonial and Postcolonial Geographies*. New York: Guilford Press.

Bond, Barbara. 1983. "Maps Printed on Silk." *The Map Collector*. 22: 10–13.

———. 1990. "Cartographic Curiosity and Intrigue: The Story of MI9's Escape and Evasion Maps, 1939–1945." Paper presented at the International Map Collectors' Annual Symposium, Imperial War Museum, London. 16 June.

———. 2009. "Escape and Evasion Maps in World War II and the Role Played by MI9." *The Ranger: Journal of the Defense Surveyors' Association*. 2 (19): 28–32.

Bond, Fred. 1996. "Obituary: Phyllis Pearsall." *The Independent*, independent.co.uk/news/obituaries/obituary-phyllis-pearsall-1312265.html.

Bowman, John S., ed. 1995. "Kelley, Florence (Molthrop)." *Cambridge Dictionary of American Biography*. Cambridge, MA: Cambridge University Press.

Bown, Stephen R. 2000. "Shanawdithit's Story." *Mercator's World*. 5 (5): 27–31.

Brandenberger, A.J. 1993. "The Role of Women in Surveying and Mapping." Paper submitted by the United Nations Secretariat at the Fifth United Nations Regional Cartographic Conference for the Americas. New York (January): Item 6(d) of the provisional agenda.

———. 1997. "The Status of Women in the Surveying and Mapping World." *Geomatica*. 51 (2): 153–56.

Brayton, Jennifer. 2003. "The Meaning and Experience of Virtual Reality." Unpublished Ph.D. Dissertation. Department of Sociology, University of New Brunswick, Fredericton, NB.

Breckenridge, Sophonisba P. 1944 "Kelley Florence." In Harris E. Starr, ed. *Dictionary of American Biography* (21), Supplement One. New York: Scribner's and Sons.

British Cartographic Society. 2000. *Freelance Directory: A Directory of Freelance Cartographers*. London: Royal Geographical Society. 4 September. 15 pp. Copy in files of author.

Brooker-Gross, Susan R. 2000. "Semple, Ellen Churchill." In *American Women Writers: A Critical Reference Guide from Colonial Times to the Present*. Farmington Hills, MI: St. James Press.

Brooks, Cheri. 2001. "French Collection: Cartographic Treasures of the Bibliothèque Nationale de France." *Mercator's World*. 6 (4): 25–29.

Brown, Lloyd A. 1949. *The Story of Maps*. New York: Bonanza Books.

Brown, Nina. 2009a. "Florence Kelley: Slums of the Great Cities Survey Maps, 1893." CSISS Classics, http://www.csiss.org/classics/content/35.

———. 2009b. "Ellen Churchill Semple: The Anglo-Saxons of the Kentucky Mountains, 1901." CSISS Classics, http://www.csiss.org/classics/content/24.
———. 2009b. "Ellen Churchill Semple: The Anglo-Saxons of the Kentucky Mountains, 1901." CSISS Classics. http://www.csiss.org/classics/content/24.
Buchanan, Roberta, Anne Hart, and Bryan Greene, eds. 2005. *The Woman Who Mapped Labrador: The Life and Expedition Diary of Mina Hubbard.* Montreal and Kingston: McGill-Queen's University Press.
Buehler, Elisabeth. 2001. *Frauen und Gleichstellungs Atlas Schweiz.* Zurich: Seismo Verlag.
———. 2002. *Atlas Suisse des femmes et l'égalité.* Zurich: Seismo Verlag.
Bulson, Eric. 2010. "Explorers in Space." *Times Literary Supplement.* (3 Sept.): 19.
Buttenfield, Barbara P. 1998. "Looking Forward: Geographic Information Services and Libraries in the Future." *Cartography and GIS.* 25 (3): 161–71.
———. 2007. "The Goals, Cost, Appearance of CU 101 are at Issue." *The Silver and Gold Record Staff Newspaper at the University of Colorado.* 10 May, https://www.cu.edu/sg/messages/5600.html.
———. 2011. "Babs Homepage." http://www.colorado.edu/geography/babs/ webpage.
———, L.V. Stanislawski, and C.A. Brewer. 2010. "Multiscale Representations of Water: Tailoring Generalization Sequences to Specific Physiographic Regimes." GIScience Short Paper Proceedings, http://aci.ign.fr/2010_Zurich/genemr2010_submission_11.pdf.
CaGIS (Cartographic and Geographic Information Society). 2008. *ACSM Bulletin.* (June): 11.
Campbell, Tony. 1995. "Helen Margaret Wallis (1924–1995)." *Imago Mundi* 47: 185–92.
———. 2004. "Wallis, Helen Margaret." *Oxford Dictionary of National Biography.* Oxford: Oxford University Press, http://press.ocforddnb.com/view/printable/57148.
Cannon, Elizabeth. 1999. "Women in Science and Engineering Activities in Canada." *Progress and Perspectives: An Affirmative Action in Surveying and Mapping.* (July–Aug.): 3–4.
Caplan, Paula J. 1993. *Lifting a Ton of Feathers.* Toronto: University of Toronto Press.
Cappelen Antikvariat. 2009. http://images.google.ca.
Cappon, Lester, Barbara B. Petchenik, and John H. Long. 1976. *Atlas of Early American History: The Revolutionary Era, 1760–1790.* Princeton, NJ: Princeton University Press.

REFERENCES

Carrington, David K., and Richard W. Stephenson. 1978. *Map Collections in the United States and Canada: A Directory.* 3rd ed. New York: Special Libraries Association.

Carson, James P. 1987. *Finite and Infinite Games.* New York: Ballantine.

———. 1997. "In Memoriam: Richard E. Dahlberg (1928–1996)." *Cartography and Geographic Information Systems.* 24 (2): 117.

Arto Cartographic Perspectives. 2005. Special Issue. Journal of the North American Cartographic Information Society, Department of Geography, University of Minnesota, Duluth, MN.

Centrogeo. 2010. "Carmen Reyes Guerrero." Bulletin published by the Centro de Investigacion en Geografia y Geomatica.

Chancy, Clancy. 1996. "In Favor of 19th Century Maps: Capturing the Turmoil of Change." *Mercator's World.* 1 (1): 72–76.

Charles, Maria. 2011. "What Gender Is Science?" *Contexts.* 10 (2): 22–28.

Charmaz, Kathy. 1991. *Good Days, Bad Days: The Self in Chronic Illness and Time.* New Brunswick, NJ: Rutgers University Press.

Chekin, Leonid S. 1999. "Easter Tables and the Pseudo–Isidorean Vatican Map." *Imago Mundi.* 51: 13–23.

China Center. 2009. "Building U.S.-China Bridges: China Center Annual Report 2008–2009." Minneapolis, MN: University of Minnesota–Twin Cities.

Clarke, Colin. 1985. "Geographical Perspectives on Social and Economic Change in Mexico and Latin America." *Area.* 17 (2): 176–79.

Coastal Studies Institute. 2007. Louisiana State University School of the Coast and Environment, http://www.csi.lsu.edu/history.asp.

Cohen, Aaron I. 1981. *International Encyclopedia of Women Composers.* New York: Bowker.

Cole, Susan. 1996. "IMCoS: A Society for People Who Like Maps." *Mercator's World.* 1 (1): 37, 91.

Commission de la Lune. 1967. "17. Commission de la Lune: Sessions 1 and 2." *Transactions of the International Astronomical Union.* Vol. 13B. Prague: 103–5.

Cook, Ellen Piel. 1993. "The Gendered Context of Life: Implications for Women's and Men's Career-Life Plans." *Career Development Quarterly.* 41: 227–37.

Cook, Karen Severud. 2005. "Obituary: Arthur H. Robinson (1915–2004)." *Imago Mundi.* 57 (Part 2): 195–97.

Cooksey, Gloria. 2002. "Semple, Ellen Churchill (1863–1932)." *Women in World History: A Biographical Encyclopedia.* Waterford, CT: Yorkin Publications.

Cosgrove, Denis. 1999. "Introduction: Mapping Meaning." In Denis Cosgrove, ed. *Mappings.* London: Reaktion.

REFERENCES

———. 2000. "Review of *Sphaerae Mundi: Early Globes as the Stewart Museum*, by E.H. Dahl and J.-F. Gauvin.'" *Cartographica*. 37 (1): 61–62.

Crampton, Jeremy W. 2010. *Mapping: A Critical Introduction to Cartography and GIS*. Chichester, UK: Wiley-Blackwell.

Crane, Nicholas. 2002. *Mercator: The Man Who Mapped the Planet*. London: Weidenfeld and Nicolson.

Crawley, Aidan. 1985. *Escape from Germany: The Methods of Escape Used by RAF Airmen during the Second World War*. London: Her Majesty's Stationery Office.

DAAD (German Academic Exchange Service). 2009a. "Prof. Dr.-Ing. Liqiu Meng: Geoinformatics." *Research in Germany: Land of Ideas*, http://www.research-in-germany.de/researcher-portraits.

———. 2009b. "We Also Want to Attract Internationally Distinguished Scientists as Teachers." *Auswärtiges Amt*, http://www.auswaertiges-amt.de.

Dahl, Edward H., and Jean-François Gauvin. 2000. *Sphaerae Mundi: Early Globes at the Stewart Museum*. Montreal: Septentrion and McGill-Queen's University Press.

Dancocks, Daniel G. 1983. *In Enemy Hands: Canadian Prisoners of War, 1939–45*. Edmonton, AB: Hurtig Publishers.

Davidson, James West, and John Rugge. 1997. *Great Heart: The History of a Labrador Adventurer*. Montreal and Kingston: McGill-Queen's University Press.

de Haan, David Bierens. 1883. *Bibliographie néerlandaise historique-scientifique des Ouvrages importants dont les auteurs sont nés aux 16ième, 17ième et 18ième siècles sur les Sciences mathématiques et physiques avec leurs applications*. Rome: Imprimerie des Sciences mathématiques et physiques.

Dekker, Elly, and Peter van der Krogt. 1993. *Globes from the Western World*. London: Zwemmer.

Demoor, Maryse. 2000. "Women and Science: Some Facts, Some Impressions." *Newsletter No. 3/00 of the Task Force on Under-Represented Groups in Surveying*. International Federation of Surveyors: 4–5, http://www.fig.net/pub/tf/unrep/200003/newsletter200003.htm.

Dent, Borden D. 1996. *Cartography: Thematic Map Design*. Dubuque, IA: Wm. C. Brown.

Department of Geography. 2010. "Pro-Chancellor, University of Plymouth: Awarded an MBE for Service to Higher Education." University of Leeds, www.geog.leeds.ac.uk/fileadmin/downloads/school/alumni/ALUMNI2010.pdf.

DESA. 2007. "The Geographical Name Game." *Department of Economic and Social Affairs, United Nations*. 11 (8): 4–6, http://www.un.org/esa/desa/desaNews/v11no8/feature/.htm.

REFERENCES

Design Museum. 2006. "Phyllis Pearsall: Map Designer (1906–1996)", http://designmuseum.org/design/phyllis-pearsall.

Dictionary of American Biography. 1936. New York: Scribner's Sons.

Diderot, Denis. 1959a. *Diderot Pictorial Encyclopedia of Trades and Industry; Manufacturing and the Technical Arts in Plates Selected from L'Encyclopédie, ou Dictionnaire Raisonné des Sciences, des Arts et des Métiers of Denis Diderot.* Vol. 1. New York: Dover.

———. 1959b. *Diderot Pictorial Encyclopedia of Trades and Industry; Manufacturing and the Technical Arts in Plates Selected from L'Encyclopédie, ou Dictionnaire Raisonné des Sciences, des Arts et des Métiers of Denis Diderot.* Vol. 2. New York: Dover.

Dietz, Mary L., Robert Prus, and William Shaffir. 1994. *Doing Everyday Life: Ethnography as Human Lived Experience.* Toronto: Copp Clark Longman.

Dodd, Dianne E., and Deborah Gorham, eds. 1994. *Caring and Curing: Historical Perspectives on Women and Healing in Canada.* Ottawa: University of Ottawa Press.

Dodge, Jefferson. 2006. "Avery: UCB Will Be Doing 'Dance' in Allocating 100 Lines." *Silver and Gold Record: Staff Newspaper at the University of Colorado.* 23 March, https://www.cu.edu/sg/messages/4893.html.

Doel, Ronald E., Tanya J. Levin, and Mason K. Marker. 2006. "Extending Modern Cartography to the Ocean Depths: Military Patronage, Cold War Priorities, and the Heezen-Tharp Mapping Project, 1952–1959." *Journal of Historical Geography.* 32 (3): 605–26.

Domosh, Mona, and Joni Seager. 2001. *Putting Women in Place: Feminist Geographers Make Sense of the World.* New York: Guilford.

Doyle, David. 1999. Email from David Doyle, National Geodetic Survey, Washington, D.C., to Lenora Sleep, University of New Brunswick, 9 December.

Drazniowsky, Roman. 1979. "The American Geographical Society's Collection." In Helen Wallis and Lothar Zögnar, eds. *The Map Librarian in the Modern World.* Munich: K.G. Saur.

Drygulski Wright, Barbara, ed. 1987. *Women, Work, and Technology: Transformations.* Ann Arbor, MI: University of Michigan Press.

Dubreuil, Lorraine, ed. 1993. *World Directory of Map Collections.* 3rd ed. Munich: K.G. Saur.

Duncker, Dieter R., and Helmut Weiss. 1983. *Het Hertogdom Brabant in kaart en prent.* Knokke, Belgium: Mappamundi.

Dymon, Ute J., and Margit Kaye. 1999. "Maps and Women." *Meridian.* 15: 5–9.

Earth Institute News Archive, Columbia University. 2006. "Remembered: Marie Tharp, Pioneering Mapmaker of the Ocean Floor," http://www.earth.columbia.edu/news/2006/story08-24-06.php.

Ebert, John, and Katherine Ebert. 1974. *Old American Prints for Collectors*. New York: Scribner's and Son.

Ellis Magazine. 2009. Fall. Pittsburgh, PA, http://www.theellisschool.org.

Etzkowitz, Henry, Carol Kemelgor, Brian Uzzi, with Michael Neushatz. 2000. *Athena Unbound: The Advancement of Women in Science and Technology*. New York: Cambridge University Press.

Fairclough, Roger H. 1974, "Review of '*My Head Is a Map: Essays and Memoirs in Honour of R.V. Tooley*, eds. Helen Wallis and Sarah Tyacke." *Geographical Journal*. 140 (3): 513–14.

Finnie, Ross, Marie Lavoie, and Maud-Catherine Rivard. 2001. "Women in Engineering: The Missing Link in the Canadian Knowledge Economy." *Education Quarterly Review*. 7 (3). Statistics Canada—Catalogue no. 81-0033.

Firth, Edith G. 2000. "Gwillim, Elizabeth Posthuma (Simcoe)." *Dictionary of Canadian Biography Online*, http://www.biographi.ca/009004-119.01-e.php?BioId=37540.

Fitzsimons, D., and E. Turner. 2006. "The Changing Look of Maps within Geography Journals: Proceedings of AutoCarto." Vancouver, WA.

Foghorn. 2005. Newsletter of the Friends of Geography, University of California-Los Angeles. 33 (Fall), www.geog.ucla.edu/downloads/alumni/fall2005.pdf.

Foster, Natt. 1997. "Content Analysis of 27 Atlases of the World." Prepared for Will C. van den Hoonaard. Summer. Unpublished. 10 pp.

Foster, Robert. 1991. "Survey for Women in Surveying." *Progress and Perspectives: An Affirmative Action in Surveying and Mapping*. (Mar.–Apr.): 7.

Fowler, Marian Little. 1977. "Portrait of Elizabeth Simcoe." *Oral History*. 69: 79–100.

Fox, Margalit. 2006. "Marie Tharp, Oceanographic Cartographer, Dies at 86." *New York Times* (26 August), http://www.nytimes.com/2006/08/26/obituaries/26tharp.html.

Freitag, Ulrich. 2004. "Laudatio Ingrid Kretshmer, 2004." Presentation at the Vienna University, 26 November, http://homepage.univie.ac.at/regina.schneider/events/FK2004/laudatio.html..

Geographers' A–Z Maps. 2006? "Our History," http://www.a-zmaps.co.uk/?nid=39.

Gerstein, M., M. Lichtman, and J.U. Barokas. 1988. "Occupational Plans of Adolescent Women Compared to Men: A Cross-Sectional Comparison." *Career Development Quarterly*. 36: 222–30.

Giddens, Anthony. 1991. *Introduction to Sociology*. New York: W.W. Norton.

Gillispie, Charles Coulston, ed. 1959b. *A Diderot Pictorial Encyclopedia of Trades and Industry: Manufacturing and the Technical Arts in Plates Selected from*

REFERENCES

'L'Encyclopédie, ou Dictionnaire Raisonné des Sciences, des Arts et des Métiers of Denis Diderot. Vol. 2.' New York: Dover Publications.

Gilmartin, Patricia. 1981. "Influences of Map Context on Circle Perception." *Annals of the Association of American Geographers.* 71 (2): 253–58.

———. 1985. "The Cued Spatial Response Approach to Macro-scale Cognitive Maps." *Canadian Geographer / Le Géographe canadien.* 29 (1): 56–59.

———. 2011. "Personal blog: About Me." http://patgilmartinaboutme.blogspot.com.

Glaser, Barney, and Anselm Strauss. 1967. *The Discovery of Grounded Theory.* Chicago: Aldine.

Grace's Guide. 2008. "Geographia," http://www.gracesguide.co.uk/wiki/Geographia.

Greed, Clara H. 1991. *Surveying Sisters: Women in a Traditional Male Profession.* London: Routledge.

———. 1994. *Women and Planning: Creating Gendered Realities.* London: Routledge.

Grenacher, F. 1970. "The Woodcut Map: A Form-Cutter of Maps Wanders through Europe in the First Quarter of the Sixteenth Century." *Imago Mundi.* 24: 31–41.

Greulich, Gunther. 2006. "The Sky Is the Limit: GPS and Its Roots." *ACSM Bulletin.* October: 22–28.

Hamicz, Edin. 2001. "Simpler Maps Put Women on the Right Track." *Sunday Times* (London) 4 March: Section 1: 8.

Hardie, Martin. 1906. *English Coloured Books.* New York: G.P. Putnam's Sons.

Hargitai, Henrik I. 2004. "Planetary Maps: Visualization and Nomenclature." *Cartographica.* 41 (2): 149–64.

Harley, J. Brian. 2001. *The New Nature of Maps: Essays in the History of Cartography.* Baltimore, MD: Johns Hopkins University Press.

Harley, J. Brian, and David Woodward, eds. 1994. *The History of Cartography: Cartography in the Traditional East and Southeast Asian Societies.* Vol. 2, Book 2. Chicago: University of Chicago Press.

Harley, J.B., and David Woodward, eds. 1987. *The History of Cartography: Cartography in Prehistoric, Ancient, and Medieval Europe and the Mediterranean.* Vol. 1. Chicago: University of Chicago Press.

Harris, Elizabeth M. 1975. "Miscellaneous Map Printing Processes in the Nineteenth Century." In David Woodward, ed. *Five Centuries of Map Printing.* Chicago: University of Chicago Press.

Hartley, Sarah. 2002. *Mrs P's Journey: The Remarkable Story of the Woman Who Created the A–Z Map.* New York: Simon and Schuster.

Heald, Claire. 2006. "From Aaron Hill to Zoffany St." *BBC News. UK Magazine,* http://news.bbc.co.uk.

REFERENCES

Hebbert, Michal. 2000. "Singing Streets of London: The Biennual Eila M.J. Campbell Lectures." Birkbeck College, University of London.1 March, http://www.bbk.ac.uk/events/campbell/hebbert.

Heller, Nancy G. 1991. *Women Artists: An Illustrated History.* New York: Abbeville Press.

Herbert, Francis. 2009. "[MapHist] Escape and Evasion Maps of WWII," http://mailman.geo.uu.nl/pipermail/maphist/2009-June/014127.html.

Helmfrid, Staffan. 1996. *National Atlas of Sweden: The Geography of Sweden.* Vällingby, Sweden: Sveriges Nationalatlas, KartCentrum.

Helmholtz Association. 2011. "Prof. Liqiu Meng," http://www.helmholtz.de/print/en/about_us/organisation/senate/members_of_the_senate/p.

Hermann, Michael, and Margaret Pearce. 2010. "'They Would Not Take Me There': People, Places, and Stories from Champlain's Travels in Canada 1603–1616." *Cartographic Perspectives.* 66: 41–46.

Herzig, R. and Jarausch, H. 2003. "Development of Cartographic and Spatial Comprehension by Children: An Investigation about Barbara Petchenik Children's Map Competition." http://cartography.tuwien.ac.at/ica/documents/ICC_proceedings/ICC2003/Papers/057.pdf.

Hessels, J.H., ed. 1969. *Epistulae Ortelianae.* Osnabruck: Otto Zeller.

Hinterhuber, Hans H., and Monika Stumpf. 1990. "Human Resource Management in Italy." In Rüdiger Pieper, ed. *Human Resource Management: An International Comparison.* Berlin: Walter de Gruyter.

Hodgkiss, Alan G. 1981. *Understanding Maps: A Systematic History of Their Use and Development.* Folkestone, UK: Dawson.

Hoogvliet, Margriet. 1996. "The Mystery of the Makers: Did Nuns Make the Ebstorf Map?" *Mercator's World.* 1 (6): 16–21.

———. 1998. Personal Communication with author. E-mail, 25 August.

Howley, J.P. 1915. *The Beothucks or Red Indians: The Aboriginal Inhabitants of Newfoundland.* Cambridge, MA: Cambridge University Press (Reprinted Coles Canadiana Collection, Toronto 1974).

Hsu, Mei-Ling. 1972. "The Role of Projections in Modern Map Design." *Cartographica.* 18 (2). Monograph 27: 151–86.

Hudson, Alice C. 1989. "Pre-Twentieth Century Women Mapmakers." *Progress and Perspectives: An Affirmative Action in Surveying and Mapping.* (Nov.–Dec.): 2, 4, 8.

———. 1999a. "Pre-Twentieth Century Women in Cartography: Who Are the Groundbreakers?" In C. Peter Keller, ed. *Conference Proceedings of the 19th International Cartographic Conference, Ottawa, Canada, 14–21 August 1999.* Ottawa, ON: Organizing Committee for Ottawa ICA 1999.

———. 1999b. "Research Note: Women in Cartography." *Meridian.* 15: 29–30.

REFERENCES

———. 2000. "Canadian Women in Maps: Looking for the Pioneers." *Association of Canadian Map Libraries and Archives Bulletin.* 107: 29–30.

———, and Mary McMichael Ritzlin. 2000a. "Preliminary Checklist of Pre-Twentieth-Century Women in Cartography." *Cartographica.* 37 (3): 3–8.

———, and Mary McMichael Ritzlin. 2000b. "Checklist of Pre-Twentieth-Century Women in Cartography." *Cartographica.* 37 (3): 9–24.

Huffman, Nikolas H. 1997. "Charting the Other Maps: Cartography and Visual Methods in Feminist Research." In John Paul Jones, Heidi J. Nast, and Susan M. Roberts, eds. *Thresholds in Feminist Geography: Difference, Methodology, Representation.* Lanham, MA: Rowman and Littlefield.

Huq-Hussain, Shahnaz, Amanat Ullah Khan, and Janet Momsen. 2006. *Gender Atlas of Bangladesh.* Dhaka: Geographical Solutions Research Centre.

Hutchings, Roger. 2004. "Blagg, Mary Adele (1858–1944)." *Oxford Dictionary of National Biography.*

Ilcan, Suzan. 2002. "Review of Draft Manuscript." Email to author. 26 December.

Innis, Mary Quayle. 1965. *Mrs. Simcoe's Diary.* New York: St. Martin's Press.

International Association of Chinese Professionals in Geographic Information Science (ISCP). *Newsletter.* 2007 (August).

International Cartographic Association *Newsletter.* International Cartographic Association, 1999–2011.

International Cartographic Association. 1995. *Directory of Women in Cartography, Surveying, and G.I.S.* Ottawa, ON: ICA Secretariat.

International Cartographic Association Commission on Gender and Cartography. 1999. "Report of ICA Commission on Gender and Cartography, 1995–1999." Distributed at the 19th ICA Conference, Ottawa, ON, August.

International Task Force on Women in Cartography. 1990. [Statement on Task Force]. March.

Jacobsen, Joyce P. 2007. *The Economics of Gender.* 3rd ed. Malden, MA: Blackwell.

Jaquette, Jane S. 2000. "Women and Democracy: Past, Present, Future." Democracy Seminar, Stanford University, 25 May, http://democracy.stanford.edu/Seminar/Jaquette_Women_Democracy.htm.

Johnson, Allan G. 1997. *The Gender Knot: Unraveling Our Patriarchal Legacy.* Philadelphia: Temple University.

Jones, John Paul, Heidi J. Nast, and Susan M. Roberts, eds. 1977. *Thresholds in Feminist Geography: Difference, Methodology, Representartion.* Lanham, MA: Rowman and Littlefield.

Joy, A.H. 1916. "Review of 'Collected List of Lunar Formations by Mary A. Blagg.'" *Astrophysical Journal.* 43: 88.

REFERENCES

Judd, Jacob. 1974. "Conference Reports—America's Wooden Age: Sleepy Hollow Restoration, April 1973." *Technology and Culture.* 15 (1): 64–69.

Kass-Simon, G., Patricia Farnes, and Deborah Nash, eds. 1993. *Women of Science: Righting the Record.* Bloomington, IN: Indiana University Press.

Keates, John S. 1981. *Understanding Maps.* London: Longmans.

Keeble, N.H. 1994. *The Cultural Identity of Seventeenth-Century Women: A Reader.* London: Routledge.

Kelley, Florence. 1970 [1895]. *Hull-House Maps and Papers.* By the residents of Hull House. New York: Arno Press.

Kerfoot, Helen. 1991. "Towards a Concise Gazetteer of Canada: Submission by the Canadian Parliament on Canadian Geographical Names, to the United Nations Group of Experts on Geographical Names." November, http://unstats.un.org/unsd/geoinfo/UNGEGN/docs/15th_gegn-WP62.pdf.

———. 2004. "Geographical Names: Some Current Issues in the Context of the United Nations." *Cartographic Journal.* 41 (2): 89–94.

Keuning, J. 1948. "Jodocus Hondius Jr." *Imago Mundi.* 5: 63–71.

King, Geoff. 1996. *Mapping Reality: An Exploration of Cultural Cartographies.* New York: St. Martin's Press.

Knowles, Leo. 2003. "She Mapped London with her Feet." *Mercator's World.* 8 (2): 16–19.

Koeman, Cor, ed. 1967. *Atlantes Neerlandici,* Vols. 1–3. Amsterdam: Theatrum Orbis Terrarum.

Koeman, Cor. 1975. "The Application of Photography to Map Printing and the Transition to Offset Lithography." In David Woodward, ed. *Five Centuries of Map Printing.* Chicago: University of Chicago Press.

———. 1980. "Review of 'Arnberger, E., and Kretschmer, I. *Enzyklopädie der Kartographie*." *GeoJournal.* 4 (4): 386–87.

Kozlowa, Anna V. 1979. "Collection of Cartographic Works in the Lenin State Library of the USSR." In Helen Wallis and Lothar Zögnar, eds. *The Map Librarian in the Modern World.* Munich: K.G. Saur.

Kretschmer, Ingrid, John Doerflinger, and Franz Wawrik, eds. 1986. *Lexikon zur Geschichte der Kartographie. Von den Anfängen bis zum Ersten Weltkrieg* [Encyclopedia of the History of Cartography. From the beginnings to World War I]. Vienna: Deuticke.

Krzywicka-Blum, Ewa. 2002 General email to members of the Commission on Gender and Cartography from its Chair. 14 August.

Lapp, Ralph. 1956. *Atoms and People.* New York: Harper.

Lassalle, D., and A.R. Spokane. 1987. "Patterns of Early Labour Force Participation of American Women." *Career Development Quarterly.* 36: 55–65.

REFERENCES

Latitudes: Newsletter of the LSU Department of Geography and Anthropology. 2002. (Summer), 2005. (Spring/Summer).

Lawrence, David M. 1999. "Mountains under the Sea: Marie Tharp's Maps of the Ocean Floor Shed Light on the Theory of Continental Drift." *Mercator's World.* 4 (6): 36–43.

Lecordix, François. 2011. *International Cartographic Exhibition Catalogue, 3–8 July 2011, Paris, France.* Paris: French Cartography Association.

Lengermann, Patricia M., and Jill Niebrugge-Brantley, eds. 1998. *The Women Founders: Sociology and Social Theory, 1830–1930: A Text/Reader.* Boston: McGraw-Hill.

Library of Congress. 2001. "Manuscript Map Drawn by Mary Van Schaack of Kinderhook, New York, 1811." Geography and Map Division Washington, D.C. Call No. G3200 1811.V2, http://memory.loc.gov/ammem/awhhtml/awgmd7/d11.html.

Livengood, R. Mark. 1998. "Review of 'Barbara G. Shortridge and James R. Shortridge, *The Taste of American Place*.'" *Western Folklore.* 57 (2/3): 200–2.

Lobeck, Armin K. 1956. *Things Maps Don't Tell Us.* New York: Macmillan.

Lohr, Nouna. 1999. "Women in Non-Traditional Roles at NRCan." *Corporate News.* 17 December, http:nrn1.nrcan.gc.ca:80/source/archive/femar96/corpnews.htm.

Louter, David. 2006. *Windshield Wilderness: Cars, Roads, and Nature in Washington's National Parks.* Seattle: University of Washington Press.

Lunar and Planetary Institute. 2010. "Lunar Images and Maps." Houston, TX, http://www.lpi.usra.edu/lunar/lunar_images.

Lutz, Alma. 1974. *Emma Willard: Pioneer Educator of American Women.* Boston: Beacon Hill Press.

Macdonald, Angie. 2008. "Phyllis and I." *Dulwich OnView.* 21 (October), http://dulwichonview.org.uk/2008/10/21/phyllis-and-i.

Macionis, John J., and Linda M. Gerber. 1999. *Sociology.* 3rd ed. Toronto: Pearson Education.

Maddrell, Avril. 2009a. "An Interview with Anne Buttimer: An Autobiographical Window on Geographical Thought and Practice, 1965–2005." *Gender, Place and Culture.* 16 (6): 741–65.

———. 2009b. *Complex Locations: Women's Geographical Work in the UK, 1850–1970.* Chichester: Wiley-Blackwell.

Madej, Ed. 2001. *Cartographic Design Using ArcView GIS.* Albany, NY: OnWord Press.

Mandell, Mel. 1996. "World War II Maps for Armchair Generals: Civilians Make a Major Market for War and Postwar Maps." *Mercator's World.* 1 (4): 42–45.

REFERENCES

Mapping Science Committee. 1990. "Spatial Data Needs: The Future of the National Mapping Program." Commission on Physical Sciences, Mathematics and Resources, National Research Council. http://www.nap.edu/catalogue.php?record_id=9616.

Margary, Harry. 1995. "*Imago Mundi* Saved by Eila Campbell." *Imago Mundi*. 47: 8–9.

Mark, David M. 1997. "The History of Geographic Information Systems: Invention and Re-Invention of Triangulated Irregular Networks (TINS)." *GIS/LIS '97 Annual Conference and Exposition Proceedings, Cincinnati, Ohio, October 1997*: 267–72.

Martin, Geoffrey J. 1994. "In Memoriam: Richard Hartshorne, 1899–1992." *Annals of the Association of American Geographers*. 83 (3): 480–92.

Mason, Karen. 2009. "Pro-Chancellor awarded MBE." 17 June. http://www.plymouth.ac.uk/pages/view.asp?page=26782.

Massey, Doreen, and John Allen. 1984. *Geography Matters! A Reader*. Cambridge, UK: Cambridge University Press.

McCorkle, Barbara. 2008. "My Early Years with the SHD: Peoples and Places." *Terrae Incognitae*. Society for the History of Discoveries. http://www.sochistdisc.org/mccorkle_early_years.htm.

McCormack, Thelma. 1991. "Women in Engineering: Revolution or Counter-Revolution?" Paper prepared for Conference on Women in Engineering sponsored by the Canadian Committee on Women in Engineering. 21–23 May. University of New Brunswick, Fredericton, NB.

McEwan, Cheryl. 2000. *Gender, Geography and Empire: Victorian Women Travellers in West Africa*. Burlington, VT: Ashgate.

McGaw, Judith A. 1982. "Women and the History of American Technology." *Signs*. 9 (2): 798–828.

McManis, Douglas R. 1996. "Leading Ladies at the AGS (American Geographical Society)." *Geographical Review*. 86 (2): 270–77.

McMaster, Robert, and Susanna McMaster. 2002. "A History of Twentieth-Century American Academic Cartography." *Cartography and Geographic Information Science*. 29 (3): 305–21.

McPherson, Bea S. 1993. "Female Army Mapmakers: 50 Years Reunion." *Progress and Perspectives: An Affirmative Action in Surveying and Mapping*. (July–Aug.): 1–2.

Mead, W.R. 1995a. "Obituaries: Helen Wallis." *The Geographical Journal*. 161 (2): 240–41.

———. 1995b. "Eila as Geographer." *Imago Mundi*. 47: 7–8.

REFERENCES

Meng, Liqiu. 2010. "Curriculum vitae," http://129.187.175/5lfkwebsite/index.php?id=25.

Menninger, Karl. 1930. *The Human Mind.* Garden City, NY: Garden City Publishing.

Mersey, Janet E. 1990. *Colour and Thematic Map Design: The Role of Colour Scheme and Map Complexity in Choropleth Map Communication.* Published as a *Cartographica* Monograph, No. 41. Toronto: University of Toronto Press.

Miller, Michael. 1993. "Women in Mapmaking." *The Portolan.* (Spring): 7.

Mills, Sara. 1994. "Knowledge, Gender, and Empire." In Alison Blunt and Gillian Rose, eds. *Writing Women and Space: Colonial and Postcolonial Geographies.* New York. Guilford Press.

Möbius, Helga. 1982. *Woman of the Baroque Age.* Montclair, NJ: Abner Schram. Trans. Barbara Chruscik Beedham.

Modelski, Andrew M. 1979. "List of Officially Published Works of the Geography and Map Division [of the United States Library of Congress]." In Helen Wallis and Lothar Zögnar, eds. *The Map Librarian in the Modern World.* Munich: K.G. Saur.

Monk, Janice. 1990. "Interview with Mei-ling Hsu." 4 May. 21 pp. transcript in author's possession.

———. 1998. "The Women Were Always Welcome at Clark." *Economic Geography.* (Special Issue): 14–30.

———. 2003. "Women's World at the American Geographical Society." *Geographical Review.* 93 (2): 237–57.

———. 2004. "Presidential Address: Women, Gender, and the Histories of American Geography." *Annals of the Association of American Geographers.* 94 (1): 1–22.

———. 2006. "Changing Expectations and Institutions: American Women Geographers in the 1970s." *Geographical Review.* 96 (2): 259–77.

Monmonier, Mark. 1977. *Maps, Distortion, and Meaning.* Washington, DC: Association of American Geographers.

———. 1996. *How to Lie with Maps.* Chicago: University of Chicago Press.

Morrison-Low, A.D. 1991. "Women in the Nineteenth-Century Scientific Instrument Trade." In Marina Benjamin, ed. *Science and Sensibility: Gender and Scientific Enquiry, 1780–1945.* Oxford: Basil Blackwell.

NACIS (North American Cartographic Information Society). 2005. "University Staff Cartographers and Cartographic Laboratories in North America," http://www.nacis.org/index.cfm?x=17.

Nash, Catherine. 1994. "Remapping the Body/Land: New Cartographies of Identity, Gender, and Landscape in Ireland." In Alison Blunt and Gillian Rose, eds.

REFERENCES

Writing Women and Space: Colonial and Postcolonial Geographies. New York: Guilford Press.

Nelson, Lise, and Joni Seager, eds. 2005. *A Companion to Feminist Geography.* Malden, MA: Blackwell.

NVK *Membership Directory: 1996–1997.* Annual Report of the Netherlands Cartographic Society.

Ochberg, R. 1988. "Life Stories and the Psychological Construction of Careers." In D.P. Adams and R.L. Ochberg, eds. *Psychobiography and Life Narratives.* Durham, NC: Duke University Press.

OIP (Office of International Programs). 2009. *OIP Staff News.* Minneapolis, MN: University of Minnesota-Twin Cities, http://www.international.umn.edu/intranet/staff_news/2009/June 1–5.html.

Olin, Kelly. 1988. "Discussions with Surveying Mothers." *Progress and Perspectives: An Affirmative Action for Surveying and Mapping.* (Summer): 3–4.

Olson, Judy M. 1997. "Presidential Address–Multimedia in Geography: Good, Bad, Ugly, or Cool?" *Annals of the Association of American Geographers.* 87 (4): 571–78.

———. 2003. "Cartography 2003." *Cartographic Perspectives.* 47: 4–12.

———. 2005. "A Tribute to Our Founding Editor: Arthur H. Robinson, 1915–2004." *Cartography and Geographic Information Service.* 33 (2).

———. 2011. "Map Typologies: 20th Usage Revisited." Presentation to the 25th Conference of the International Cartographic Association, Paris, 7 July.

———, and Cynthia A. Brewer. 1997. "An Evaluation of Color Selections to Accommodate Map Users with Color-Vision Impairments." *Annals of the Association of American Geographers.* 87 (1): 103–34.

Oxford Cartographic Conference. 2000. "Delegates List." 31 August. 6 pp. Mimeo. Copy in files of author.

Palfrey, Dale Hoyt. 2000. "Canadian Club Focuses on Chapala Cyber-Atlas." *Guadalajara Reporter.* 25 February, http://guadalajarareporter.com/news-mainmenu- 82/lake-chapala-mainmenu-84/10818-canadian-club-focuses-on-chapala-cyber-atlas.html.

Papp-Váry, Árpád. 1998. *Képes történelmi atlasz [Illustrated Historical Atlas].* Budapest: Cartographia.

———. 2000. *Irodalomtörténeti atlasz [Atlas of Literary History].* Budapest: Cartographia.

Pastoureau, Mireille. 1984. *Les atlas français XVI–XVII siècles.* Paris: Bibliothèque Nationale.

REFERENCES

Pavlovskaya, Mariana. 2009. "Feminism, Maps, and GIS." In R. Kitchin and N. Thrift, eds. *International Encyclopedia of Human Geography*. Maryland Heights, MO: Elsevier.

———, and Kevin St. Martin. 2007. "Feminism and GIS: From a Missing Object to a Mapping Subject." *Geography Compass*. 1 (3): 583–606.

Pearsall, Phyllis. 1990. *From Bedsitter to Household Name: The Personal Story of A–Z Maps*. Sevenoaks, Kent, UK: Geographers' A–Z Map Company.

Pelletier, Monique. 1995. "Eila and the International Conferences on the History of Cartography." *Imago Mundi*. 47: 9.

Penny, Virginia. 1863. *The Employment of Women: A Cyclopaedia of Woman's Work*. Boston: Walker, Wise, and Company.

Pickles, John. 1992. "Texts, Hermeneutics, and Propaganda Maps." In Trevor J. Barnes and James S. Duncan, eds. *Writing Worlds: Discourse, Text, and Metaphor in the Representation of Landscape*. London: Routledge.

———, ed. 1995. *Ground Truth: The Social Implications of Geographic Information Systems*. New York: Guilford Press.

Pillsbury, Richard. 1998. "Review of 'Barbara G. Shortridge and James R. Shortridge, *The Taste of American Place*.'" *Geographical Review*. 88 (4): 604–5.

Plymouth University Newsletter (United Kingdom).

Pognon, Edmond. 1979. "Les Collections du Département des Cartes et Plans de La Bibliothèque Nationale de Paris." In Helen Wallis and Lothar Zögnar, eds. *The Map Librarian in the Modern World*. Munich: K.G. Saur.

Postman, Neil. 1995. *The End of Education: Redefining the Value of School*. New York: Knopf.

Potter, Jonathan. 1999. "Curiosities." In *Collecting Antique Maps: An Introduction to the History of Cartography*. rev. ed. London: Jonathan Porter.

Pressl, Harald. 2009. "Deutsch ist das Problem" ["German Is the Problem"]. *Time* magazine. 2 November.

Progress and Perspectives: An Affirmative Action for Surveying and Mapping. W.J.W. Straight, ed. Dunkirk, NY. Vol. 1 (1988) to present.

Prouse, A. Robert. 1982. *Ticket to Hell via Dieppe: From a Prisoner's Wartime Log, 1942–1945*. Toronto: Van Nostrand Reinhold.

Pruitt, Evelyn L. 1979. "The Office of Naval Research and Geography." *Annals of the Association of American Geographers*. 69: 103–8.

Pye, Norman. 1991. "Ordnance Survey: Bicentenary Symposium." *Geographical Journal*. 157 (3): 349–51.

Raji, Saraswati, Peter Atkins, and Janet G. Townsend. 2000. *Atlas of Women and Men in India*. New Delhi: International Books.

Ramaswamy, Sumathi. 2001. "Maps and Mother Goddesses in Modern India." *Imago Mundi.* 53: 97–114.

Ravenhill, William. 1996. "Obituary: Helen Margaret Wallis, 1924–1995." *Transactions, Institute of British Geographers* NS 21: 299–301.

Reitinger, Franz. 1999. "Mapping Relationships: Allegory, Gender, and the Cartographical Image in Eighteenth-Century France and England." *Imago Mundi.* 51: 106–30.

Reyes, Carmen. 2005. "National Data Infrastructures: Sharing Geospatial Information as an Asset for Environmental Public Policies." E/CONF.96/I. Eighth United Nations Cartographic Conference for the Americas, New York.

———. 2006a. "Cybercartography from a Modeling Perspective." In D.R.F. Taylor, ed. *Cybercartography: Theory and Practice.* Maryland Heights, MO: Elsevier.

———. 2006b. "Cybercartography and Society." In D.R.F. Taylor, ed. *Cybercartography: Theory and Practice.* Maryland Heights, MO: Elsevier.

———. 2006c. "Technology and Culture in Cybercartography." In D.R.F. Taylor, ed. *Cybercartography: Theory and Practice.* Maryland Heights, MO: Elsevier.

———, Liliana Guerrero, and Levi Lopez. 1996. "Redistritación electoral 1996: el diseño de una solución" [Electoral Redistricting 1996: the design of a solution]. In Mario Alejandro Carrillo Luvianos, Sergio de la Vega Estrada, and Alejandra Toscano Aparicio, eds. *Imagen electoral de México (1980–2002)* [Electoral Redistricting 1980–2002]. Universidad Autónoma Metropolitana (Unidad Xochimilco, División de Ciencias Sociales y Humanidades).

———, Fraser Taylor, Elvia Martinez, and Fernando Lopez. 2006. "Geo-Cybernetics: A New Avenue of Research in Geomatics?" *Cartographica.* 41 (1): 7–20.

Ring, Betty. 1993. *Girlhood Embroidery: American Samplers and Pictorial Needlework, 1650–1850.* New York: Alfred Knopf.

Ristow, Walter W. 1975. "Lithography and Maps, 1796–1850." In David Woodward, ed. *Five Centuries of Map Printing.* Chicago: University of Chicago Press.

———. 1980. "Eliza Colles: America's First Female Map-Engraver." *Map Collector.* 10: 14–17.

———. 1985. *American Maps and Mapmakers: Commercial Cartography in the Nineteenth Century.* Detroit: Wayne State University Press, 1985.

Ritter, Michael. 2001. "Seutter, Probst and Lotter: An Eighteenth-Century Map Publishing House in Germany." *Imago Mundi.* 53: 130–35.

———. 1986. "The Role of Women in the Development of Cartography." *AB Bookman's Weekly.* 9 June: 2709–13.

———. 1989. "Women's Contributions to North American Cartography: Four Profiles." *Meridian.* 2: 5–16.

———. 1990a. "Women's Contributions to North American Cartography: Mary Biddle." *Progress and Perspectives: An Affirmative Action in Surveying and Mapping.* (Mar.–Apr.) and (July–Aug.): 6.

———. 1990b. "Women's Contributions to North American Cartography: Elizabeth Simcoe." *Progress and Perspectives: An Affirmative Action in Surveying and Mapping.* (May–June) and (July–Aug.): 6.

———. 1990c. "Women's Contributions to North American Cartography: Emma Hart Willard." *Progress and Perspectives: An Affirmative Action in Surveying and Mapping.* (July–Aug.): 4.

———. 1993. "School Marm to Author: Nineteenth Century Women Geographers." *The Portolan.* (Fall): 12–15.

Ritzlin, Mary McMichael. 1999. "Sweeping the Skies: Some Celestial Ladies of the 17th–19th Centuries." *Meridian.* 15: 11–21.

Robinson, Arthur H. 1960. *Elements of Cartography.* New York: John Wiley and Sons.

———. 1975. "Mapmaking and Map Printing: The Evolution of a Working Relationship." In David Woodward, ed. *Five Centuries of Map Printing.* Chicago: University of Chicago Press.

———. 1982. *Early Thematic Mapping in the History of Cartography.* Chicago: University of Chicago Press.

———. 1989. "Cartography as an Art." In D.W. Rhind and D.R.F. Taylor, eds. *Cartography Past, Present and Future.* New York: Elsevier.

———. 1994. "B.B. Petchenik (1939–1992)." *Imago Mundi.* 46: 175–76.

——— and Barbara Bartz Petchenik. 1976. *The Nature of Maps: Essays Towards Understanding Maps and Mapping.* Chicago: University of Chicago Press.

Robinson, Jennifer. 1994. "White Women Researching/Representing 'Others': From Antiapartheid to Postcolonialism?" In Alison Blunt and Gillian Rose, eds. *Writing Women and Space: Colonial and Postcolonial Geographies.* New York: Guilford Press.

Ronan, Colin A. 1981. *The Shorter Science and Civilisation in China: An Abridgement of Joseph Needham's Original Text. Vol. 2.* Cambridge, UK: Cambridge University Press.

Ross, Robert. 1996. "The Lure of Old Maps." *Mercator's World.* 1 (1): 42–45.

Roth, Robert E. 2010. "Interview with a Celebrity Cartographer: Cindy Brewer." *Cartographic Perspectives.* 66: 91–101.

Royal Canadian Geographical Society. 2011. "Camsell Medal," http://www.rcgs.org/awards/camsell_medal_winner_camsell2010.asp.

Royal Geographical Society. 1979. "Meetings: Session 1978–1979." *Geographical Journal.* 145 (3): 522–29.

Rundstrom, Robert A. 1991. "Mapping, Postmodernism, and Indigenous People and the Changing Direction of North American Cartography." *Cartographica*. 28 (2): 1–12.

Rysted, Bengt. 1995. "Current Trends in Electronic Atlas Production." *Cartographic Perspectives*. 20: 5–11.

——. 2007. "Helen Kerfoot." ICA *News*. December: 6.

Saigal, Anil. 1987. "Women Engineers: An Insight into their Problems." *Engineering Education*. (December): 194–96.

School of Geography. 2010. "Barbara Bond: Pro Chancellor, University of Plymouth." *Alumni Newsletter*. University of Leeds, Faculty of Environment. 3 (April): 10.

Schweik, Joanne. 1994. "Revamping Education for 2000 and Beyond." *Progress and Perspectives: An Affirmative Action in Surveying and Mapping*. (Nov.–Dec.): 1–2.

Scullen, Wendy. 1990. "Women in Surveying in Australia: The Road Ahead." Paper presented at the XVIX International Federation of Surveyors Congress in Helsinki, Finland: 87–95.

Seager, Joni, and Ann Olson. 1986. *Women in the World: An International Atlas*. (Now in its 4th edition as *The Penguin Atlas of Women in the World*, 2009.)

Selley, Lea. 2000. "Advancements in Cartographic Design at Statistics Canada." Joint Conference of the Association of Canadian Map Libraries and Archives and the Canadian Cartographic Association, Edmonton, May 30–June 2.

Semple, Ellen Churchill. 1897. "The Influence of the Appalachian Barrier upon Colonial History." *Journal of School Geography*. Vol. 1.

——. 1931. *The Geography of the Mediterranean: Its Relationship to Ancient History*. New York: AMS Press.

——, and Clarence Jobes. 1903. *American History and Its Geographic Conditions*. Boston: Houghton Mifflin.

Sheffield, Suzanne Le-May. 2006. *Women and Science: Social Impact and Interaction*. New Brunswick, NJ: Rutgers University Press.

Shingareva, K.B., Irina P. Karachevtseva, and Elena V. Cherepanova. 2003. "Geostatistical Analysis of the Planetary Nomenclature Database as a Method of Extraterrestrial Territories Mental Exploration," icaci.org.

Shingareva, K.B., Irina P. Karachevtseva, and Elena V. Cherepanova. 2007. "Geostatistical Analysis of the Planetary Nomenclature Database as a Method of Extraterrestrial Territory Mental Exploration." Proceedings of the 23rd International Cartographic Conference, Moscow, Russia (Planetary Cartography Session) http://planetcarto.wordpress.com/projects/planetary-nomenclature.

Shirley, Rodney. 1996. "The Face of the Maker." *Mercator's World*. 1 (4): 14–19.

REFERENCES

Shoffey, Valerie. 1999. Email from Valerie Shoffey, Executive Secretary of Canadian Association of Geographers, to Lenora Sleep, University of New Brunswick, 11 November.

Shortridge, Barbara G. 1987. *Atlas of American Women.* New York: Macmillan.

———, and James R. Shortridge, eds. 1998. *The Taste of American Place: A Reader on Regional and Ethics Foods.* Lanham, MD: Rowman and Littlefield.

Siekierska, Eva. 1984. "Towards an Electronic Atlas." *Cartographica.* 21 (2 & 3). Monograph 32–33, Auto-Carto Six: Selected Papers, D.H. Douglas, ed.: 110–20.

———. 1994. "Gender and Cartography: A Background Document for the 13th United Nations Cartographic Conference for Asia and the Pacific, held in Beijing, China, 1994." Mimeo. 9 pp (in possession of author).

———. 1995. "From the Task Force on Women in Cartography to the ICA Working Group on Gender and Cartography: What We Learned." Presented at the 17th International Cartographic Conference, 10th General Assembly of ICA. (3–9 Sept.)

———. 1997. Email from Eva Siekierska, Chair of the ICA Commission on Gender and Cartography, to Natalie Foster. Summer.

———, and Costas Armenakis. 1999. "Territorial Evolution of Canada: An Interactive Multimedia Cartographic Presentation." In W. Cartwright, M.P. Paterson, and G. Gartner, eds. *Multimedia Cartography.* Berlin: Springer.

———, and William McCurdy. 2008. "Internet-Based Mapping for the Blind and People with Visual Impairment." In Michael Peterson, ed. *International Perspectives on Maps and the Internet.* Berlin: Springer.

———, Linda O'Neil, and Donna Williams. 1991. "The Participation of Women in the International Cartographic Association (ICA): Report and Recommendations." Ottawa, Canada: ICA Task Force on Women and Cartography.

———, and D.R.F. Taylor. 1991. "Electronic Mapping and Electronic Atlases: New Cartographic Products for the Information Era—The Electronic Atlas of Canada." *CISM Journal ACSGC.* 45 (1): 11–21.

Siltanen, Janet, and Michelle Stanworth. 1984. *Women and the Public Space: A Critique of Sociology and Politics.* London: Hutchinson.

Simonton, Dean Keith. 1999. *Origins of Genius: A Psychology of Science.* Cambridge, MA: Cambridge University Press.

Sinitsyna, Anna, and Kira Shingareva. 2006. "Megacity Transport." *Geo: Connexion.* 1 December. www.highbeam.com.

Skelton, R.A. Papers. Various dates. Memorial University of Newfoundland, St. John's, NF, Canada.

REFERENCES

Sklar, Lawrence. 1995. *Physics and Chance: Philosophical Issues in the Foundations of Statistical Mechanics.* Cambridge, UK: Cambridge University Press.

Slocum, Terry A. 1999. *Thematic Cartography and Visualization.* Upper Saddle River, NJ: Prentice Hall.

Smith, Nancy Johnson. 1997. "Adapting Grounded Theory Methods to Interpret the General Story Line in Young Women's Life Histories Regarding their Occupational Aspirations." Qualitative Analysis Conference: Interdisciplinary Perspectives. Toronto: OISE. (5–8 Aug.)

Smith, Terence R. 1998. "Educational Activities," http://www.alexandria.ucsb.edu/ public-documents/annual_report98/node/76.html.

Statistics Canada. 1996. *1996 Census: Detailed Occupation by Sex, Canada.* Catalogue No. 93F0027XDB96007, http://www.statcan.ca/english/census96/mar17/ occupa/table1/t1p00c.htm.

———. 1998. *1996 Census: Population 20 Years and Over with University Certificates or Degrees by Major Field of Study and Sex, Showing School Attendance, for Canada, 1996 Census (20% Sample Data).* Catalogue No. 93F0028XDB96009, http://www.statcan.ca/english/census96/Apr14.

———. 2010. "University Degrees, Diplomas and Certificates Awarded." *The Daily.* Available on CANSIM: Table 477-0014, http://www.statcan.gc.ca/ daily-quotidien/100714/dq100714b-eng.htm.

Stauffer, David McNeely. 1907. *American Engravers upon Copper and Steel, Part 1.* New York: Burt Franklin.

Stefoff, Rebecca. 1992. *Women of the World: Women Travelers and Explorers.* New York: Oxford University Press.

Steward, Henry. 2001. "And Great Fun It All Is: Or How I (Almost) Became a Feminist Cartographer." *Cartouche.* 42: 5–7.

Stewart, Roger. 2011. "In the Family Way," http://stewartcollection.googlepages .com/ Inthefamilyway-mapmakingdynasty08090.pdf.

St. Martin, Kevin, and Marianna Pavlovskaya. 2009. "Ethnography." In N. Castree, D. Demeritt, D. Liverman, and B. Rhoads, eds. *Companion to Environmental Geography.* Oxford: Blackwell.

Stone, Lawrence. 1977. *The Family, Sex and Marriage in England, 1500–1800.* London: Weidenfeld and Nicolson.

Straight, Wendy. 1990. "Patricia Caldwell Lindgren." *Progress and Perspectives: An Affirmative Action for Surveying and Mapping.* (Mar.–Apr.): 5.

———. 1991a. "Communiqués." *Progress and Perspectives: An Affirmative Action in Surveying and Mapping.* (May–June): 6.

———. 1991b. "Survey for Women in Surveying." *Progress and Perspectives: An Affirmative Action in Surveying and Mapping.* (May–June): 3.

———. 1993. "Female Army Mapmakers: 50 Years Reunion." *Progress and Perspectives: An Affirmative Action in Surveying and Mapping.* (July-Aug.): 1–2.

———. 1997. "Phyllis Pearsall Dies." *Progress and Perspectives: An Affirmative Action in Surveying and Mapping.* (Jan.–Feb.): 4.

Strauss, David. 2002. "Review of 'Ewen A. Whitaker. Mapping and Naming the Moon: A History of Lunar Cartography and Nomenclature'." *Isis.* 93 (2).

Suárez, Thomas. 1997. *The Art of Maps: Of Mortals and Myth: The Human Figure on Antique Maps.* Eugene, OR: Aster Press.

Takeda, Yuko, Reiko Kinoshita, and Takashi Nakazawa. 2007. *Gender Atlas of Japan.* [in Japanese] 2007. Tokyo: Akashi.

Tang, Joyce. 2006. *Scientific Pioneers: Women Succeeding in Science.* Lanham, MD: University Press of America.

Task Force on Women in Cartography. 1989. *Minutes from the First Meeting, 21 August, in Budapest.* Mimeo. 2 pp (in possession of author).

Teixeira, Marco A.P., and William B. Gomes. 2000. "Autonomous Career Change among Professionals: An Empirical Phenomenological Study." *Journal of Phenomenological Psychology.* 31 (1): 78–96.

Terenzoni, A.M. Piras. 1990. "Women in the Surveyor's Profession." Paper presented at the XVIX International Federation of Surveyors Congress, June 10–19, in Helsinki, Finland: 293–99.

Tharp, Marie. 1999. "Marie Tharp Bio: Mary Sears Woman Pioneer in Oceanography Award at the Woods Hole Oceanographic Institute," http://www.whoi.edu/sbl/liteSite.do?litesiteid=9092&articleId=13407.

Thrower, Norman J.W. 1969. "Annals Map Supplement Number Eleven: Taiwan Population Distribution, 1965." *Annals of the Association of American Geographers.* 59 (3): 611–12.

———. 1972. *Maps and Man.* Englewood Cliffs, NJ: Prentice-Hall.

———. 1996. *Maps and Civilization: Cartography in Culture and Society.* University of Chicago Press, Chicago.

———. 1999. "Tribute to Helen Margaret Wallis, 1924–1995." *Meridian: A Journal of the Map and Geography Round Table of the American Library Association.* 15: 31–32.

Tierney, Helen. 2008. "Objects through Time … Statement of Significance: Convict Sandstock Bricks." *NSW Migration Heritage Centre.* Fairfield City Museum and Gallery. http://www.migrationheritage.nsw.gov.au/exhibition/objects throughtime/convict-sandstock-bricks.

Tooley, Ronald Vere. 1978. "Women in the Map World." *Map Collector.* 4 (Sept.): 16–17.

Traub, Valerie. 1997. "Mapping the Global Body." Paper presented at Conference, Landscapes: Maps, Texts, and the Construction of Space, 1500–1700. Queen Mary and Westfield College, London. (July.)

Tyacke, Sarah. 1978. *London Map-Sellers, 1660–1720*. Tring, Hertfordshire, UK: Map Collector Publications.

Tyner, Judith. 1969. "Early Lunar Cartography." *Surveying and Mapping*. 4 (Sept.): 583–96.

———. 1973. *The World of Maps and Mapping*. New York: McGraw-Hill.

———. 1974. "Persuasive Cartography: An Examination of the Map as a Subjective Tool in Cartographic Communication." Unpublished Ph.D. dissertation, Department of Geography, University of California–Los Angeles.

———. 1976. *California Patterns on the Land*. 5th ed. Palo Alto, CA: Mayfield.

———. 1982. "Persuasive Cartography." *Journal of Geography*. 81 (4): 140–44.

———. 1987. "Interactions of Culture and Cartography." *History Teacher*: 20 (4): 455–64.

———. 1992. *Introduction to Thematic Cartography*. Englewood Cliffs, NJ: Prentice-Hall.

———. 1994. "Geography through the Needle's Eye: Embroidered Maps and Globes in the Eighteenth and Nineteenth Centuries." *Map Collector*. 66: 3–7.

———. 1996. "The World in Silk: Embroidered Globes of Westtown School." *Map Collector*. 74: 11–14.

———. 1997. "The Hidden Cartographers: Women in Mapmaking." *Mercator's World*. 2 (6): 46–51.

———. 1999. "Millie the Mapper and Beyond: The Role of Women in Cartography Since World War II." *Meridian*. 15: 23–28.

———. 2001. "Following the Thread: The Origins and Diffusion of Embroidered Maps." *Mercator's World*. 6 (2): 36–41.

———. 2006. "A Day with Norman J.W. Thrower." *Cartographic Perspectives*. No. 55 (Fall): 7–15.

———. 2010. *Principles of Map Design*. New York: Guilford.

van den Hoonaard, Deborah. 1997. "Identity Foreclosure: Women's Experiences of Widowhood as Expressed in Autobiographical Accounts." *Ageing and Society*. 17: 533–51.

van den Hoonaard. Willem. 1843. *Korte Aardrijkskundige Beschrijving der Stad Rotterdam, In Lesjes voor de Jeugd, met een Topographische kaartje* [Brief Geographical Description of the City of Rotterdam, in Lessons for the Youth, with a Small Topographical Map]. Rotterdam: Van Mensing and van Westrenen.

van den Hoonaard, Will C. 1994. "Do City Maps Show Gender Bias?" *Daily Gleaner*. Fredericton, NB (23 June): 7.

REFERENCES

———. 1997. *Working with Sensitizing Concepts: Analytical Field Research.* Thousand Oaks, CA: Sage.

———. 1999. "Map Worlds: A Conceptual Framework for the Study of Gender and Cartography." In C. Peter Keller, ed. *Conference Proceedings of the 19th International Cartographic Conference, Ottawa, Canada, 14–21 August 1999.* Ottawa, ON: Organizing Committee for Ottawa ICA, 1999.

———. 2000a. "Mapping a Conference: A Participant-Observation Analysis of a Cartographers' World." *Association of Canadian Map Libraries and Archives Bulletin.* (107): 23–28.

———. 2000b. "'What's a Nice Sociologist Like You Doing in a Place Like This?' A Sociological Exploration of the World of Cartographers." Joint Conference of the Association of Canadian Map Libraries and Archives, the Cartographic Association of Canada, and the Western Association of Map Libraries, Edmonton. 31 May–4 June.

———. 2000c. "Getting There without Aiming at It: Women's Experiences in Becoming Cartographers." *Cartographica.* 37 (3): 47–60.

———. 2001a. "The Courtship of Commerce" *Posisjon.* [GeoForum: Scandinavian Organization for Geographic Information] 8 (6): 18–19.

———. 2001b. "Mapping a Conference: A Participant-Observation Analysis of a Cartographers' World." GeoForum, Summer School for Cartography, Tromsø, Norway. 20 August.

———. 2002a. "The Courtship of Commerce: Community and Competition at Cartographers' Commercial Exhibits." 19th Annual Qualitative Analysis Conference, McMaster University, Hamilton, ON. 23–25 May.

———. 2002b. "'We Are Good Ghosts!' Orientations and Expectations of Women Cartographers." Annual Meetings of the Canadian Cartographic Association. Wilfrid Laurier University, Waterloo, ON. 27 May.

———. 2009a. "Cartography in the Family Way in the 16th and 17th Centuries." *Cartouche: Newsletter of the Canadian Cartographic Association.* Summer (74): 6–7.

———. 2009b. "Kirstine Colban: Norway's First Woman Cartographer." *Cartouche: Newsletter of the Canadian Cartographic Association.* Winter (76): 7–9.

van der Krogt, Peter. 1996. "From 'Atlas' to Atlas." *Mercator's World.* 1 (1): 61–63.

van Ee, Patricia Molen. 2001. "American Women: Introduction." Library of Congress Geography and Map Division, http://memory.loc.gov.

van Egmond, Marco. 2002. "The Secrets of a Long Life: the Dutch Firm of Covens and Mortier (1685–1866) and Their Copper Plates." *Imago Mundi.* 54: 67–86.

Varley, John. 1953. "John Rocque: Engraver, Surveyor, Cartographer and Mapseller." *Imago Mundi V*: 83–85.

REFERENCES

Verner, Coolie. 1975. "Copperplate printing." In David Woodward, ed. *Five Centuries of Map Printing*. Chicago: University of Chicago Press.

Walker, J. Jesse. 2006. "In Memoriam: Evelyn Lord Pruitt, 1918–2000." *Annals of the Association of American Geographers*. 96 (2): 432–39.

Wallis, Helen, 1954. "The Exploration of the South Sea, 1519–1644: A Study of the Influence of Physical Factors, with a Reconstruction of the Routes of the Explorers." Unpublished D.Phil. thesis, Geography, St. Hugh's College, Oxford.

———. 1979a. "Walter W. Ristow." In Helen Wallis and Lothar Zögnar, eds. *The Map Librarian in the Modern World*. Munich: K.G. Saur.

———. 1979b. "Map Librarianship Comes of Age." In Helen Wallis and Lothar Zögnar, eds. *The Map Librarian in the Modern World*. Munich: K.G. Saur.

———. 1989. "Historical Cartography." In D.W. Rhind and D.R.F. Taylor, eds. *Cartography Past, Present and Future*. New York: Elsevier.

———. 1994. "Eila Muriel Joyce Campbell, 1915–1994." *Geographical Journal*. 160 (3): 361.

———. 1995. "Eila Campbell and Her Societies." *Imago Mundi*. 47: 9–10.

———, and M.H. Edney. 2003. "Cartography." In Ivor Grattan-Guinness, ed. *Companion Encyclopedia of the History and Philosophy of the Mathematical Sciences*. Baltimore, MD: Johns Hopkins University Press.

Warhus, Mark. 1997. *Another America: Native American Maps and the History of Our Land*. New York: St. Martin's Press

Warren, Carol B., and Jennifer Kay Hackney. 2000. *Gender Issues in Ethnography*. 2nd ed. Thousand Oaks, CA: Sage.

Watts, Ruth. 2007. *Women in Science: A Social and Cultural History*. London: Routledge.

Whitaker, Ewen A. 1999. *Mapping and Naming the Moon: A History of Lunar Cartography and Nomenclature*. New York: Cambridge University Press.

Whitby, Barbara. 2005. *The Last of the Beothuk: A Canadian Tragedy*. Canmore, AB: Altitude Publishing.

Wilford, John Noble. 2000. *The Mapmakers*. New York: Knopf.

Williams, David R. 2005. "Soviet Lunar Missions," http://nssdc.gsfc.nasa.gov/planetary/lunar/lunartimeline.html.

Winearls, Joan. 1979. "Map Collections and Map Librarianship in Canada: Review and Prospects." In Helen Wallis and Lothar Zögnar, eds. *The Map Librarian in the Modern World*. Munich: K.G. Saur.

Winter, Keith. 1975. *Shananditti: The Last of the Beothuks*. North Vancouver, BC: J.J. Douglas.

Wirth, Linda. 2002. *Breaking through the Glass Ceiling: Women in Management*. Geneva: International Labour Office.

REFERENCES

Wood, Denis. 1992. *The Power of Maps.* New York: Guilford.
Woodfield, Ruth. 2000. *Women, Work, and Computing.* New York: Cambridge University Press.
Woodward, David. 1975. *Five Centuries of Map Printing.* Chicago: University of Chicago Press.
———. 1977. *The All-American Map: Wax Engraving and Its Influence on Cartography.* Chicago: University of Chicago Press.
———, and G. Malcolm Lewis, eds. 1998. *The History of Cartography: Cartography in the Traditional African, American, Arctic, Australian, and Pacific Societies.* Vol. 2, Book 3. Chicago: University of Chicago Press.
WorldView Antique Maps. 2002. "A Website specializing in Maps of Africa and Sea Charts," http://www.worldviewmaps.com.
Zhou, Yu. 2010. "A Tribute to Mei-Ling Hsu." *China Geography Speciality Group Newsletter.* (Winter): 3–4.

READING SUGGESTIONS ON FEMINIST CARTOGRAPHY IN ADDITION TO ABOVE REFERENCES
(with thanks to Dr. Jennifer J. Johnson, Thorneloe College at Laurentian University)

Dando, C. 2007. "Riding the Wheel: Selling American Women Mobility and Geographic Knowledge." ACME. 6 (2): 174–210.
———. 2010. "'The Map Proves It': Map Use by the American Woman Suffrage Movement." *Cartographica.* 45 (4): 221–40.
Elwood, S. 2009. "Geographic Information Science: New Geovisualization Technologies—Emerging Questions and Linkages with GIScience research." *Progress in Human Geography.* 33 (2): 256–63.
———. 2010. "Geographic Information Science: Emerging Research on the Societal Implications of the Geospatial Web." *Progress in Human Geography.* 34 (3): 349–57.
Kwan, M. 2002. "Introduction: Feminist Geography and GIS." *Gender, Place & Culture: A Journal of Feminist Geography.* 9 (3): 261–62.
———, and T. Schwanen. 2009. "Quantitative Revolution 2: The Critical (Re)Turn." *Professional Geographer.* 61 (3): 283–91.
Pavlovskaya, M.E. 2002. "Mapping Urban Change and Changing GIS: Other Views of Economic Restructuring." *Gender, Place & Culture: A Journal of Feminist Geography.* 9 (3): 281–89.
Pavlovskaya, M. 2006. "Theorizing with GIS: A Tool for Critical Geographies?" *Environment & Planning A.* 38 (11): 2003–20.
Piper, Karen. 2002. *Cartographic Fictions:Maps, Race and Identity.*Piscataway, NJ: Rutgers University Press.

Richards, P.L. 2004. "'Could I but Mark Out My Own Map of Life': Educated Women Embracing Cartography in the Nineteenth-Century American South." *Cartographica*. 39 (3): 1–17.

Rocheleau, D., B. Thomas-Slayter, and D. Edmunds. 1995. "Gendered Resource Mapping: Focusing on Women's Spaces in the Landscape." *Cultural Survival Quarterly*. 18 (4): 62–68.

Schuurman, Nadine. 1999. "Chapter 1 Introduction: Theorizing GIS, Inside and Outside." *Cartographica*. 36 (4): 7.

———. 2009. "Critical GIscience in Canada in the New Millennium." *Canadian Geographer*. 53 (2): 139–44.

———, and K. Mei-Po, 2004. "Guest Editorial: Taking a Walk on the Social Side of GIS." *Cartographica*. Spring: 1–3.

Schwanen, T., and M. Kwan. 2009. "'Doing' Critical Geographies with Numbers." *Professional Geographer*. 61 (4): 459–64.

Copyright Acknowledgements

The author and the publisher are grateful to a number of copyright holders for permission to reprint materials used in this book.

Figure 3.2: Konrad Miller facsimile of Ebstorf map. Formerly Business, editorial and circulation offices of *Mercator's World*, 845 Willamette St., Eugene, OR 97401, USA. Phone (541) 345-3800.

Figure 3.3: The Engravers' Map Atelier. Reprinted with permission of Dover Publications, Inc., from *A Diderot Pictorial Encyclopedia of Trades and Industry*, Vol. 2 (New York: Dover Publications, 1959), Plate 380.

Figure 4.1: Elizabeth Ann Goldin. Reprinted with permission of Cooper-Hewitt, National Design Museum, New York, NY.

Figure 4.2: "Little girls" with paper destined for printing. Reprinted with permission of Dover Publications, Inc., *A Diderot Pictorial Encyclopedia of Trades and Industry*, Vol. 2 (New York: Dover Publications, 1959), Plate 368.

Figure 4.3: Two women file burrs. Reprinted with permission of Dover Publications, Inc., *A Diderot Pictorial Encyclopedia of Trades and Industry*, Vol. 2 (New York: Dover Publications, 1959), Plate 370.

COPYRIGHT ACKNOWLEDGEMENTS

Figure 4.4: Woman composes line of text. Reprinted with permission of Dover Publications, Inc., *A Diderot Pictorial Encyclopedia of Trades and Industry*, Vol. 2 (New York: Dover Publications, 1959), Plate 373.

Figure 4.5: Eighteenth-century map seller's shop. Reprinted with permission of William L. Clements Library, University of Michigan.

Figure 4.6: Title page, Mary Ann Rocque, *A Set of Plans and Forts in America*. Reprinted with permission of The Edward E. Ayer Collection, The Newberry Library, 60 W. Walton St., Chicago, IL 60610-7324. Tel. 312-943-9090.

Figure 4.7: Elizabeth Mount's globe. Reprinted with permission of Yale University Garvan Collection, Suzanne Warner, Service Representative, Rights and Reproductions, Yale University Art Gallery, 201 York Street, P.O. Box 208271, New Haven, CT 06520. Tel. (203) 432-0630, e-mail suzannewarner@yale.edu.

Figure 4.8: Kirstine Colban's first known map. Reprinted with permission of Avdeling for fag og forskning, Nasjonalbiblioteket, Oslo, Norway.

Figure 4.9: Perspective map of Kabelvåg, by Colban. Reprinted with permission of Perspective Maps, Lofotmuseet, Lofoten Islands, Norway.

Figure 4.10: Church of Vågan, by Colban. Reprinted with permission of Perspective Maps, Lofotmuseet, Lofoten Islands, Norway.

Figure 4.11: Shanawdithit, last survivor of the Beothuk. Reprinted with permission of The Collection of the Newfoundland Museum (Found in Warhus, 1997: 88) / Newfoundland Provincial Archives, Newfoundland Museum, P.O. Box 8700, St. John's, NL A1B 4J6. Tel. (709) 729-2329, e mail: Kwalsh@mail.gov.nf. Or: Provincial Archives Penny Holden, Chief Curator, 285 Duckworth St., St. John's, NL A1C 1G9. Tel. (709) 729-3065, fax: (709) 729-2179).

Figure 4.12: Red Indian Lake: Shanawdithit depicts the capture of her aunt. From Brown, 2000: 28. Reprinted with permission of Cambridge University Press, 40 West 20th Street, New York, NY 1001-4221. Tel. (212) 924-3900, fax (212) 691-3239.

Figure 6.1: Marie Tharp. Photo by Bruce Gilbert. Reprinted with permission of The Earth Institute, Columbia University, New York, NY.

COPYRIGHT ACKNOWLEDGEMENTS

Figure 6.2: *World Ocean Floor Panorama.* Copyright © Marie Tharp 1977/2003. Reproduced by permission of Marie Tharp Maps, LLC, 8 Edward Street, Sparkill, New York 10976.

Figure 7.1: Regina de Almeida in group shot. Photo provided by kind permission of Regina Araujo de Almeida, 10 June 2011.

Index

"3M Girls," 91

Aanrud, Roald, 65, 67, 311n17
Aboriginal people, 202. *See also* First Nations
Academy of Travel and Tourism, Advisor Board of (Brazil), 166
Académie royale des Sciences, 44, 45
Ackworth Boarding School, 48, 63
ACSM. *See* American Congress on Surveying and Mapping
ACSM Bulletin, 4
Adams, A., 6, 270
Adams School Teachers' Association, 83
Addams, Jane, 77
Affirmative action, 245
Africa, 129, 266
Age of Reason, 46
Ageism, 252
Ainslie, Mrs., 55
Ajtay, Agnes, 320n9
Akin, Mrs., 310n6

Almeida, Regina Arauijo de, 139, 164, 203, 226, 278, 303, 321n13
Amazonia, 165
American Association for Geodetic Surveying, 234, 245, 321n17
American Cartographer, The, 115, 119, 121, 122
American Cartographic Association, 103, 104, 115, 122, 124, 245
American Congress on Surveying and Mapping, 91, 94, 99, 100, 101, 114, 124, 130, 226, 233, 244, 245
American Folklore Society, 118
American Geographical and Statistical Society, 235
American Geographical Society, 5, 80, 83, 107, 236, 321n17
American Geophysical Union, 97
American Revolutionary War, 114
American School in Guatemala City, 160
American Society for Photogrammetry and Remote Sensing, 130

INDEX

American War of Independence, 147
Amersfoort, 233
Anderson, Jacqueline, 126, 130
Angel, Karen Nattanaelsen, 66
Anglée (factory), 310n7
Annals of the Association of American Geographers, 122
Anthropogeography, 80
Anzaldúa, G., 7
Apollo missions, 151, 316n12
Aporta, Claudio, 277
Appalachian Barrier, 79
Archival Research Task Force, 148
Arctic, 106, 127, 128, 131, 132, 133, 134
Ardener, S., 6
Armenakis, Costas, 133
Arnberger, Erik, 152, 153, 157, 317n15
Artimo, Kirsi, 228, 321n13
Art worlds, 8
Aschenberner, Peter, 152, 316n13
Association for the Study of Food and Society, 118
Association of American Geographers, 104, 108, 109, 115, 118, 120, 122, 313n13, 320n10
Association of Canada Lands Surveyors, 321n17
Association of Canadian Map Libraries/Archives, 130, 289, 290, 321n17
Association of GIS (in Mexico), 162
astronomers, women, 306n7
Atkins, Peter, 10
Atlantic Ocean, 97
Atlas(es), 9; editing of, by gender, 231–32; exhibit of in Paris, 231–32; Hungarian, 9; school, 62; tactile ~, of Canada, 133
A to Z Maps, 180, 278
Auringer Wood, Alberta, 98, 113, 115, 116, 124, 127, 129–31, 154, 286, 303, 313n16, 314n19
Australia, 225, 266, 320n9

Austria, 151, 225; folk culture, 139
Austrian Geographical Society, 153
Auto-Carto 5, 100
Awards, 80, 95, 100, 101, 108, 109, 119, 122, 126, 129, 131, 134, 136, 141, 142, 148, 153, 155, 156, 159, 163, 166, 315n32, 316n8

Baber, M.A., 5
Bagge, Oluf Olufsen, 66
Bagrow, L., 33, 38, 41, 43, 143, 174–75
Bahrain, 321n13
Baker, J.N.L., 145
Ballantyne, B., 242, 246, 248, 253, 259
Baltimore Regional Planning Council, 99
Barbara Petchenik Children's Map Competition, 19, 113, 115, 126, 130, 229, 303
Barber, P., 143, 144, 147, 148, 174, 175, 199, 315n5, 315n7
Barckley, P., 10
Barker, H., 52
Barokas, J.U., 169
Barton, C., 97, 312n2
Bassa, Gizella, 321n13
Bavaria, 158
Baxendall, J., 85
Beatus of Liebana, 30
Beaujot, R., 191
Beaver, Carol, 13, 140, 154, 321n12
Becker, H.S., 8, 14
Beeck, Anna, 36, 43
Beeck, Barents, 41, 42, 43
Beijing, 227
Belgravia (London), 87
Benson, Mina. *See* Hubbard, Mina
Bentzen, Billie L., 320n7
Beothuk(s), 68, 69–72
Bercy, daughter of, 49
Berkeley, Lady, 64
Berman, Mildred, 4, 80, 311n5
Bermingham, Elizabeth, 64

358

INDEX

Berne City and University Library, 49
Bernleithner, Ernst, 153, 179
Bert. *See* Bertius (van der Keere) family
Bertall, 278
Bertin, J., 210
Bertius family, 37, 38, 42. *See also* Kaerius, Elisabeth
Berton, Pierre, 312n8
Bibliothèque Nationale de France, 144, 235, 242
Biddle, Mary, 51, 309n5; William, 51
Bidwell, John, 99
Bingham, M.T., 4
Binney, Dr., 310n6
Blachut, Theodor, 132
Blackhawk Middle School, 125
Blaeu House, 34, 37, 38, 40–41; Cornelis, 40; Joan, 40; maps and atlases, 41, 43; printing presses, 309n15; Willem J., 40, 308n5
Blagg, Mary Adela, 76, 84–86, 281, 302; Charles John, 85; lunar crater named for, 86
Blair, R., 49
Blumberg, D.R., 77, 311n2
Blumer, H., 14
Blunt, A., 7, 73, 81, 87, 88
Bond, Barbara, 130, 139, 154–57, 303, 315n6, 317n16, 317n17, 317n18
Bond, Fred, 88, 89, 312n12
Bonsall, Caroline Bartram, 77
Booth, Charles, 77
Boston, 48, 50
Boundaries, walking on, 199–202
Bouréche, Julie, 60
Bowman, J., 311n2
Bracket, H.V., 310n6
Brandenberger, A.J., 13, 218, 265
Brauen, Glenn, 277
Brayton, J., 280
Brazil(ian), 139, 165, 203, 278, 321n13; ~ Society of Cartography, 166
Breckenridge, S., 78, 311n2

Brewer, Cynthia, 120, 126, 315n32
Britain, 63, 76
British Admiralty, 139; Directorate of Military Survey, 154; Ministry of Defence, 156; National Committee for Geography, 142
British Association for the Advancement of Science, 142
British Cartographic Society, 139, 147, 148, 156, 255
British Columbia, 160
British Committee for Map Information and Catalogue Systems, 147
British Library, 139, 145, 146, 199, 235; Map Library, 173
British Museum, 145, 146
Brooker-Gross, S.R., 79, 80, 81, 311n3, 311n4
Brooklyn Training School for Nurses, 81
Brooks, C., 235, 242
Brouwer, Anna Catherine, 49
Brown, Lloyd A., 38, 39, 52, 308n6, 308n13, 311n3
Brown, Nina, 77, 78, 79, 311n2, 311n3
Brown, Stephen R., 311n20
Brunswick, 30
Brush, Stephen G., 210
Brussels, 233
Buchanan, R., 82, 312n8
Buehler, E., 10
Bulson, E., 274
Buttenfield, Barbara, 103, 105, 124–26, 279, 303
Buttimer, Anne, 142
Buys, Catherine, 42

Caissie, Linda, 294
Caldwell, Patricia, 103, 104, 116, 122–24, 130, 189, 276, 303
Calgary, 225
California Map Society, 124

359

INDEX

Campbell, Eila, 139, 140–45, 154, 174, 203, 225, 281, 302
Campbell, Tony, 145, 146, 147, 148, 149, 175, 203, 315n7
Canada, 81, 104, 160, 193, 205, 206, 219, 225, 233, 266, 311n18, 311n6; Northern, 276; Western, 68
Canada, proportion of women surveyors, compared to US, 319n1; context of women cartographers, 126–27; women cartographers in, compared to United States, 80
Canada Aeronautics and Space Institute, 321n17; Centre for Mapping, 226; Committee on Bibliographic Control of Cartographic Materials, 130; Department of Energy, Mines and Resources, 226, 320n10; Department of Natural Resources, 126, 243, 276; Hydrographic Service, 133; Institute for Surveys and Mapping, 320n10; Permanent Committee on Geographical Names, 128; Space Agency, 133
Canadian Association of Geographers, 233
Canadian Cartographic Association, 136, 188, 287, 289, 290, 320n10
Canadian Institute of Geomatics, 136, 321n17
Canadian National Committee for the ICA, 130
Canadian Polish Heritage Society, 132
Canadian Society for the Study of Names, 129
Canadian Women and Geography Newsletter, 4
Cannon, E., 245
Caplan, P., 242
Cappon, Lester, 114
Careers of women: non-linear, 175–78, 182
Caribbean, 266

Caro, Clara, 321n13
Carson (physician), 72
Carson, James, 210
Carte de Tendre, 278
Carter, J., 105, 313n10
Cartesian view of space, 277
Cartograms, 78
Cartographer(s): expectations, by women, 204, 207; falling into occupation of, 178–79; first interest by women, 170–72; Flemish, 37; formative years of women, 172–73; lifespan of, 34; occupational culture of, 18, 170; who is?, 15, 18–19; young women, 195
Cartographic laboratories, 233
Cartographic metaphors, 277; symbolization, 110
Cartographic organizations, Netherlands and UK, 234–35; gender and regional, 232–33
Cartographic Perspectives, 120, 122
Cartography, academic, 80, 92, 102–5, 276; allegorical, 278; automated, 133; commercialization of, 44; as corporate pursuit, 279; critical, 279, 283; and feminism, 273, 277, 283 (*see also* feminism/-ist); golden age of, 33, 43; historical, 139; as immature science, 200, 204; internationalizing, 73; lunar, 111, 112, 281; and masculinity, 273–75, 283; as occupation, 18, 170; oceanic, 93; one of the most remarkable achievements in, 95; persuasive, 25, 111, 112, 137, 203, 278; planetary, 151; Polish, 132; positivist vision of, 273–75; scientific, 45, 274; seafloor, 95; use of narratives to teach, 200; Western, 26
Cartography and Geographic Information Society, 122
Cartography, houses of, 270

Cartography Speciality Group (Association of American Geographers), 103, 104, 232
CartoPhilatelic Society, 13
Cartouche, 289
Carvel and Bowles (firm), 47
Catholic Church, 29
Catskills, 81
Central America, 266
Centro de Investigación en Geografía y Geomática, 159
CentroGeo, 162
Champlain Institute, 320n10
Chancy, C., 76, 308n6
Charles, Maria, 267
Charmaz, Kathy, 179
Chekin, 307n4
Chicago, 75, 77
Chicago Geographic Society, 5
Child(ren), 61–62, 77, 113, 166, 189; care of, at conferences, 243–44
Children's National Map Competition, 136
China, 10, 109, 139, 159, 157, 321n13
Chinese cartography, 110; princess map maker, 11
Cholera, map of, 76
Christiania, 67
Christian history, 31
Clarke, Dawne, 163
Clarkson, M., 51
Clawson, Mary G., 93, 98–102, 130, 303; David, 99
Clayton, Keith, 128
Clio, 308n5
Cloisters, 30
Cnobbari, family, 42, 309n19
Coastal Society, 108; Coastal Studies Institute, 107
Cohen, A., 273
Coindé, Madame, 61
Colban, Kirstine, 47, 64, 65–68, 271, 311n17
Cold War, 281
Cole, S., 39, 308n6
Coleman, David, 286,
Coles, Eliza, 50, 51, 270
Colet Girls' School, 145
Colius family, 39
Colombia, 321n13
Colonization, 45, 72, 279
Colouring/colourists, 33, 35, 36, 39, 47, 274
Commerce, courtship of, 230
Committee on the Status of Women in Geography, 108
Congress of Cartographic Information Specialists Associations, 130
Continental drift, 95, 98
Conzen, Michael, 119
Cook, Capt., voyages of, 147
Cook, Ellen P., 169
Cook, Karen, 104, 170
Cooksey, Gloria, 79, 80, 311n3
Cool. *See* Colius family
Cooperation vs. competition, 204
Coopers, Lybrand and Deloitte (UK), 320n9
Copper engraving/plates, 22, 23, 33, 274
Cormack, William E., 71, 72
Cornell, Sarah Sophia, 61
Coronelli World League of Friends of the Globe, 153
Cosgrove, Denis, 18, 62
Coulsdon South Station, 83
Covens, Amelia, 41; Johannes, 41. *See also* 's-Gravensande, Amelia
Covens and Mortier (firm), 49, 309n4
Crampton, J., 120
Crane, N., 26
Crawley, 48, 309n3
Creutzberg, N., 77, 311n1
Crowd sourcing, 15, 276
Cruz, Britta, 117
Crystal Palace, 63
Cyber-cartography, 159, 166; -feminists, 280
Cybernetic theory, 162

INDEX

Cyberspace, 278
Czech Republic, 321n13; Czechoslovakia, 225

Dahl, E., 62
Dahlberg, R., 105, 112
Danckerts family, 41
Dancocks, D., 48, 309n3
Danet, Guillaume, 53
Darby, Clifford, 143, 144
Davidson, James W., 312n8
deFer, Nicholas, 49, 53; Marguerite-Geneviève, 53; Marie-Anne, 53
de Haan, 308n12
de Hondt. *See* Hondius
de Jode, Gerard, 40, 309n19; Paschina, 309n19
Dekker, Elly, 63, 311n16
Demasduit, 71
Demographic data, spatial representation of, 78
Demoor, Maryse, 206, 243, 283
Dent, Borden D., 186
Deny, Jeanne, 49; Martial, 49
de Scudéry, Mme., 278
Design Museum (London), 86
Desmaretz, Marie, 52
Developing vs. developed countries: status of women in, 218
de Wit, Frederick, 40, 41
Diderot: *Pictorial Encyclopedia*, 35, 53, 54
Dietz, M.L., 12
Digital Earth Conference, 290
Digital mapping data, 98, 100; digitalization, 186; pioneer in, 98
Divorce, impact of, on map ateliers, 34
Dixon, Judith, 320n7
DMTI Spatial, 126
Dodd, D.E., 273
Dodge, J., 126
Doel, R.E., 95
Doerflinger, John, 152
Domosh, M., 6
Doyle, D., 234

Drake, F., voyages of, 147
Drazniowsky, R., 235, 236
Dubreuil, Lorraine, 235, 236, 237
Dunbar, Moira, 127
Duncan, James, 88
Duncker, 37, 38, 43, 49, 308n5, 308n6
Durham, Don, 100
Durval, Ulla, 320n9
Dutch East India Company, 40
du Val, Madame, 52; Pierre, 52
Duval, Pierre, 55
Dymon, Ute, 63, 64
Dynamic/quantitative geomorphology, 107

Earth & Ocean Research, 320n10
Eastern Europe, 266
Ebert, John and Katherine, 36
Ebstorf map, 30, 32, 270, 307n2; image, 32
École Polytechnique (Montreal): massacre at, 248
Ecology, 79
Edney, M.H., 76, 77, 311n1
Education, public, 59; involving maps, 58–60
Electronic atlas, 133
Elg, Margereta, 321n13
Elgin, Stephen, 210
Ellis, Harold, 83
Ellis, Richard H., 130
Ellison, George, 82
Ellis School, 124
Elwood, Sarah, 315n32
Ende (Spanish nun), 30
Endonyms, 129
Enfranchisement movement, 83
Engels, Friedrich, 77
England, 36, 47, 59, 63, 193, 203
engravers, 35, 49–51
Enlightenment, 45, 47
Environmental determinism, 79
Epidemics, 76
Equality, 239, 279; disavowing, 247,

259–65, 268; mainstreaming, 246, 248–52, 268; observing, 247, 257–59, 268; parallelism in, 247, 252–57, 268; pathways through, 246–65, personal experience, 262–63
Equity compulsion, 245
Equity relapse, 243
Esselte International, 320n9
Etzkowitz, H., 273
EuroGeographics, 233
EUROGI, 232
Evelyn L. Pruitt Lecture Series, 109
Ewing, Maurice, 97

Fairclough, R., 146, 315n7
Fairholt, F.W., 36
Falk-Plan, 1
Family: balance with work, 191–96, 204
Farmer, John, 61; Silas, 61, 310n14
Farnes, Patricia, 3
Farrar, Virginia, 11, 47. *See also* Farrer
Farrell, Barbara, 126
Farrer, John, 63, Nicholas, 63; Virginia, 63–64. *See also* Farrar
Fédération Internationale des Gèometres (FIG). *See* International Federation of Surveyors
Federation of University Women, 142
Feindt, Mary, 255
Feminist/feminism, 5, 6, 7, 13, 86, 160, 249, 250, 259, 262, 268, 269, 270, 273–79, 280, 281, 282, 283; cartography, 277; critique, 279; cyber-, 280; visualities, 277
Field Enterprise Education Corporation, 114
FIG. *See* International Federation of Surveyors
Finland, 225, 321n13
Finnesset, 66
Finnie, Ross, 206
First Nations, 7, 134
Firth, Edith G., 69, 311n18
Fitz, Ellen Eliza, 63

Fitzsimons, D., 120
Flanders, 26, 42, 49
Fofana, Mohamed Hassimiou, 321n13
Foley, John E., 211
Ford Motor Company, 100
Foreign and Commonwealth Office, 155
Foster, Natt, 9
Foster, Robert, 112, 245, 246
Foundation for Science and Technology, 147
Fowler, Marian Little, 311n19
Fox, Margalit, 95, 174, 312n2
France, 41, 44, 49, 53, 59, 225, 277
Frankfurter, Felix, 77
Franklin, Ursula, 280
Fredericks van der Linden, Aaltje, 43
Freelance work, 256
Freitag, Ulrich, 152, 153, 316n13
French Revolution, 45–46
Fussey, I., 319n3

Ganeau, widow of, 53
Gaspé Peninsula, 202
Gastaldi, Giacomo, 147
Gauvin, François, 62
Gävle, 243
Gender: and atlas editing, 231–32; and cartographic organizations, 232–35; comparing developing and developed countries, 265–68; rejection concept of, 263
Gender and Cartography (ICA), 132, 134, 160, 162, 220, 224–29, 253, 263, 289, 290, 320n8
Gender and Geography Newsletter, 4
Gender knot, 281–83
General Memorial Hospital of New York, 81
Geo-cybernetics, 159, 162, 163
Geo-data generalization, 139
Geodesy and geomatics vs. surveying engineering, 305n3
GeoForum, Summer School for Cartography (Tromsø), 290

INDEX

Geographers, women, 306n7
Geographers A–Z Map Company, 87
Geographia (firm), 87
Geographical names, standardization of, 127
Geographical Names Section, 128
Geographic Informations Systems in Mexico City, 320n9
Geographic Perspectives on Women Newsletter, 4
Geographic Society of Chicago, 80
Geography, 59, 98; education/teachers, 73, 306n7; human, 76, 79–81; remembered, 124
Geography and Map Groups of the Special Libraries Association, 236
Geological Survey of Canada, 83, 127, 128
Geomatica, 136
George Rivers, 82, 84
Georgian Bay, 202
Geovisualization, 102
Gerber, Linda M., 13
German Cartographical Society, 153
Germany, 22, 52, 57, 59, 79, 157, 159, 225
Gerstein, M., 169
Gervase of Tilbury, 32
Ghent, 38
Giddens, Anthony, 58
Gillispie, Charles C., 35
Gilmartin, Patricia, 102, 103, 105, 116–17, 303
GIS, 18, 102, 107, 120, 125, 280
Glaser, Barney, 14
Global positioning systems, 93, 100; invention of, 123
Global rift system, 97
Globes: rising importance of, 58; making, 60–63
Glynne (or Glin), Richard, 55
Goldin, Elizabeth Ann, 48
Goldman, Rabbi Yosef, 13
Gomes, William B., 183
Google Earth, 15
Goos family, 37–38, 42, 52, 309n19

Gorham, Deborah, 273
Gosson, Claire, 127
GPS. *See* Global positioning systems
Graf, Will, 117
Great Britain, 254, 255
Great Exhibition, 63
Greece, 321n13
Greed, Clara, 5, 86, 87, 259
Greene, Bryan, 82, 312n8
Greenland, Arnold, 100
Greenpeace, 78
Grenacher, F., 22
Greulich, G., 123, 276, 314n22
Grierson, Jane, 56
Griffin, Ian, 89
Grønland Asylum, 67
Gross, Phyllis. *See* Pearsall, Phyllis
Grounded Theory, 14
Guarani people, 165; image, 165
Guerrero, L., 162
Guild, 44, 54, 275
Guinée, 321n13
Guschina, Natalia, 320n9
Gutenberg, J., 33
Gwich'in Social and Cultural Institute, 277

Hackney, Jennifer Kay, 182
Hakluyt Society, 142, 147
Haldeman, Prof., 310n6
Hall, Alice, 310n6
Hamicz, 14
Hamilton (Ontario), 84
Hamilton, Roxanna, 61
Hamilton Club of Brooklyn, 83
Hammond (firm), 90
Hanham, Sue, 242, 246, 248, 253, 259
Haraway, Donna, 280
Hardcastle, J.A., 84
Hardie, M., 35, 36
Hargitai, H., 151, 316n10, 316n11
Harley, J.B., 11, 12, 25, 31, 112
Harris, E., 23, 24
Harsthorne, Richard, 80
Hart, Anne, 82, 83, 84, 312n12

INDEX

Hartley, S., 312n13
Hartshorne, Richard, 104
Harzell, Mary E., 236
Haussard, E., 36
Heald, Claire, 87, 88, 89, 312nn12–14
Heap, Abigail, 51
Heavenly objects, naming of, 316n10
Hebbert, Michael, 141
Heezen, Bruce, 95, 97, 98
Hegstad, Sneibulf, 311n17
Heist, Daniel, 120
Helen Wallis Memorial Lecture, 156
Heller, Nancy G., 30
Helmfrid, Staffan, 277
Helmholtz Association, 159
Herbert, Francis, 156
Herlihy, Elisabeth M., 91
Herlin, Olga, 50
Hermann, Michael, 202
Herrade, of Landberg, 30, 31; of Hohenbourg, 30
Herzig, R., 115
Hessels, J.H., 309n16
Himalayas, 172, 318n4
Hinterhuber, Hans H., 243
Hodgkiss, A.G., 90, 312n15
Holland, 42. *See also* Netherlands
Hollar, Mrs., 53; Wenceslaus, 53
Home Intelligence Division (UK), 89
Hondius family, 34, 35, 36–39, 42, 43, 308n10. See also Kaerius (van der Keere) family
Hoogvliet, M., 30, 32, 35, 270, 307n2
Howley, James, 72
Hoya, Mutsuko, 321n13
Hsu, Mei-Ling, 103, 104, 109–11, 302
Hubbard, Leonidas, 81, 82, 94
Hubbard, Mina, 76, 81–84, 270, 302; image, 83
Hudon, Alice, 3, 4, 10, 17, 36, 39, 40, 42, 43, 49, 50, 51, 52, 55, 56, 59, 60, 61, 63, 64, 81, 270, 305n2, 308n6, 309n19, 312n7
Huffman, N.H., 6, 270, 318n2
Huguenots, 55

Hull House, 77, 78
Hungary, 255, 276, 321n13
Huq-Hussain, Shahnaz, 10
Hure, Sebastian, 53
Hus (or Huée), Anne, 53
Husbands, role of, 193
Hutchings, Roger, 85, 312n10
Hutchinson, Ebenezer, 61
Hydrography, and gender, 232

Ibraheem, Seema Ahmed, 321n13
ICA conferences, 130, 181; exhibits, 229–31, 238, 276. *See also* International Cartographic Association
ICA Working Groups and Commissions, 13, 122, 126, 132, 133, 134, 149, 151, 160, 162, 166, 218, 222–29, 253, 263, 290, 320n8. *See also* Gender and Cartography (ICA)
Identifying moments, 179
IIT Research Institute, 100
Ilcan, Suzan, 46, 250
Imago Mundi, 60, 139, 140, 141, 143–44, 149, 153, 281, 285
Imhof, Eduard, 277
Imperial Public Library (Russia), 235
Indian Subcontinent, 266
India Office Records Map Collection, 235
Indigenous populations, 165. *See also* First Nations
Individualism, belief in, when dealing with sexism, 259–60
Inequality, pathways through, 246–65
Innis, Mary Quayle, 311n19
Institut(e), of British Geographers, 145; Cartogràfic de Catalunya, 144; of Cartography and Reproduction Technology (U of Vienna), 316n14; Nacional de Informacion Estadistica, Geografica e Informatica, 161; National de Géographie, 277; of Space Researches at the Academy of Science (Moscow), 149–50

365

Interactive Global Change Encyclopedia, 133
Intergraph Corporation, 320n10
Intermarriage, 11, 37
International Association of Academies, 84
International Association of Chinese Professionals in Geographic Information Science, 159
International Astronomical Union, 85, 86, 150, 151, 281
International Cartography Association, 13, 113, 115, 122, 134, 140, 142, 143, 146, 147, 148, 166, 173, 220–32, 288; International Secretariat, 320n10; loss of women in organizational executives, 220–22, 238; proportion of women in scientific sessions at, 238. *See also* ICA
International Conferences on the History of Cartography, 143, 153
International Congress of Onomastic Sciences, 129
Internationale Coronelli Gesellschaft, 147
International Federation of Library Associations, 148
International Federation of Surveyors (FIG), 100, 101, 244, 251, 263; Gender and Surveying Working Group, 228
International Geographic Information Foundation, 124
International Geographical Union (IGU), 124, 142
International Hydrographic Organisation, Antarctic Commission, 155
International Map Collectors' Society, 142, 148, 156
International Society for Photogrammetry and Remote Sensing, 232
Inuit, 134, 310n9. *See also* Arctic

Isolines, 310n10
Israel, Nico, 144
Italian National Council of Surveyors, 233

Jaillot, 49
Janssonius, family, 37, 38, 43, 307n5, 309n19
Japan, 321n13; proportion of women surveyors in, 319n3
Jaquette, Jane, 243
Jarausch, H., 115
Jenks, George, 102, 103, 105, 117, 118, 124, 313n7
Jerusalem, 30, 31
Jesuit mapping of China, 147
Jobes, C., 79
Jodocus, Gerard, 40; Pashina, 40
Johnson, Nancy, 170
Johnston, Allan G., 281, 282
Jones, John Paul, III, 318n2
Jorge L. Tamayo Center for Research on Geography and Geomatics (CentreGeo), 162
Joy, A.H., 84
Judd, J., 63
Jun, Yang, 320n9

Kabelvåg, image of, 65
Kaerius (van den Keere) family, 37, 38, 39, 42. *See also* Bertius family
Kansas, 99
Karachevtseva, I.P., 151
Kartografiya, 320n9
Kartographischen Nachrichten, 153
Kass-Simon, G., 3
Kaye, M., 63, 64
Keates, John, 210
Keeble, N.H., 52
Kelley, Florence, 75, 76–78, 272, 277, 302
Kelnhofer, Fritz, 152
Kemelgor, Carol, 273
Kerfoot, Helen, 126, 127–29, 276, 302, 314n24, 314n27

INDEX

Keuning, J., 37, 308n10
Khan, U., 10
Kindergarten: Norway's first, 67
King, Geoff, 11, 25, 26, 32, 275
Kinoshita, R., 10
Kishimoto, Mme., 226
Kitikmeot Heritage Society, 277
Kniffen, Fred, 108
Knowledge, transforming, 279–81; Knowledge Integration for Sustainable Development Program (Canada), 203
Knowles, Leo, 86, 88, 89, 199, 312n12
Koeman, Cor, 22, 38, 144, 152, 307n6, 308n6, 315n4
Kozlowa, Anna, 235
Kremer. *See* Meracator
Kretschmer, Ingrid, 139, 140, 151–54, 179, 225, 303
Krzuwicka-Blum, Ewa, 228, 321n13
Kudronovska, Olga, 140, 225
Kwan, Mei-Po, 315n32

Labrador, 76, 82, 84, 270
Laidler, Gia, 277
Lakota, 159
Lamont Geological Observatory, 96, 98
language, gendered, in history, 5
Lapp, Ralph, 114
Lassalle, D., 169
Latin America, 176, 265
Laurie and Whittle, firm of, 47
Laussedat, Aimé, 90, 312n15
Lavoie, Marie, 206
Lawrence, David, 95, 98, 175
Lawson, Helen E., 310n6
Lea, Anne Fitz (or Fitch), 55; Philip, 55
Lebreton *fils,* firm, 52
Lecordix, François, 231
Le Gay, Fanny, 47
LeGear, Clara Egli, 236
Lengermann, 273
Lenin State Library, 235
"Leo Belgicus" (map), 42, 308n14

Levin, J., 95
Lewis, Samuel, 48
Li, Li, 321n13
Library Association, 147
Library of Congress, 130
Lichtman, M., 169
Liebenberg, Elri, 320n9
Lindgren, Caldwell, Patricia, 154. *See also* Caldwell, Patricia
lithography, 22, 35–36, 45, 47, 49, 309n4
Livengood, H. Mark, 118
Lobeck, A., 25, 275
Lofoten Islands, 66; image, 65
Lofotmuseet, 67, 311n17
Lohr, Nouna, 246
Lohrmann, W.G., 85
London, 54, 77, 86, 180
Long, Letitia A., 102
Lopez, L., 162
Los Rios Community College, 123
Lotter, Euphrosina, 57; Tobias Conrad, 57. *See also* Seater, Anna Sabina
Louis XIV, 44, 45
Louis-Philippe, King, 66
Louter, D., 50
Lovelace, Lady, 11
Low Countries (Lands), 33, 34, 35, 49, 56
Lunar, 150; image, 86. *See also* Moon
Lunar Commission, International Astronomical Union, 85
Lunar Nomenclature (IAU), 150, 316n12
Luna satellites, 150
Lundemo, Edel, 320n9
Lutz, Alma, 60
Luxton, Karen, 320n7
Lynam, Edward, 143

Macdonald, Angie, 312n12
MacDonald Setwiller, 320n9
Macionis, J.J., 13
Mackenzie Delta, 128
Maddrell, Avril, 5, 89, 142, 174, 235
Madej, Ed, 186

INDEX

Magna Carta of women's education, 59
Magnus, Albert, widow of, 43
Maillart, Jeanne S., 49; Philippe Joseph, 49
Maine, 206
Makkonen, Kirsi, 320n9
Mandell, Mel, 90
Mann, Diane, 127
Map archives, archivists, 218, 232, 235–37
Map ateliers, 11, 15, 29, 33–34, 43, 72, 274, 275; images, 36
Map Collector, The, 42, 148
Map Curators' Group of the British Cartographic Society, 130
Mapmaking, companies, 276; satellite, 276
Mappaemundi, 31, 307n1
Mapping, alternatives in, 277
Mapping Maids, 12, 17, 89–91, 93
Map projection, 110
Maps(s) approaches to: democratization of, 275–77; not for walls, 188–89; Picassos on the wall, 188; psychology of, 120; purpose of, 24; reconciling beauty and usefulness of, 186–90; social activism through, 77; subjectivity of, 25, 275
Map(s), distribution of, 54
Map(s) illiteracy, 176
Map(s) libraries, ~ians, ~ianships, 16–17, 91, 131, 140, 218, 232, 235–37; and archivists by gender, 237; Map Library (British Museum), 281; in 19th c., 236; in North America vs. Europe vs. Latin America, 236; in United Kingdom, 145
Map(s) and technology, 190; digitizing, 93, 280; first use of aerial photos, 90; isarithm, 109; making relevant for users, 188
Map(s), military influence in, 254
Map(s), publishers and sellers of, 51–57
Map(s) (specific) *A to Z*, 76, 86, 87–88; of cholera epidemic, 76; International, 320n9; Louisiana, 53; Mars, 150; moon, 150; Phobos, 150; "T-O," 29, 30; Venus, 150
Map(s), typography of, 123
map(s) (types), action, 71; birchbark, 68; cadastral, 116; cognitive, 117; cross-stitched, 309n3; embroidered, 47–48, 73, 112, 113, 306n7, 309n1; escape, 156; European, 29; for the blind, 124; for children, 113, 114, 164, 278; for colourblind, 120; for visually impaired, 164; on handkerchief, 47, 48, 309n3; as icosahedron, 24; jigsaw, 47; lunar, 139; perspective, 66, 67; power, 25, 80; silk, 112; tactile, 164, 278; textile, 111; thematic, 76, 77; image of thematic map, 78
Map trades, 7, 46–47
Map workshops. *See* Map ateliers
Map worlds, concept of, 7; internationalization of, 9; social organization of, 217–18
Marcotte, Louise, 225
Margary, Harry, 144, 315n5
Margot, Helena, 321n13
Marino, Jill, 105
Mark, David, 107
Marker, M.K., 95
Martin, Geoffrey, 80
Martinez, Elvia, 162
Marx, Karl, 77
Mary Troy Seminary (Kinderhook, NY), 59, 60
Mason, Karen, 317n17
Massey, Doreen, 6
Maverick, Maria A., and Emily, 310n6
Mayer, Maskelyne, 52; Johann Tobias, 52
Mbyá tribe, 165
McCleary, George, 119
McCorkle, Barbara Backus, 148

McCormack, Thelma, 242, 245, 263, 280
McCurdy, William, 133
McDonnell, Porter W., 211
McElhanny Group, 320n10
McEwan, Cheryl, 306n7
McGaw, Judith, 305n1
McKay, Ed, 210
McManis, Douglas R., 80, 311n3
McMaster, Robert, and Susanna, 102, 104, 313n4, 313n6
McPherson, Bea, 12, 17
Mead, George Herbert, 14
Mead, W.R., 143, 145, 315n5, 315n7
Medicine, women in history of, 273
Men, older, 264; criticizing women's making maps, 186
Meng, Liqiu, 139, 157–59, 275, 303; source of name, 317n20
Menninger, Karl, 114
Mentor(ship), 209–10; comparing women and men, 212–13; men, 210; personal, 210; qualities of, 211–12
Menzel, D., 150
Mercator, 26, 29, 36, 38, 307n5
Meridian Lab, 125
Meridian, 4
Merkel, Angela, 159
Mersey, Janet, 127, 134–36, 303
Methodist, 81; Men's Social Union, 82
Methodology of *Map Worlds*, 14, 285–95; documents, 285–86; international field research, 294–95; interviews, 290–93, 297–99, participant observation, 287–90, vignettes, 286–87
Metropolitan London Police, 88
Mexico/Mexican, 52, 139, 159, 160, 203, 278, 321n13; Association of Geographic Information Systems, 162; Ministry of Education, 160
Miekkavaara, Leena, 140, 225
Mikhailov, A., 150
Military/militaries, 230, 275–76, 279, 283; Mapping Agency, 94; Map-Making Girls, 91 (*see also* "3M Girls"); Map-making Corps, 91
Military University of Information Engineering (Zhengzhou), 157
Miller, Konrad, 3, 30, 31, 32, 307n4
Millie the Mapper, 12, 76, 276
Mills, C. Wright, 3
Mills, Sara, 69, 81
Miss Roullet's Academy, 60
Möbius, Helga, 41, 44, 46, 52, 53, 56, 57, 278
Modelski, Andrew, 130, 236
Moen, Marit, 311n17
Molyneux, Emery, 147
Momsen, Janet, 10
Monasteries, 30
Monk, Janice, 5, 80, 107, 109, 110, 111, 234, 311n5, 313n11, 313n13, 313n14
Monmonier, Mark, 25, 112, 210, 275, 286
Montanus (van den Berghe) family, 37, 38
Montargis, 310n7
Moon, 84, 85, 150; Exploration Project, 151; map of, 63. *See also* Selonography, Lunar
Morentdort family, 43, 309n19. *See also* Moretus family
Moretus, family, 39, 40. *See also* Morentdort family
Morrison, Joel, 119
Mortier family, 41, 43, 52, 57–58. *See also* 's-Gravensande, Amelia
Moscow Megacity Road and Transport Complex, 151
Mounsey, Helen, 320n9
Mount, Elizabeth, 63; globe by, 62
Mount School, 48
Mueller, Fritz, 132; Karl, 85
Mushi, Anna, 321n13
Music, women in history of, 273

Nakazawa, T., 10
Nanjing Geoinformatics Conference, 159

INDEX

Nash, C., 3, 6
Naskaupi, 82, 84
Nast, J., 318n2
National Atlas of Sweden, 277
National Center for Geographic Information and Analysis, 125
National Geodetic Survey (United States), 233
National Geographic Society, 104
National Geophysical Data Center, 24
National Maritime Museum, 148
National Science Foundation Committee, 103
National Society of Professional Surveyors, 231, 232, 244, 253
Naylor, Mrs., 88
Needham, Joseph, 11
Nelson, Lise, 6
Netherlands, 42; ~ Cartographic Society, 234, 238. See also Holland; Low Countries
Newberry Library, 113, 114, 313n16
Newfoundland, 68, 69
New Zealand, 253; women surveyors in, 259
Nhandeva tribe, 165
Nichols, Susan, 126
Niebrugge-Brantley, Jill, 273
Nissen, Ole Hartvig, 67; Nils Kristian, 68; Per Schelderup, 67
Nobel Peace Prize, 123
North America, 93, 226
North American Cartographic Association, 136
North American Cartographic Information Society, 103, 104, 125, 130, 210
Northwest Territories, 128
Norway, 47, 65–68, 232, 243, 255, 288, 290, 311n17
Norwegian Geographic Survey, 68; Association for Cartography; Geodesy, Hydrography and Photogrammetry, 320n10;
Mapping Authority, 320n9, 320n10
Nunavut, 133

Ocean Floor Panorama (image), 97
Ochberg, R., 182
Office of Strategic Services, 91, 94, 104
Olin, Kelly, 244, 255
Olson, Ann, 10
Olson, Judith, 103, 104, 105, 109, 111, 119–22, 124, 130, 199, 225, 226, 278, 303, 313n13
O'Neil, Linda, 134, 218, 219, 226, 288, 320n6, 321n12
Ontario Heritage Foundation, 84; Geographic Names Board, 129
Open Geospatial Consortium, 162
Open Street Map, 16, 276
Oram, Elizabeth, 63
Orr, Alfred, 87
Ortel. See Ortelius family
Ortelius family, 34, 35, 39–40; maps, 39; title page of *America*, 307n5. *See also* Colius family
Ottens family, 41, 43
Ottoson, P., 50
Owen, Robert, 67
Oxford Cartographic Conference, 234, 255

Palfrey, Dale Hoyt, 162, 163
Pan-American Institute of Geography and History (Organization of American States), 161, 163
Panouse, Jacqueline, 49
Papp-Váry, Árpád, 9
Paris Salon, 46, 52, 310n6
Pastoureau, Mireille, 52
Pathways, female, 239, 241ff., 282, 283
Pavlovskaya, Marianna, 5, 277, 279
Pearce, Margaret, 202
Pearsall, Phyllis, 76, 86–89, 180, 199, 278, 302, 312n12, 312n13
Peetersen, H.: widow of, 42, 309n19

INDEX

Pelletier, Monique, 144, 225, 242, 315n5
Pennsylvania, 63,113
Penny, Virginia, 51
People with Disabilities On Line, 133
Peplinski, Lynn, 277
Perry, Joanne., 236
Petchenik, Barbara, 25, 98, 100, 103, 104, 113–16, 119, 120, 130, 154, 178, 189, 210, 225, 226, 275, 303
Petchenik International Children's Map Competition. *See* Barbara Petchenik International Children's Map Competition
Peucker, T.K. *See* Poiker
Peyton, John, Jr., 70; John, Sr., 71
Philadelphia, 51, 63; Philadelphia School of Design for Women, 310n6
Phobos, mapping of, 150
Photo: interpretation, 108; ~grammetrists, 90, 91; ~graphic plates, 23; ~lithography. *See* Lithography
Piaget, Jean, 210
Pickles, J., 25, 275
Pietkiewicz, Prof., 132
Pillewizer, Wolfgang, 152, 316n14
Planetary bodies, naming practices of, 151
Plant, Sadie, 280
Plantijn, Christoffel, 40; Maria (Martina), 40. *See also* Plantin family
Plantin family, 39, 40
Platonic relationships, 278
Pognon, Edmond, 235
Poiker, T.K., 107, 125, 160, 203
Point Barrow (Alaska), 106
Poland, 52, 59, 134, 228, 321n13
Portuguese discovery of Australia, 147
Postman, Neil, 210
Potter, Jonathan, 39
Poverty, 77, 78
Power mapping, 80
Prescott, Dorothy F., 140, 225

Pressl, Harald, 158, 159
Printers and printing presses, 23–24, 33, 275
Private sector vs. public, 193, 254
Privatization, 255–56, 283
Probst family, 57
Professional Geographer, The, 108
Progress and Perspectives, 4, 202, 305
Prouse, A. Robert, 49, 309n3
Pruitt, Evelyn, 12, 103, 105–9, 172, 191, 203, 234, 275, 302
Pruitt National Fellowship for Dissertation Research, 108
Ptolemy, 343; map by, 30, 37
Public Archives of Canada, 235
Public sector vs. private, 193, 254
Publishers, 275
Pugh, John, 128,
Putnam, William, 106
Pye, Norman,156

Quakers, 77; School in Westtown, 62; School, in Maryland, 48; (Oakwood) School, in Poughkeepsie, 48; Quaker schools vs. non-Quakers schools, 309n2

Raji, S., 10
Ramaswamy, S., 60
Rand McNally, 90, 118, 320n10
Ratajski, Lech, 132
Ratzel, Friedrich, 79, 311n4
Ravenhill, William, 140, 145, 146, 315n7
Ravenstein, Helga, 225
Ravn, Nils Frederik, 310n10
Rayburn, Alan, 128
Readhouse, Miss, 63
Reagan, Ronald, 147
Red Indian Lake (NL): image of, 71
Reitinger, Franz, 278
Relationships: importance of in careers, 190–91; Platonic, 278
Remote sensing, 98, 102, 105, 108; term coined by E. Pruitt, 105

Renaissance, 29
Renwick, Jane, 61
Research Institute of Surveying and Mapping, Beijing, 320n9
Reyes, Carmen, 139, 159–64, 203, 278, 303, 320n9, 321n13
Richardson, Diane, 127
Ring, Betty, 306n7
Ristow, Walter W., 22, 51, 61, 236, 310n14
Ritter, Michael, 57
Ritzlin, Mary McMichael, 3, 4, 10, 17, 35, 36, 38, 39, 40, 42, 43, 49, 50, 51, 52, 53, 54, 55, 56, 59, 60, 61, 63, 64, 81, 270, 286, 305n2, 306n7, 308n6, 308n7, 309n5, 309n16, 309n19, 310n11, 310n12, 311n19, 312n7
Rivard, Maude-Catherine, 206
Roberts, Susan M., 318n2
Robijn, Jacobus, 38, 42
Robinson, Arthur, 22, 23, 25, 44, 46, 47, 76, 94, 103, 104, 105, 109, 110, 115, 119, 135, 146, 153, 172, 189, 210, 275, 307n7, 313n7, 319n2
Robinson, Elisha, 61
Robinson, Jennifer, 313n16
Rocque (née Bew), Mary Ann, 55, 56; John, 55
Roedean (Girls' Boarding School), 87
Ronan, Colin A., 11
Roosevelt, Franklin D., 90
Rose, Gillian, 7, 73, 81
Rosien, W., 32
Ross, Robert, 49, 55
Roth, Robert E., 120, 315n32
Rotz, Jean, 147
Rowley, Diana, 127
Rowson's Academy, Mrs., 48
Roy, Laura, 322n5
Royal Astronomical Society: first woman Fellow, 86
Royal Canadian Air Force, 48, 309n3
Royal Canadian Geographical Society, 128, 129
Royal College of Physicians (London), 72
Royal Geological Society, 102
Royal Geographical Society (London), 79, 141, 142, 143, 146, 147, 148, 156, 235, 316n8
Royal Historical Society, 147, 148
Royal Institute of Chartered Surveyors, 319n3
Royal Society of Canada: first woman Fellow, 127
Royal United Services Institution, 147
R.R. Donnelly & Sons, 114, 320n10
Rugge, John, 312n8
Rundstrom, Robert A., 310n9
Russell, Richard, 108
Russia/Russians, 52, 149, 151, 235, 321n13. *See also* Soviets
Ryhiner Collection, 49
Rysted, Bengt, 128, 314n27

's-Gravensande, Amelia, 41, 4
Saigal, Anil, 246
Sandy Point (NL), 7
Sanson, Hubert Alexis, 49; Marie-Angelique, 52; Michelle, 52; Nicholas, 52
Sartain, Emily, 310n6
Scandinavia, 255, 267
Schalk, Agnes Coentgen, 49
Schenk, Pieter, 39, 40, 41; wife of, 40
Schweik, Joanne, 206, 321n12
Science and technology: women in history of, 273
Scientific-instrument makers, women, 306n7
Script, non-Gothic, 29
Scull, Nicholas, 51
Scullen, Wendy, 241, 245
Seager, Joni, 6, 10
Seater, Anna Sabina and Matthäus, 57. *See also* Lotter, Euphrosina; Probst family
Second shift, 192
Seegers Linken, 39. *See also* Verhoeven family

Seghers. *See* Seegers
Seile, Anna, 55
Selley, 188
Selonography, 84. *See also* Moon
Semple, Ellen Churchill, 4, 75, 79–81, 273, 302, 311n5
Semple Collegial School for Girls in Louisville, 79
Senefelder, Aloys, 46
Senex, John, 55; Mary, 55
Seutter-Probst-Lotter House, 57
Severud, Karen, 119
Sexist advertising, 231
Shaffir, William, 12
Shanawdithit, 68, 69–72, 270, 311n20; image of, 70
Shaw, Sandra, 320n9
Sheffield, S. Le-May, 4
Shelesnyak, M.C., 106
Sherman, John, 103, 104, 105, 124, 313n7
Sherrill, Elizabeth, 59
Shingareva, Kira, 149–51, 139, 281, 302
Shirley, Rodney, 38, 43, 307n5
Shoffey, Valerie, 233
Shortridge, Barbara, 103, 105, 117–19, 203, 278, 303; James, 118
Siekerska, Eva, 13, 126, 127, 131–34, 171, 178, 180, 189, 203, 218, 219, 225, 226, 227, 275, 276, 288, 303, 314n24, 320n6, 320n8, 321n12
Siemer, Julia, 286
Siltanen, Janet, 6
Simcoe, Elizabeth, 68–69, 311n18; John Graves, 68
Simms, Attorney General, 72
Simonton, Dean Keith, 94
Sinitsyna, Anna, 151
Sisterhood, 263
Sitting Bull, Chief, 159
Skelton, R.A., 145, 146, 173
Sklar, Lawrence, 311n2
Sleep, Lenora, 294
Slocum, Terry A., 186
Slums, survey of, 77

Smith, Nancy J., 170, 182
Smith, Terrence R., 125
Smythe, Kerry, 320n9
Snow, John, 76
Social activism/reform, 76, 77
Social organization, study of, 281–82
Social processes, study of, 281
Social Sciences and Humanities Research Council of Canada, 288
Social structures, being uncritical of, 260–61
Société de Géographie, 147, 235
Society for the History of Discoveries, 142, 148; of Antiquaries, 147; of Automotive Engineers, 115; of Cartographers, 142; for Nautical Research, 142, 147; for University Cartographers, 142; of Woman Geographers, 108, 124, 126, 142, 148, 191
Sociological imagination, 3; perspectives, 13
Sociology, 2, 281–82; women in history of, 273
Someran, Joannis A. (widow of), 43
South Africa, 321n13
South America, 266
South Pacific: exploration of, 281
Soviet Union/Soviets, 151, 225, 281; portrayal of, in American textbooks, 279. *See also* Russia
Space, masculinist view of, 277
Spain, 52
Spatial metaphors, 6
Special Libraries Association (Geography and Map Division), 130
Spilburg, John, 47
Sputnik, 97
Staffmaecker, Anna, 37
Stanislawski, L.V., 126
Stanolind Oil and Gas Co., 96
Stanworth, M., 6
State, the, 44, 45, 275
Stationers' Company, 54, 55

INDEX

Stauffer, David, 50, 51, 310n6
Stefoff, Rebecca, 306n7
Stephenson, Richard W., 130
Steward, Henry, 2, 115, 141, 149, 286
Stewart, Roger, 37
Stine. *See* Colban, Kirstine
St. John's (NL), 71
St. Luke's Guild, 39
St. Martin, K., 5, 277, 279
Stone, Lawrence, 46
Stoughton, Herbert, 211
Straight, Wendy, 86, 90, 91, 94, 123, 124, 202, 255, 286, 314n22
Strande, Kari, 311n17
Strauss, Anselm, 14
Strauss, David, 151, 316n12
Strutt, Joseph, 49–50
Stumpf, Monika, 243
Suarez, Thomas, 307n5
Suchan, Trudy, 315n32
Surveying: and women, 245; surveying engineering vs. geodesy and geomatics, 305n3
SWECO, 157
Sweden, 50, 157, 243, 255, 321n13; topography in, 50
Swedish Armed Forces, 158; Cartographic Association, 320n10; Landsurvey, 320n10; Royal Institute of Technology, 157; State Topographic Services, 255
Switzerland, 77, 225, 232, 277
symbolic interactionism, 14, 203, 247

Taiwan, 109, 110
Takeda, Yuko, 10
Tamayo, J.L., 159,
Tancred, Peta, 6, 270
Tang, Joyce, 4, 93, 94
Tanzania, 321n13
Tapley, Charles, 245
Taylor, Evan, 143; Eva, 235; Fraser, 77, 115, 132, 133, 159, 161, 162, 166, 226, 286

Teaching, Maps for Children Conference, 290; use of narrative in, 200
Technology, 16–17, 21–27, 29, 190, 201, 203, 274, 280, 283; and gender, 243; redemptive, 280
Teixeira, Marco A.P., 183
Tennessee Valley Authority, 91
Terenzoni, A.M. Piras, 233
Tharp, Marie, 93, 95–98, 174, 175, 273, 302; image, 96; created akin to Copernican revolution, 95
Thomas, W.I., 247
Thrower, Norman J.W., 10, 11, 30, 34, 35, 37, 43, 58, 63, 103, 104, 110, 112, 123, 148, 306n6, 310n10, 313n7, 315n7
Tiebout, Cornelius, 51
Tierney, Helen, 128, 314n27
Timation, 123
Tobler, Waldo, 130
Tooley, R.V., 3, 29, 30, 33, 307n1, 307n3
Topography, 45
Toponyms, 129, 276, 310n9
Tournachon, Gaspard Félix ("Nadar"), 90, 312n15
Townsend, Janet G., 10
Traub, Valerie, 307n5
Triangulated, 44; ~ irregular networks (TIN), 107
Tuberculosis, 72
Turkey, 321n13
Turner, H.H., 85, 120
TWIC—Tall Women in Cartography, 116
Twyman, Paul, 22
Tyacke, Sarah, 54, 55
Tyler, Judith, 3, 12, 19, 25, 47, 48, 55, 59, 60, 62, 63, 64, 90, 91, 94, 102, 103, 104, 109, 123, 186, 203, 210, 275, 278, 286, 302, 306n7, 309n1, 309n2, 312n16, 313n7

Ulugtekin, Necla, 321n13

INDEX

UNESCO: Division of Earth Sciences, 320n9
Ungava Bay, 82, 84
Unilever, 128
United Kingdom, 6, 140, 160, 192, 225, 234, 238, 319n3; Hydrographic Charting and Marine Sciences, 155
United Nations, Group of Experts on Geographical Names, 127, 128, 277; Law of the Sea Conference, 95
United States, 50, 52, 59, 90, 160, 205, 206, 225, 226, 233, 255; Air Force, 100, 123; Army, 100; Census Bureau, 119; Central Intelligence Agency, 123; Coast and Geodetic Survey, 91, 106; Defense Mapping Agency, 99, 125, 154; Department of Defense, 99; Department of State, 320n9; Department of Transportation, 119; Geological Survey, 91, 125; Geospatial-Intelligence, 101; Library of Congress, 95, 99, 235, 236; Marine Corps, 100; Marine Geography Committee, 108; National Committee on Lunar Mapping and Nomenclature, 150; National Geospatial-Intelligence Agency, 101; Navy, 91, 100, 105, 123; Navy, did not permit women, 97; Navy, highest-ranking woman scientist, 108; Office of Global Navigation, 101; Office of Naval Research, 106, 107, 108; Office of Strategic Services, 94, 104; Strategic Planning Division, 101; Supreme Court, 77; War Department, 91
United States: compared to Canada, 80; women cartographers in, 61; proportion of women surveyors, compared to Canada, 319n1
Universities/colleges in Africa: South Africa, 320n9

Universities/colleges in Asia: Taiwan Normal U., 109
Universities/colleges in Canada: Calgary, 206; Carleton, 134, 126, 161, 166, 226, 286; Concordia, 126; Guelph, 135; McGill, 106; Memorial of Newfoundland, 130, 286; Mount Allison, 134; New Brunswick, 126, 228; Ottawa, 126; Regina, 286; Simon Fraser, 125, 160, 164; Waterloo, 132
Universities/colleges in Central and South America: Metropolitan Autonomous University of Mexico, 160; National Autonomous University of Mexico, 160; São Paulo, 164, 166, 278
Universities/colleges in Europe: Birkbeck College, 141, 320n9; Dresden, 149; Eötvös Loránd, 320n9; ETH-Z, 132, Gävle, 157; Hanover, 157; King's College, Universties London, 127; Leeds, 154; Leiden, 37; Leipzig, 79; London, 140, 141, 281; London School of Economics, 128; Moscow, 149; Moscow State, 149; Oxford, 80; Oxford Observatory, 85; Plymouth, 155; Portsmouth Polytechnic, 148, 316n8; Sorbonne, 87; St. Hugh's College, Oxford, 145; Technical University of Munich, 157, 158, 275; Technology, Otakaari, 320n9; University College, Dublin, 142; Utrecht, 144; Victoria, 117; Vienna, 125, 151, 152; Warsaw, 132; Zurich, 77
Universities/colleges in the United States: California, Santa Barbara, 125, 130: California–Los Angeles, 103, 106, 112, 123; Chicago, 96, 107; Clark, 80, 115, 124, 149, 286, 311n5; Colorado, 80; Colorado–Boulder, 125, 126; Columbia, 95, 96, 98;

375

INDEX

Davidson College (NC), 147, 148, 316n8; Kansas, 99, 102, 103, 105, 117, 118, 124, 125; Louisiana State, 105, 107, 108; Michigan, 96, 107, 129, 130; Minnesota, 80, 110; Ohio State, 104, 315n32; South Carolina, 102, 117; Washington, 102, 103, 104, 107, 123, 125, 315n32; Wellesley College, 80, 84; Wisconsin, 80, 102, 103, 104, 107, 109, 118, 119; Wisconsin–Madison, 119, 125, 134
USSR, 151
Uzzi, Brian, 273

Vågan (Kjerkvågen), Church of, 66; image, 67
Valck, family, 39, 41, 42
Valsamaki, Maria, 321n13
van den Berghe. *See* Montanus family
van den Hoonaard, Deborah K., 180, 318n7, 318n3
van den Hoonaard, Will C., 29, 185, 186, 276, 288, 306n1, 308n6, 308n8, 311n17, 318n1, 318n6, 321n16, 321n2
van den Keere, Coletta, 11, 37, 38; Pieter, 308n14. *See also* Kaerius (van den Keere) family
van den Scott, Lisa-Jo, 309n3
van der AA, Pieter, 41
van der Krogt, Peter, 38, 63, 308n6, 308n11, 311n16
van Ee, Patricia Molen, 6
van Egmond, 41, 49, 309n4
van Keulen family, 39, 42, 309n15
van Meurs, Jacob, widow of, 43
van Offenberg, Francina, 38
Vanova, Jarmila, 321n13
van Peetersen, Henrick, widow of, 43
Van Schaack, Mary, 60
van Waesberger family, 37, 38
Varley, John, 56
Venus, mapping of, 150
Vereschaka, Tamara, 321n13
Verhoeven family, 39, 40. *See also* Seegers, Linken

Verner, Coolie, 307n5
Verseyl, Elizabeth, 38, 39. *See also* Visscher, Elizabeth
vignettes of women cartographers: role of, 272
"Vinland" map, 146
Visschers family, 36, 38
Visualities, feminist, 277
Visualization of Urban Archetypes (Canada), 203
Vrients family, 40, 309n19

Waage, T.C., xvi
Wackeren Hondt, den, 37
Waldorf, Susan, 105
Waldseemüllers, Martin, 95
Walker, J. Jesse, 106, 107, 108, 172, 313n11
Wallace, Dillon, 81, 82
Wallis, Helen, 20, 76, 77, 139, 140, 142, 143, 144, 145–49, 154, 173, 199, 225, 235, 236, 281, 302, 311n1, 315n3, 315n5
Warburg Institute (London), 290
Warren, Carol B., 182
Watrous, Mrs. S.I., 61
Watt, Ruth, 4
Wawrik, Franz, 152
Wealth: as source of longevity, 34
Weapons system developer, 100
Wegener, Alfred, 97
Weiss, Helmut, 37, 38, 43, 49, 308n5, 308n6
Welch, Grace, 126
Wells, John, 141
Wergeland, Henry, 67
Western Europe, 266
Westerstee Beeck, Anna van, 41, 42
Westtown globes, 62
Weyerstraet, Elizée, 38; Jodocus Janszoon, 38
Whitaker, Ewen A., 150, 151, 286, 316n9, 316n12
Whitby, Barbara, 69, 71, 270, 311n20
W.H. Smith and Son, 88

Widow(s), 34, 37, 38, 41, 42, 43, 44, 51, 52, 53, 54, 55, 56, 61, 88, 180, 191, 269, 271, 309n18, 309n19
Wilding, Faith, 280
Wilford, John N., 95
Willard, Emma, 59–60, 310n11
Williams, David, 150; Donna, 127, 134, 218, 219, 226, 288, 320n6, 321n12
Wilson, Alice, 127; James, 63
Winearls, Joan, 235
Winter, Keith, 71, 331n20
Wirth, Linda, 243
Wischnewetzky, Lazare, 77
Women, cartographic, in various contexts: 10, 30, 81; Canadian context of, 126–27, in United States, 61; comparing United States and Canada, 80; in France, 49; in New Zealand, 259
Women, cartographic activities of: creators of atlases, 9, 59; explorers, 81; making globes, 63; schoolteachers, 58; rise of, 89; forbidden to be apprenticed, 57; nursing next generation of, cartographers, 213–14
Women "firsts," 50, 63, 76, 80, 86, 91, 101, 106, 109, 115, 127, 135, 139, 142, 143, 145, 150, 156, 158, 159, 225, 235, 236, 241, 242, 270, 275, 281; difficult to establish "first," 113
Women, participation of cartographic, in: academia, 182, 206; architecture, 6; cartography and related fields, 218–19; engineering, 206; an European institute, 207–8; geography, 4; geomatics, 206; science/scientific fields, 3, 206; technical fields, 206
Women, proportion of: compared to men in engineering, 214; as surveyors (Canada and US), 319n1, as surveyors in Japan, 319n3; in various organizations, 218–20; in American Society of Professional Geographers, 94; in cartographic organizations, 103, 104; in American Congress on Surveying and Mapping, 94; not in government service, 140; absence in Scientific Cartography, 46
Women, self-perceptions of women, cartographic: challenges in organizations, 241; obstacles, 4; expectations of, in careers, 185; orientations of, in careers, 185; see selves as intruders, 173–75
Women's Bureau of the United States Department of Labor, 91
Women's Forum (American Congress on Surveying & Mapping), 244–45
Women travellers, Victorian, 306n7
Wonders, Lillian, 225
Wood, Chuck, 85; Denis, 25, 275
Wood, Clifford, 131
Woodbridge, William Channing, 59
woodcutting, 22
Woodfield, Ruth, 4, 280, 283, 321n1
Woods Hole Oceanographic Institute, 129, 312
Woodward, David, 11, 22, 23, 31, 115, 119, 135, 314n20
Woolridge, Sidney, 143
work and family, 191–96, 204
World Bank, 161
World Ocean Floor Panorama, 98; image, 97
World War I, 83, 85, 276
World War II, 17, 48, 72, 88, 89–91, 102, 140, 156, 174, 210, 276, 309n3
Wright, Barbara D., 305n1
Wright, Dawn, 315n34

Yonge, Ena L., 236
Yukon Territory, 128

Zaroutskaja, Irene, 225
Zegers. *See* Seegers
Zhou, Yu, 109, 110, 313n13
Zierer, Clifford, 103
Zond 3, 150

377

www.ingramcontent.com/pod-product-compliance
Lightning Source LLC
Chambersburg PA
CBHW060450030426
42337CB00015B/1532